I0035076

WSN and IoT

Nowadays, all of us are connected through a large number of sensor nodes, smart devices, and wireless terminals. For these Internet of Things (IoT) devices to operate seamlessly, the Wireless Sensor Network (WSN) needs to be robust to support huge volumes of data for information exchange, resource optimization, and energy efficiency. This book provides in-depth information about the emerging paradigms of IoT and WSN in new communication scenarios for energy-efficient and reliable information exchange between a large number of sensor nodes and applications.

WSN and IoT: An Integrated Approach for Smart Applications discusses how the integration of IoT and WSN enables an efficient communication flow between sensor nodes and wireless terminals and covers the role of machine learning (ML), artificial intelligence (AI), deep learning (DL), and blockchain technologies which give way to intelligent networks. This book presents how technological advancement is beneficial for real-time applications involving a massive number of devices and discusses how the network carries huge amounts of data allowing information to be communicated over the Internet. Intelligent transportation involving connected vehicles and roadside units is highlighted to show how a reality created through the intelligent integration of IoT and WSN is possible. Convergence is discussed and its use in smart healthcare, where only through the intelligent connection of devices can patients be treated or monitored remotely for telemedicine or telesurgery applications. This book also looks at how sustainable development is achieved by the resource control mechanism enabling energy-efficient communication.

A wide range of communication paradigms related to smart cities, which includes smart healthcare, smart transportation, smart homes, and intelligent data processing, are covered in the book. It is aimed at academicians, researchers, advanced-level students, and engineers who are interested in the advancements of IoT and WSN for various applications in smart cities.

Prospects in Smart Technologies

Series Editor
Mohammad M. Banat
Mohammad M. Banat, Jordan University of Science and Technology, Irbid, Jordan
banat@just.edu.jo
Sara Paiva, Instituto Politécnico de Viana do Castelo, Viana do Castelo, Portugal
sara.paiva@estg.ipvc.pt

Published Titles

Emerging Technologies for Sustainable and Smart Energy
Edited by
Anirbid Sircar, Gautami Tripathi, Namrata Bist, Kashish Ara Shakil, and Mithileysh
Sathiyanarayanan

CYBORG
Human and Machine Communication Paradigm
Kuldeep Singh Kaswan, Jagjit Singh Dhatterwal, Anupam Baliyan, and Shalli Rani

Emerging Technologies and the Application of WSN and IoT
Smart Surveillance, Public Security, and Safety Challenges
Edited by
Shalli Rani and Himanshi Babbar

WSN and IoT
An Integrated Approach for Smart Applications
Edited by
Shalli Rani and Ashu Taneja

For more information on this series, please visit: https://www.routledge.com/Prospects-in-Smart-Technologies/book-series/CRCPST

WSN and IoT
An Integrated Approach for Smart Applications

Edited by
Shalli Rani and Ashu Taneja

CRC Press
Taylor & Francis Group
Boca Raton London New York

CRC Press is an imprint of the
Taylor & Francis Group, an **informa** business

Designed cover image: Shutterstock

MATLAB® and Simulink® are trademarks of The MathWorks, Inc. and are used with permission. The MathWorks does not warrant the accuracy of the text or exercises in this book. This book's use or discussion of MATLAB® or Simulink® software or related products does not constitute endorsement or sponsorship by The MathWorks of a particular pedagogical approach or particular use of the MATLAB® and Simulink® software.

First edition published 2024
by CRC Press
2385 NW Executive Center Drive, Suite 320, Boca Raton FL 33431

and by CRC Press
4 Park Square, Milton Park, Abingdon, Oxon, OX14 4RN

CRC Press is an imprint of Taylor & Francis Group, LLC

© 2024 selection and editorial matter, Shalli Rani and Ashu Taneja; individual chapters, the contributors

Reasonable efforts have been made to publish reliable data and information, but the author and publisher cannot assume responsibility for the validity of all materials or the consequences of their use. The authors and publishers have attempted to trace the copyright holders of all material reproduced in this publication and apologize to copyright holders if permission to publish in this form has not been obtained. If any copyright material has not been acknowledged please write and let us know so we may rectify in any future reprint.

Except as permitted under U.S. Copyright Law, no part of this book may be reprinted, reproduced, transmitted, or utilized in any form by any electronic, mechanical, or other means, now known or hereafter invented, including photocopying, microfilming, and recording, or in any information storage or retrieval system, without written permission from the publishers.

For permission to photocopy or use material electronically from this work, access www.copyright.com or contact the Copyright Clearance Center, Inc. (CCC), 222 Rosewood Drive, Danvers, MA 01923, 978-750-8400. For works that are not available on CCC please contact mpkbookspermissions@tandf.co.uk

Trademark notice: Product or corporate names may be trademarks or registered trademarks and are used only for identification and explanation without intent to infringe.

ISBN: 978-1-032-56689-4 (hbk)
ISBN: 978-1-032-56771-6 (pbk)
ISBN: 978-1-003-43707-9 (ebk)

DOI: 10.1201/9781003437079

Typeset in Times
by KnowledgeWorks Global Ltd.

Contents

List of Figures and Tables

Preface

This book presents the integration of the Internet of Things (IoT) and Wireless Sensor Networks (WSNs) for smart applications. It includes the emerging technologies of IoT and WSN, the latest innovations, trends, and concerns, as well as practical challenges encountered, and solutions adopted.

At present, all of us are connected through a large number of sensor nodes, smart devices, and wireless terminals. For these IoT devices to operate seamlessly, the WSNs need to be robust. To support huge volumes of data for information exchange, resource optimization and energy efficiency play an important role. The role of machine learning (ML), artificial intelligence (AI), blockchain, caching, deep learning (DL) in optimum network performance will lead to smarter applications of the future. A wide range of communication paradigms pertaining to smart applications, which includes smart cities, smart healthcare, smart transportation, and intelligent data processing, are also discussed.

This book also covers a thorough convergence of IoT and WSN for different use case scenarios. It is revealed that how a set of key enabling technologies (KET) related to network management, resource control, data sensing, decision-making, and automation can be efficiently integrated in one system. The technological advancement is beneficial for real-time applications involving a massive number of devices. The technology can be extremely useful for Industrial IoT (IIoT) and Industry 5.0. The vision of intelligent transportation which involves connected vehicles and the road side units can be made a reality through the intelligent integration of IoT and WSN. This convergence finds usage in smart healthcare where through the intelligent connection of the devices, the patients can be treated or monitored remotely for telemedicine or telesurgery applications. The resource control mechanism enables energy efficient communication for sustainable development. This book provides step-by-step guidance to readers, starting from the technological background to real-time implementation involving different use-case scenarios.

MATLAB® is a registered trademark of The Math Works, Inc.
For product information, please contact:

The Math Works, Inc.
3 Apple Hill Drive
Natick, MA 01760-2098
Tel: 508-647-7000
Fax: 508-647-7001
E-mail: info@mathworks.com
Web: http://www.mathworks.com

Acknowledgments

First and foremost, I would like to thank my husband, **Shvet Jain**, for standing beside me throughout my research career so far. He has been my motivation for continuing to improve my knowledge and move forward in my career. I dedicate this book of mine to him.

Moreover, without the prayers and best wishes of my parents, my sister, and both of my sons, none of my achievements would have been possible.

I am also thankful to the editorial team and my co-editor, **Dr. Ashu Taneja**, who really helped me in every aspect of the preparation of this book. Her prompt response and care of the literature and contributions was remarkable. She is indeed a great and responsible researcher.

I would also like to thank each and every one behind the completion of this book, who were a great help, and without their efforts, it wouldn't have been possible to publish this book.

Last, but not least, my gratitude is due to my inspiration **Dr. Archana Mantri**, Vice Chancellor, Chitkara University, Punjab, for her trust in me and for giving me the confidence to make my efforts successful in each and every field.

Editors

Shalli Rani, PhD, has been pursuing a postdoctoral fellowship from Manchester Metropolitan University, UK, since July 2022. She is a professor at Chitkara University Institute of Engineering and Technology, Chitkara University, Rajpura, Punjab, India. She has more than 18 years of teaching experience. She received an MCA degree from Maharishi Dyanand University, Rohtak in 2004, the M.Tech degree in Computer Science from Janardan Rai Nagar Vidyapeeth University, Udaipur in 2007, and a PhD degree in Computer Applications from Punjab Technical University, Jalandhar in 2017. Her main areas of interest and research are Wireless Sensor Networks, Underwater Sensor Networks, Machine Learning, and Internet of Things. She has published/accepted/presented more than 100 papers in international journals/conferences (SCI & Scopus) and edited/authored five books with international publishers. She is serving as the associate editor of *IEEE Future Directions Letters*. She served as a guest editor in *IEEE Transaction on Industrial Informatics*, Hindawi WCMC, and Elsevier IoT Journals. She has also served as reviewer in many repudiated journals of IEEE, Springer, Elsevier, IET, Hindawi, and Wiley. She has worked on Big Data, Underwater Acoustic Sensors, and IoT to show the importance of WSN in IoT applications. She received a young scientist award in February 2014 from Punjab Science Congress, Lifetime Achievement Award and Supervisor of the Year Award from Global Innovation and Excellence, 2021.

Ashu Taneja, PhD, is an associate professor at Chitkara University Research and Innovation Network (CURIN), Chitkara University, Punjab, India. She has a PhD in Electronics and Communication Engineering with 12 years of experience serving industry, teaching, education, and research. She received her Master of Engineering (M.E.) degree from Thapar University, Patiala, Punjab in 2011 and Bachelor of Technology (B.Tech) degree from Kurukshetra University, Haryana, in 2008. She has authored more than 50 publications in international journals/conferences (SCI & SCOPUS) and she has filed more than 20 patents. Dr. Taneja has also served as a reviewer in many reputable journals from IEEE, Springer, Elsevier, IET, Hindawi, and Wiley. Her research interests include Next Generation Wireless Networks, IoT, 5G, 6G, Energy Harvesting, User Selection, Antenna Selection, Intelligent Reflecting Surfaces, and Vehicular Communication.

Contributors

Mainak Adhikari
Indian Institute of Information
    Technology
Lucknow, India

Satyam Kumar Agrawal
Centre for In Vitro Studies and
    Translational Research
Chitkara School of Health Sciences,
    Chitkara University
Rajpura, India

Syed Hassan Ahmed
California State University
Long Beach, USA

Samir N. Ajani
Department of Computer Engineering
St. Vincent Pallotti College of
    Engineering and Technology
Nagpur, India

Srinivas Ambala
Pimpri Chinchwad College of
    Engineering
Pune, India

Salim Y. Amdani
Babasaheb Naik College of Engineering
Pusad, India

Ankita Sharma
Chitkara University Institute of
    Engineering and Technology
Chitkara University
Rajpura, India

Wankhede Vishal Ashok
Department of Electronics and
    Telecommunication Engineering
S.H.H.J.B. Polytechnic
Nashik, India

Rashmi Ashtagi
Department of Computer
    Engineering
Vishwakarma Institute of Technology
Pune, India

Mrunal Swapnil Aware
Computer Science and Engineering
MIT World Peace University
Pune, India

Lalit Kumar Awasthi
National Institute of Technology
Uttarakhand, India

Jyoti L. Bangare
Department of Computer Engineering
Vishwakarma Institute of Information
    Technology
Pune, India

Ali Kashif Bashir
Manchester Metropolitan University
Manchester, UK

Parul Bhanarkar
Babasaheb Naik College of
    Engineering
Pusad, India

Anup Bhange
Department of Computer Science and
    Engineering
KDK College of Engineering
Nagpur, India

Ashwin S. Chatpalliwar
Department of Industrial
    Engineering
Shri Ramdeobaba College of
    Engineering and Management
Nagpur, India

Sarika T. Deokate
Department of Computer Engineering
Pimpri Chinchwad College of
    Engineering
Pune, India

Imali Dias
Faculty of Science, Engineering, and
    Built Environment
Deakin University, Burwood Campus
Melbourne, Australia

Yatin Gandhi
Competent Softwares
Pune, India

Namrata Gawande
Department of Computer Engineering
Pimpri Chinchwad College of
    Engineering
Pune, India

Vaibhav V. Gijare
Department of Electronics and
    Telecommunication
School of Electrical Engineering
MIT Academy of Engineering,
    Alandi (D)
Pune, India

Krishan Kumar Goyal
Faculty of Computer Application
R.B.S. Management Technical Campus
Agra, India

Lipika Gupta
Department of Electronics and
    Communication Engineering
Chitkara University Institute of
    Engineering and Technology
Rajpura, India

Shubham Gupta
Rayat-Bahra University
Mohali, India

Abhishek Hazra
National University of Singapore
Singapore

Sushmita Sunil Jain
Centre for In Vitro Studies and
    Translational Research
Chitkara School of Health Sciences
Rajpura, India

Vinit Khetani
Cybrix Technologies
Nagpur, India

Saurabh Manoj Kothari
College of Engineering Design and
    Physical Sciences
Brunel University
Uxbridge, UK

Sanjivani Hemant Kulkarni
Computer Science and Engineering
MIT World Peace University
Pune, India

Shailesh V. Kulkarni
Department of Electronics and
    Telecommunication Engineering
Vishwakarma Institute of Information
    Technology
Pune, India

Kamini Lamba
Chitkara University Institute of
    Engineering and Technology
Rajpura, India

Suresh Limkar
Department of Artificial Intelligence
    and Data Science
AISSMS Institute of Information
    Technology
Pune, India

Parikshit N. Mahalle
Department of Artificial Intelligence
  and Data Science
Vishwakarma Institute of Information
  Technology
Pune, India

Jyoti Maini
Chitkara University Institute of
  Engineering and Technology
Rajpura, India

Shivani Malhotra
Department of Electronics and
  Communication Engineering
Chitkara University Institute
  of Engineering and
  Technology
Rajpura, India

Sulakshana Sagar Malwade
Department of Computer Science and
  Engineering
School of Polytechnic and Skill
  Development
MIT World Peace University
Pune, India

Deepak T. Mane
Department of Computer
  Engineering
Vishwakarma Institute of
  Technology
Pune, India

Bhushan M. Manjre
Department of Computer Science and
  Engineering
G. H. Raisoni Institute of Engineering
  and Technology
Nagpur, India

Imran Memon
Bahria University
Islamabad, Pakistan

Rais Allauddin Mulla
Department of Computer Engineering
Vasantdada Patil Pratishthan College of
  Engineering and Visual Arts
Mumbai, India

Rakhi Mutha
Amity University
Jaipur, India

Neeraj
Chitkara University School of
  Engineering and Technology
Solan, India

Shraddha Ovale
Department of Computer Engineering
Pimpri Chinchwad College of
  Engineering
Pune, India

Namita Parati
Department of Computer Science and
  Engineering,
Maturi Venkata Subba Rao (MVSR)
  Engineering College
Hyderabad, India

Mahendra Eknath Pawar
Department of Computer Engineering
Vasantdada Patil Pratishthan College of
  Engineering and Visual Arts
Mumbai, India

Sashikanta Prusty
Department of Computer Science and
  Engineering
Siksha 'O' Anusandhan (Deemed to be
  University)
Bhubaneshwar, India

Satpalsing D. Rajput
Pimpri Chinchwad College of
  Engineering
Pune, India

Shalli Rani
Chitkara University Institute of
    Engineering and Technology
Rajpura, India

Thippa Reddy
College of Information Science and
    Engineering
Jiaxing University
Jiaxing, China

Nilesh P. Sable
Vishwakarma Institute of Information
    Technology
Pune, India

Vyasa Sai
Intel Corporation
Santa Clara, USA

Parul Saini
Department of Electronics and
    Communication Engineering
Chitkara University Institute of
    Engineering and Technology
Rajpura, India

Raju M. Sairise
Yadavrao Tasgaonkar College of
    Engineering and Management
Pune, India

Mahendra Balkrishna Salunke
Department of Computer Engineering
Pimpri Chinchwad College of
    Engineering and Research
Pune, India

Farhadeeba Shaikh
Department of Computer Science and
    Engineering
School of Polytechnic and Skill
    Development
MIT World Peace University
Pune, India

Pooja Sharma
Rayat-Bahra University
Mohali, India

Gitanjali Shinde
Vishwakarma Institute of Information
    Technology
Pune, India

Shrinivas T. Shirkande
Department of Computer
    Engineering
S.B. Patil College of Engineering
Pune, India

Ashu Taneja
Chitkara University Institute of
    Engineering and Technology
Rajpura, India

Basant Tiwari
Hawassa University
Hawassa, Ethiopia

Kishor S. Wagh
Department of Computer Engineering
AISSMS Institute of Information
    Technology
Pune, India

1 Introduction to IoT and WSN

Jyoti Maini and Shalli Rani

1.1 INTRODUCTION

Internet of Things (IoT) and Wireless Sensor Networks (WSNs) are two closely related technologies that are transforming the way we collect and analyze data in a wide range of industries. IoT is a connected arrangement of various devices, vehicles, and other objects installed with different sensors and software that empower data collection and data interchange with other systems. These data can then be used to inform decision-making, enhance processes, and improve efficiency across a wide range of applications.

WSN, on the other hand, is a specific type of IoT network that is designed to collect data from a large number of small sensors spread over a wide area. These can be used to analyze a variety of environmental factors, such as temperature, humidity, and pressure, and can transmit these data wirelessly to a central hub or data collection system.

The main distinction between IoT and WSN is that IoT is a more general term that encompasses a large count of devices and applications, while WSN is a specific type of IoT network that is focused on collecting data from sensors in a wireless, distributed manner. WSNs are often used in many areas such as monitoring environmental factors, industrial automation, and smart cities, where it is important to collect data from many sensors over a wide area.

Together, IoT and WSN are revolutionizing industries such as agriculture, transportation, manufacturing, and healthcare by enabling real-time operational insights, enabling predictive maintenance, and optimizing resource utilization. The potential benefits of these technologies are vast, and as they are progressive, they will have an increasingly crucial role in developing the future of the global economy.

1.1.1 INTERNET OF THINGS

IoT is described as the interconnection of physical objects, home appliances, vehicles, and other products that incorporate electronics, software, sensors, and connectivity that allow these items to connect and exchange data through the Internet (Figure 1.1). IoT devices encompass everything from smartwatches and fitness trackers to home security systems, thermostats, and even cars. IoT allows these devices to communicate with each other and with other systems, providing new ways to monitor and control various aspects of daily life [1].

The potential benefits of IoT include increased efficiency, improved safety, and reduced costs across various industries, from manufacturing and agriculture

DOI: 10.1201/9781003437079-1

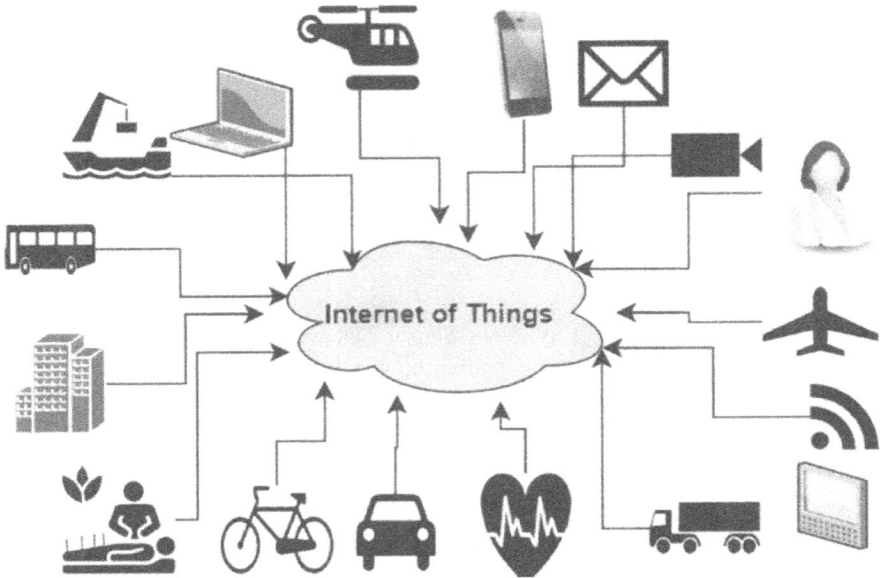

FIGURE 1.1   Internet of Things.

to healthcare, transportation, and smart homes. However, IoT also raises concerns about potential abuse and unintended consequences, as well as privacy and data security.

The adoption rate of IoT devices has increased rapidly over the last few years and is expected to continue to succeed remarkably. According to Statista's report, the number of IoT devices worldwide will reach to approx. 76 billion by 2025, from 27 billion in 2019 [2]. This growth can be attributed to several factors, including:

1. *Advancements in technology:* The rapid development of new technologies such as low-power wireless networks, miniaturized sensors, and cloud computing have made it easier and more affordable to deploy IoT devices.
2. *Cost reduction:* The cost of IoT devices has decreased significantly over the years, making them more approachable to an ample range of industries and applications.
3. *Increased connectivity:* The widespread availability of high-speed Internet and the increasing adoption of 5G technology have made it easier for IoT devices to communicate and share data.
4. *Business benefits:* IoT devices offer numerous benefits to businesses, including improved efficiency, productivity, and cost savings, which have encouraged their adoption.
5. *Consumer demand:* The growing demand for connected devices and smart home technologies has also contributed to the increased adoption of IoT devices.

It will become increasingly important to address concerns such as security, privacy, and interoperability to ensure that these devices are used safely and responsibly, as the number of IoT devices continues to grow.

1.1.1.1 Key Features of IoT

Here are few of the features of IoT [3, 4]:

- *Connectivity*: IoT devices are often connected to the Internet to be able to exchange data and interact with other devices.
- *Sensors and actuators*: These devices often are embedded with sensors that collect data about various factors such as temperature, environment, etc. as well as actuators that can control and respond to that environment.
- *Data analytics*: IoT devices generate large amounts of data that can be analyzed to generate insights and improve decision-making, optimize processes, automate tasks, and enhance overall productivity and efficiency.
- *Machine learning (ML) and artificial intelligence (AI)*: IoT can incorporate ML and AI algorithms to automate tasks and improve efficiency.
- *Interoperability*: IoT devices are designed to work together and with other systems, allowing for seamless integration and interoperability, which means you can connect anything to your communications infrastructure and global information.
- *Inconsistent or non-uniformity*: IoT system is based on various networks and hardware platforms. These can communicate with other devices or service platforms with varying sizes and functionality over various networks, meaning IoT devices are inconsistent.
- *Dynamic and self-adapting*: IoT devices need to adapt dynamically depending on the situation. For example, a camera can capture data depending on the lighting conditions. It automatically switches to night mode or day mode. It is a self-adaptive technique.
- *Security*: IoT devices must be secure to protect against cyberattacks and ensure the privacy of user data.
- *Scalability*: IoT systems can be scaled up or down to accommodate changes in demand and the addition of new devices and applications.

1.1.1.2 Layers of IoT Architecture

The IoT architecture typically consists of several layers that work together to enable the connectivity and functionalities of IoT devices and systems [5]. Here is a brief overview of each layer:

- *Perception layer*: This layer consists of the sensors and various physical devices that collect data about the environmental parameters, such as temperature, light, sound, or motion.
- *Network layer*: This layer connects devices and sensors to each other and to the Internet, typically using wireless protocols such as Wi-Fi, Bluetooth, or Zigbee.

- *Middleware layer*: This layer handles the interaction between the devices and the applications that use the data they generate. It also provides services such as data storage, messaging, and security.
- *Application layer*: This layer consists of the software applications and services that analyze and use the data generated by the devices. Examples of applications include smart home systems, healthcare monitoring tools, or industrial control systems.
- *Business layer*: Under this layer, the business models and processes are included that drive the development and deployment of IoT solutions. It also includes the governance and management of the data generated by IoT systems.

The architecture of IoT is designed for collecting, processing, and analyzing data which has been generated by various physical devices and sensors, enabling the development of applications and services that improve efficiency, security, and quality of life across various industries and applications.

Here is a simplified diagram of the architecture of IoT (Figure 1.2), showing the different layers and their interconnections:

1.1.1.3 Building Blocks of IoT Architecture

IoT architecture, shown in Figure 1.3, consists of four building blocks. These are (1) Sensors, (2) IoT gateways and frameworks, (3) Cloud server system, and (4) Mobile applications.

1. *Sensors*: Sensors obtain information from the environment or specific locations and are ubiquitous. For example, an IoT gateway allows a temperature sensor to send room temperature. Sensors can collect a wide range of data, including location, weather and environmental conditions, machinery in operation, engine maintenance information, human body information, and vehicle vital signs.
2. *IoT Gateways and frameworks*: The network of sensor nodes is linked to the Internet through gateways. It is an entrance or a path to the Internet

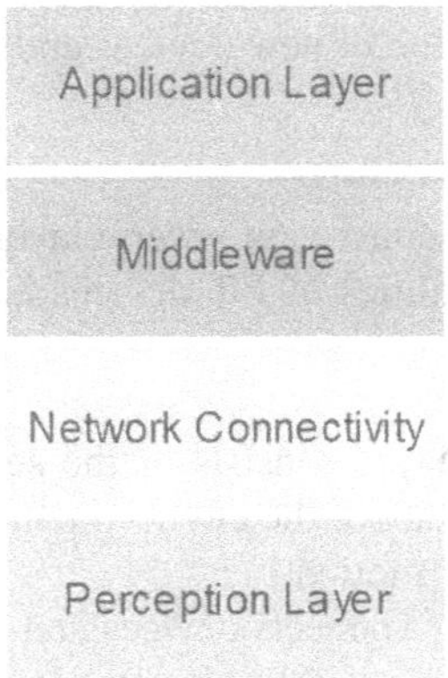

FIGURE 1.2   Layers of IoT architecture.

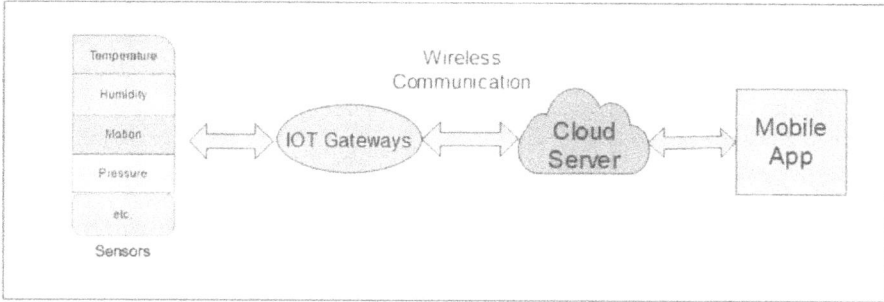

FIGURE 1.3 IoT architecture.

for all the things and devices that we want to interact with. It collects data from sensors and sends it to the Internet infrastructure to establish connectivity.

3. *Cloud server system*: The data transmitted over the gateway is securely processed and stored in cloud servers, i.e. in data centers. The processed data is then utilized to execute intelligent actions that make all of our gadgets intelligent. In the cloud, all analysis and decision-making are done by keeping the user's convenience in mind.

4. *Mobile applications*: Thanks to an easy-to-use smartphone application, customers can remotely manage and monitor devices such as interior thermostats and car engines. These applications deliver critical information directly from the cloud on your smartphone or tablet. Utilizing the following analytics, the information is displayed to the customer in the form of graphs, bar charts, and pie charts, so that users can easily understand the scenario. We may pass the signals to guide the sensors to change their default settings via our mobile application. For example, adjust the default temperature of AC and many other default values.

1.1.2 WSN – INTRODUCTION

WSN is a special kind of network consisting of a large number of tiny, low-power, inexpensive, and autonomous sensor nodes that are connected to each other with wireless facilities. These sensor nodes are equipped with various capabilities such as acquisition, processing, and communication that allow them to collect and transmit data from the physical environment to a central location [6].

WSNs are typically used for monitoring and controlling physical environments, such as temperature, humidity, pressure, light, sound, speed, direction and motion, pollutant levels, chemical concentration, vibrations, and many more such conditions. They are mostly used in different applications areas, such as environmental monitoring, industrial automation, home automation, healthcare monitoring, and military surveillance.

WSNs are characterized by their ability to function in harsh and remote environments, low power consumption, self-organization, and self-healing capabilities.

However, WSNs also face several challenges, such as limited bandwidth, limited energy, and limited processing capabilities, as well as security and privacy concerns [7].

Therefore, WSNs provide an economical and flexible solution for collecting and transmitting data from physical environments, enabling real-time tracking and control of various applications.

1.1.2.1 Features of WSN

WSNs are designed to have several features that enable them to perform their intended functions [8]. Here are some of the key features of WSNs:

1. *Low-power consumption*: WSNs are designed to consume minimal power, allowing the nodes to operate for extended periods of time on limited energy resources such as batteries.
2. *Small size*: The nodes in a WSN are typically small, allowing them to be deployed in large numbers and in tight spaces.
3. *Wireless communication*: The nodes in a WSN communicate with each other using wireless communication protocols, allowing them to be deployed in areas where wired communication is not practical or feasible.
4. *Self-organization*: WSNs are designed to self-organize, allowing the nodes to automatically establish and reconfigure their network connections.
5. *Autonomous operation*: The nodes in a WSN are typically autonomous and can operate independently of human intervention.
6. *Low cost*: The nodes in a WSN are typically low cost, allowing them to be deployed in large numbers and over a wide area.
7. *Fault tolerance*: WSN is designed to be fault tolerant and can continue to function even if some nodes fail or are damaged.
8. *Data processing and analysis*: WSNs are designed to perform data processing and analysis locally on the nodes, allowing them to filter and compress the data before transmitting it to the central location.

These features of WSNs enable them to be used in a large variety of applications, from environmental monitoring to industrial automation, and from healthcare to military surveillance.

1.1.2.2 Types of WSN

There are several types of WSNs that can be classified based on their applications, communication protocols, power source, and topology [6, 9]. Here are some common types of WSNs:

1. *Environmental sensor networks*: An environmental sensor network is a system of interconnected sensors designed to monitor and collect data about various aspects of the environment. These sensors are typically deployed in different locations to gather information on parameters such as temperature, humidity, air quality, noise levels, pollution levels, and other environmental factors. The sensors in the network continuously measure and record data, which is then transmitted wirelessly or through a wired connection to a

central server or data repository for analysis. The collected data can be used to monitor and assess environmental conditions in real-time, detect patterns, identify trends, and make informed decisions regarding environmental management and resource allocation. Environmental sensor networks find applications in various fields, including climate monitoring, urban planning, agriculture, industrial monitoring, and disaster management. They play a crucial role in improving environmental sustainability, enhancing public health and safety, and supporting evidence-based decision-making to address environmental challenges.

2. *Terrestrial WSNs (TWSNs)*: These include many wireless sensor nodes, which can be used as unstructured or structured. Nodes are randomly distributed in unstructured mode staying within the allocated destination area [6, 9]. In pre-planned or structured mode, optimized placement, grid placement, 2D and 3D placement models are considered [6]. These networks are above ground and can get activated through solar cells. Energy can be saved by minimizing latency, optimizing routing, and using low duty cycle operation. TWSNs find applications in diverse fields, including environmental monitoring, agriculture, infrastructure monitoring, wildlife tracking, and disaster management. They provide real-time or near real-time data about the monitored parameters, allowing for better understanding, analysis, and decision-making.

3. *Underground WSNs (UWSNs)*: The entire network is underground, as they are effectively used to monitor conditions underground. However, a terrestrial sink node is used to transfer information to the base station. Challenges in UWSNs include limited power supply, harsh underground conditions, network topology planning, and communication reliability (Figure 1.4). Researchers

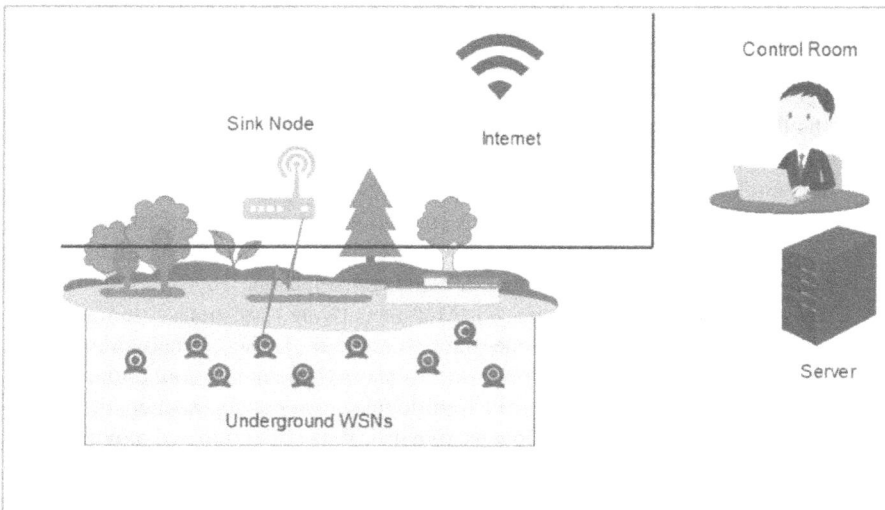

FIGURE 1.4 Underground WSNs.

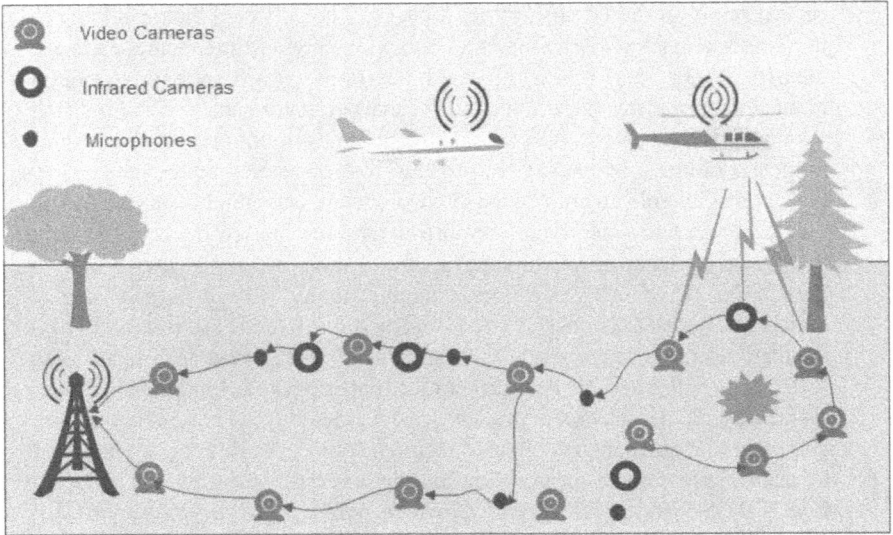

FIGURE 1.5   Multimedia WSNs.

and engineers continue to explore innovative solutions to improve the efficiency, range, and resilience of communication in these networks. These sensor networks are more expensive. The devices and machines used require proper maintenance.

4. *Multimedia WSNs* (*MWSNs*): The nodes in these networks collect information in the form of pictures, audio, and video by connecting to cameras and microphones. They can keep a check on various actions as they occur and display the events visually (Figure 1.5). These nodes are interconnected via wireless links for the purposes of data retrieval and data compression. The latest techniques of data processing and compressing are used to compress the data. The compressed audio and video data can also be sent over these sensor networks, to reduce the power consumption and high bandwidth. Research and development in MWSNs focus on improving multimedia data processing algorithms, developing energy-efficient communication protocols, and enhancing the overall network performance to enable reliable and real-time multimedia data collection and transmission.

5. *Mobile sensor networks* (*MSNs*): Mobile networks are not settled at one location but can shift from one place to another. These are networks consisting of mobile nodes equipped with sensors that can move autonomously or be controlled remotely. Unlike traditional sensor networks where the sensors are stationary, MSNs allow for dynamic data collection and monitoring capabilities in various environments. They can easily connect with your surroundings. Their main benefit is to provide better area coverage, and better channel capacity. These MSNs are more adaptable when compared to other static sensor network systems. These WSNs are used in applications

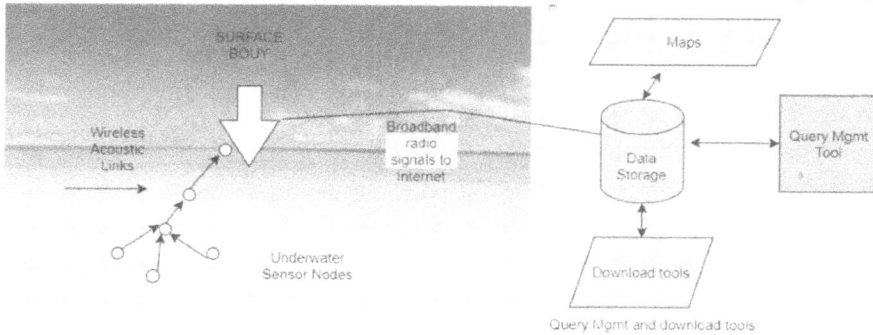

FIGURE 1.6 Underwater WSNs.

where the sensors are mobile, such as in vehicle tracking, animal tracking, and search and rescue operations.

6. *Underwater sensor networks* (*UWSNs*): The wireless underwater system consists of devices or sensor nodes installed underwater. These WSNs are used to monitor and collect data in underwater or aquatic environments such as oceans, lakes, and rivers (Figure 1.6). These vehicles are mostly used to collect information from sensor nodes. These consist of small, battery-powered sensor nodes to measure parameters such as water temperature, salinity, pressure, pH levels, dissolved oxygen, or the presence of pollutants. These nodes communicate with each other and with a surface-based sink node or base station to transmit the collected data. Long propagation delays and sensor failures are major challenges for these communication systems, and the batteries in these WSNs are limited and cannot be recharged. Therefore, various technologies have been evolved to solve this problem of energy utilization and conservation. Underwater communication in UWSNs faces challenges due to the features of the underwater medium, including high attenuation, limited bandwidth, multipath fading, and the requirement for specialized communication techniques. Acoustic waves are commonly used for long-range communication in UWSNs, although optical and electromagnetic methods can be employed for short-range communication in clear water.

1.1.2.3 How It Works

A WSN is a set of special devices called sensors that are used to monitor and record various environmental conditions and store this data in a central location.

A WSN has multiple nodes. These nodes are portable, small-sized detection stations. Each sensor node consists of a transducer or sensor, transceiver, microcontroller, and power supply. The transducer detects changes in your physical condition and produces an electrical signal. These signals are sent to a microcomputer for further processing. The microcomputer transmits signal to the receiving node and data is then sent to the computer over the Internet (Figure 1.7).

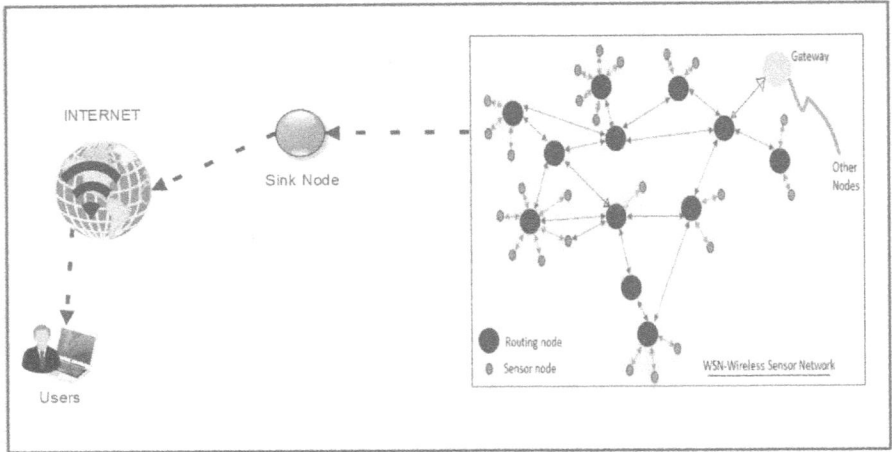

FIGURE 1.7 Wireless sensor network.

1.1.2.4 WSN Topologies

The nodes in WSN are arranged in geometric representations, called topologies [6, 9, 10]. The most common sensor network topologies are as follows (Figure 1.8):

1. Point-to-point (dedicated) network
2. Star network
3. Tree network
4. Mesh network

1. *Point-to-point (dedicated) network topology*: In dedicated topology, the nodes can communicate directly with each other. There is only one data communication path, realizing highly secure communication. Since there is no central hub, each node acts as both client and server (Figure 1.9).
2. *Star network topology*: In a star topology, communication takes place through a central hub and in contrast to point-to-point topology, direct communication between sensor nodes is not possible. Hubs act as servers and nodes act as clients (Figure 1.10).
3. *Tree network topology*: In tree topology, the central hub performs as a root node or parent node. Each node is connected to the previous higher node, and finally connects to the root node (Figure 1.11). The leading advantages of this topology are that the network is easy to expand, and errors can be easily detected. However, it has the drawback of being heavily dependent on the base cable. If the base cable cracks or breaks, the entire network collapses. The data is moved back from the leaf node to the parent node and finally to the root node.
4. *Mesh network topology*: A mesh network allows all nodes to interact directly with each other without relying on a central hub. Data can be moved from one node to another. This is the most reliable network communication

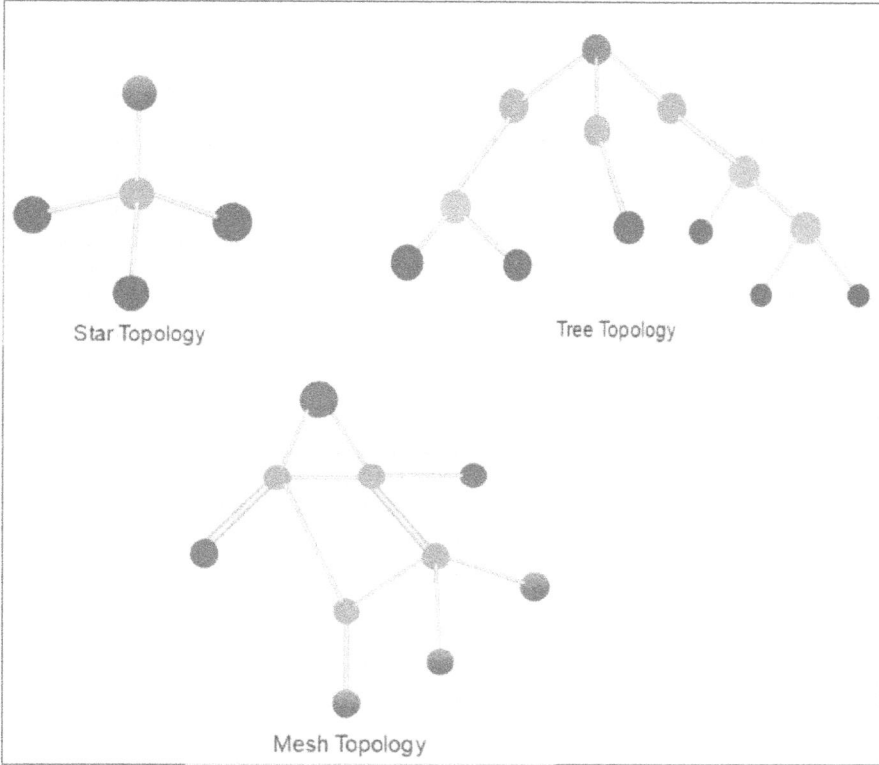

FIGURE 1.8   Wireless sensor network topologies.

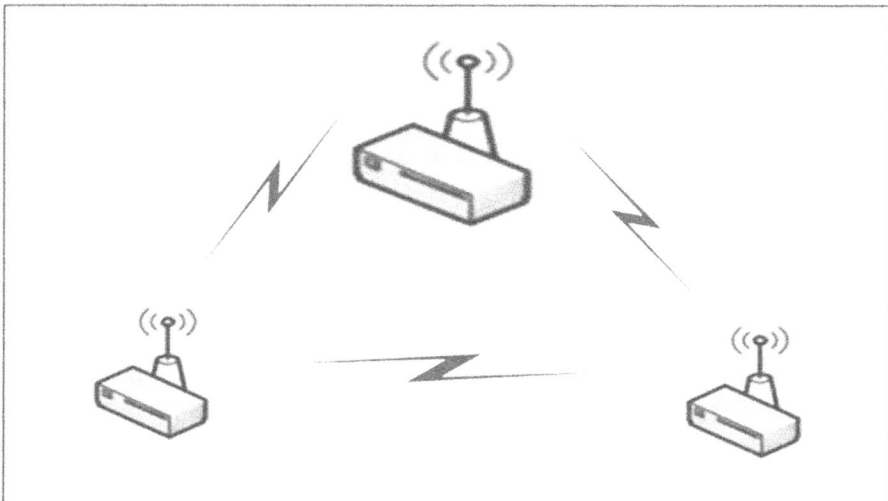

FIGURE 1.9   Point-to-point network.

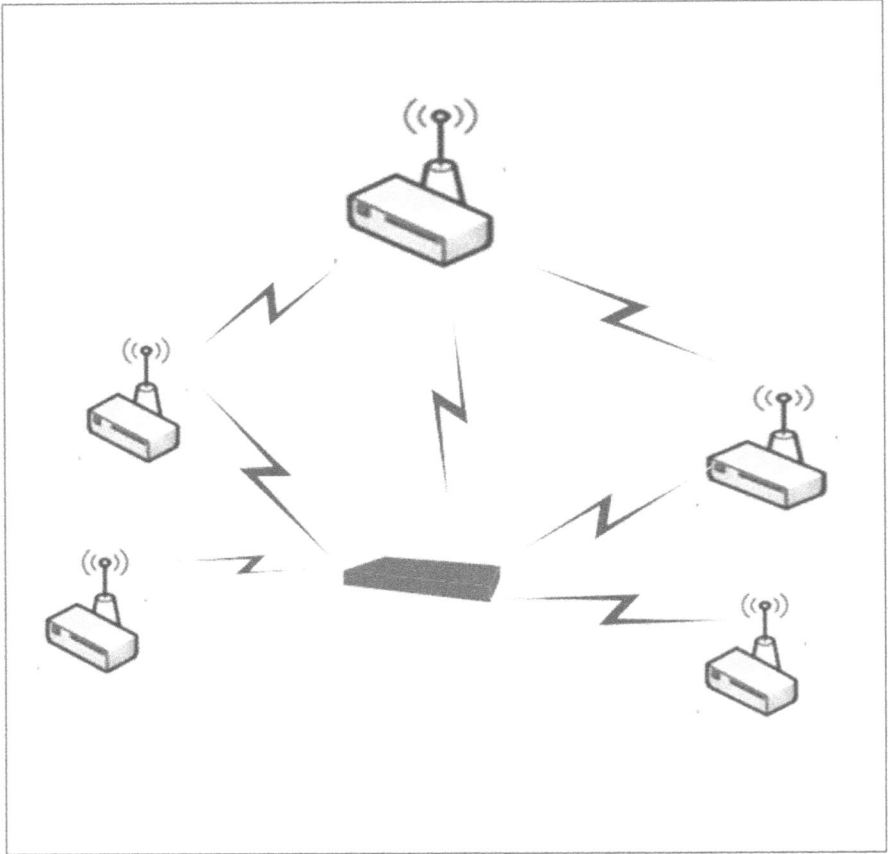

FIGURE 1.10   Star network.

structure as it has no single failure point. However, this complicated structure consumes a lot of power.

1.1.2.5   Differences between Wireless Sensor Network and Internet of Things

Table 1.1 shows the differences between WSN and IoT.

1.1.3   CONVERGENCE OF IoT AND WSN

The convergence of the IoT and WSNs is becoming increasingly common and important. This convergence involves combining the data collection capabilities of WSNs
with the communication and data processing capabilities of IoT systems [12, 13].

Combining WSNs and IoT can enable more efficient, intelligent, and scalable data
collection and analysis in a variety of applications. For example, WSNs can be used

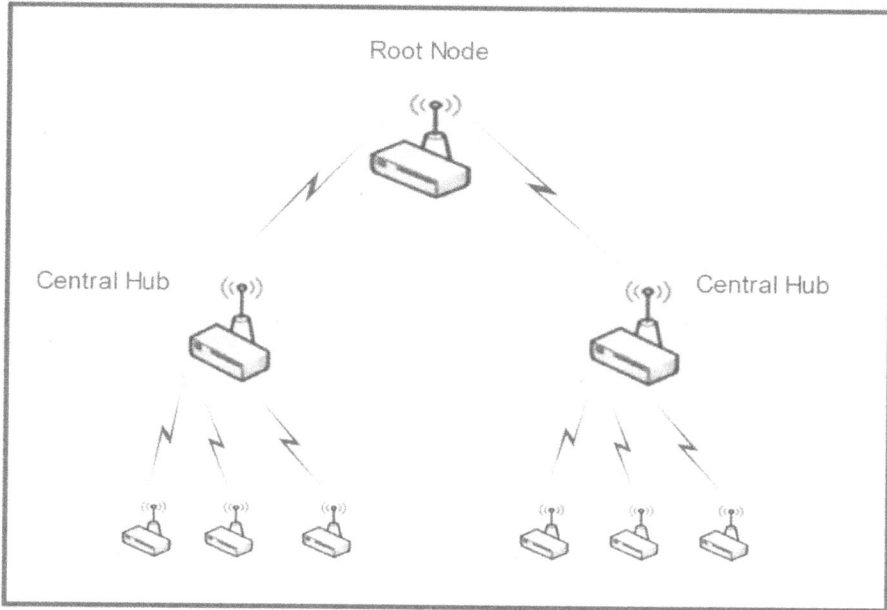

FIGURE 1.11   Tree network.

to collect data from remote or harsh environmental conditions, while IoT can provide cloud-based processing and storage capabilities for the collected data.

It is a natural evolution of both technologies. The two technologies have many similarities, including the ability to collect and transmit data from the physical environment, low power consumption, and wireless communication. The convergence of

TABLE 1.1
Differences between WSN and IoT

Wireless Sensor Network	Internet of Things
WSN is anything that connects to a wireless network to collect data or monitor the environment.	IoT means WSN + Internet + Cloud Storage + Mobile/Web Applications.
WSN consists of a network of a large number of tiny low-power devices called nodes that communicate through wireless channels for information sharing.	IoT consists of a gateway that is used to connect the devices to internetworks (using switches, routers, Access Points, etc.).
WSN follows IPv4 and it has a sink node but no gateway.	IoT follows IPv6 on the sensor network (802.15.4 MAC/PHY) and IPv4 on the internetwork part [11].
Routing protocols in WSN include, DSR, CTP, PEGASIS, LEACH, SEP, AODV, OLSR, SPIN, and TEEN.	Routing protocols in IoT include, RPL, 6LoWPAN, MQTT-SN, CoAP, AODV, BLE Mesh, and Zigbee.

these technologies has led to the creation of new applications and opportunities in various domains, including industrial automation, healthcare, environmental monitoring, and smart cities.

One of the major advantages of the convergence of IoT and WSN is the capability to integrate large-scale networks with local sensor networks. This integration enables the collection and analysis of data from multiple sources, providing a better and improved view of the physical environment. For example, a smart city network can use IoT devices and WSNs to monitor traffic flow, air quality, and noise pollution, providing a better understanding of the city's environmental conditions.

Another benefit of this combination is the ability to reduce power consumption and increase the lifespan of the network. By using WSNs for local sensing and IoT devices for data aggregation and communication, the network can be optimized for low power consumption, reducing the need for frequent battery replacements.

It also poses several challenges, such as interoperability, security, and power management. In order to fully realize the potential of this convergence, there needs to be standardization and coordination among different technologies and protocols. Additionally, security measures need to be implemented to protect the sensitive data collected and transmitted by WSNs and IoT devices. Finally, efficient power management strategies need to be developed to extend the lifespan of WSN nodes and IoT devices.

Therefore, the convergence of IoT and WSN is expected to drive innovation and create new opportunities for businesses and individuals alike. The ability to collect and analyze data from the physical environment in real-time has the potential to revolutionize or transform many different industries and domains, improving efficiency, safety, and quality of life, but it also requires careful planning, coordination, and implementation to ensure its success.

1.1.4 Sustainable Integration of IoT and WSN for Smart Applications

The integration of the IoT and WSN can enable the development of smart applications that enhance sustainability in various domains. This integration enables seamless data exchange between IoT devices and sensors, enabling real-time monitoring, control, and analysis of physical systems. Sustainable IoT is an approach to designing and implementing IoT systems in a way that prioritizes sustainability and environmental responsibility. Smart applications that use sustainable IoT technology aim to reduce energy consumption, minimize waste and environmental impact, and promote social well-being [14, 15].

Here are some key considerations for designing sustainable IoT systems for smart applications: [16]

1. *Energy efficiency*: One of the crucial considerations for sustainable integration is optimizing the energy consumption of IoT devices and sensors. WSNs typically operate on limited battery power, so energy-efficient protocols and algorithms should be employed to minimize energy consumption. Sustainable IoT systems should be designed to consume minimal power, using energy-efficient components and power management techniques and

to explore energy harvesting techniques to minimize energy consumption. This includes using low-power sensors, such as sleep modes and wake-on-demand, duty cycling, adaptive sampling, and data aggregation that can help prolong the network's lifetime, and leveraging renewable energy sources for power generation, including solar or kinetic energy, that can also be used to power IoT devices.

2. *Recycling and waste reduction*: Sustainable IoT systems should be designed to minimize waste and promote recycling. This includes using recyclable materials in IoT devices, implementing take-back programs, and designing systems that can be easily upgraded or repurposed.

3. *Data management and analytics*: Efficient data management and analytics play a vital role in sustainable IoT. IoT generates a massive amount of data from various connected devices. Sustainable IoT systems should prioritize efficient data management, minimizing the amount of data transmitted and processed to reduce energy consumption, and implement data compression techniques, edge computing, and intelligent analytics to reduce data transmission and processing. Data compression and filtering techniques can be used to minimize data transfer, and edge computing enables data processing locally to the source node, reducing the need to transfer data to the cloud, thereby conserving energy and network bandwidth. WSNs can also employ data fusion techniques to combine data from multiple sensors and eliminate redundant information.

4. *Lifecycle analysis and circular economy*: Sustainable IoT systems should be designed with a lifecycle analysis approach and to support circular economy principles, taking into account the environmental impact of the system throughout its entire lifecycle, from manufacturing to disposal. Consider factors like manufacturing, deployment, operation, and end-of-life management. Promote circular economy principles by designing devices for ease of repair, upgradeability, and recycling to minimize waste and resource depletion.

5. *Smart resource management*: Use IoT for smart resource management to optimize resource consumption and reduce waste. Monitor and control energy usage, water consumption, and other resources in real-time, enabling efficient allocation and conservation. Therefore, by leveraging IoT and WSN, real-time data can be collected, analyzed, and used to make intelligent decisions that minimize waste, improve resource allocation, and reduce environmental impact.

6. *Social impact and inclusion*: Ensure that IoT solutions contribute to social well-being and inclusivity. Consider accessibility, privacy, and security aspects to protect user data and maintain trust. Develop solutions that address social challenges and provide equitable access to smart services for all.

7. *Collaboration and partnerships*: Foster collaboration among stakeholders, including governments, industry, academia, and communities, to promote sustainable IoT initiatives. Encourage knowledge sharing, research, and development of sustainable practices and technologies.

8. *Interoperability and standardization*: IoT and WSN integration should adhere to interoperability standards to facilitate seamless communication between devices and systems from different vendors. Standardization efforts promote compatibility, reduce implementation complexity, and foster innovation within the IoT ecosystem.

9. *Connectivity and communication*: Seamless connectivity and reliable communication between IoT devices and WSN sensors are essential. Robust wireless communication protocols and networking architectures should be used to ensure efficient and uninterrupted data transmission. This enables the integration of diverse sensors and devices into a unified smart system.

10. *Standards and certification*: Adhere to industry standards and certifications that promote sustainable practices in IoT. Certification programs, such as Energy Star, EPEAT, or LEED, can provide guidelines and benchmarks for sustainable IoT design, operation, and disposal.

Therefore, sustainable IoT for smart applications has the potential to create more efficient and environmentally responsible systems that can reduce energy consumption, minimize waste, and improve the sustainability of our society. By prioritizing sustainability in IoT design and implementation, we can create a more resilient and environmentally conscious future.

1.1.4.1 Applications Areas of IoT and WSN

IoT and WSN have a wide range of applications areas across various industries. These applications cover different domains, including the industrial sector, medical sector, agriculture, smart cities, security, and emergencies, etc. A few of application areas shown in Figure 1.16 have been discussed here:

a. *Smart cities*: Building smart cities includes intelligent transportation systems [17], smart buildings, traffic jams [17, 18], waste management [19], intelligent lighting, smart parking, and city maps (Figure 1.12). This may incorporate a variety of functionalities such as AI-enabled IoT devices that can be used to track, control, and mitigate traffic congestion in smart cities [20]. Moreover, IoT will also enable the installation of intelligent street lighting that adapts to the weather. To realize an IoT smart city, it is necessary to utilize RFID and sensors. IoT for most systems such as parking meters, streetlights, sprinkler systems, etc. Many countries have plans to implement the IoT facilities.

b. *Healthcare*: As [21] explains, enhanced and automated IoT devices are assisting in the transformation of healthcare systems. Other technologies that improve a range of activities, such as communicating reports with different persons and places, maintaining records, and administering medications. It significantly improves the transformation of the healthcare industry [21]. IoT applications offer benefits to health sector as it may be broadly classified as automated data gathering and analysis, person identity and authentication, and patient, employee, and asset monitoring (Figure 1.13). Moreover, authentication and identification reduce the requirement for

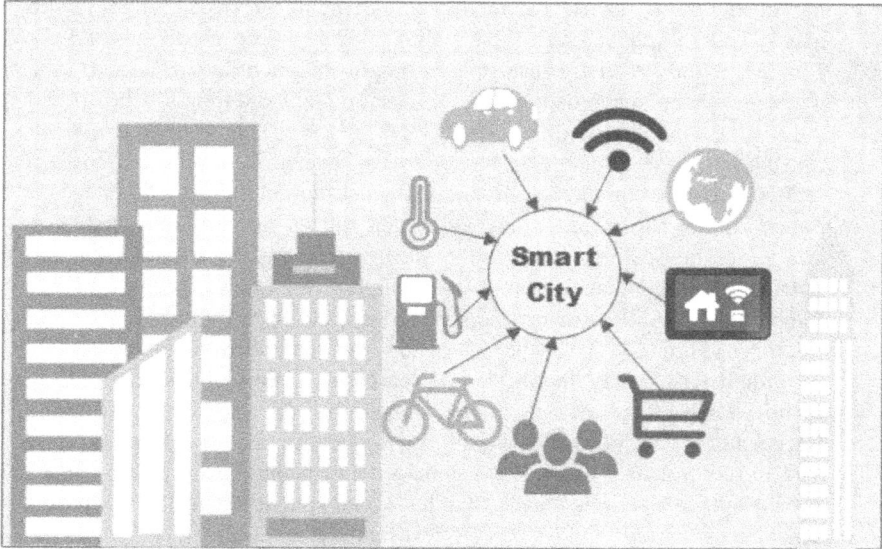

FIGURE 1.12   Smart city.

record-keeping, infant mismatch scenarios, and patient-harming incidents. Sensor devices provide the patient-centric capability for illness detection and using real-time data on patient health indicators [20]. The sensor may be applied to toothbrushes, dental Bluetooth devices, and inpatient and outpatient settings. It can also offer information after patient monitoring

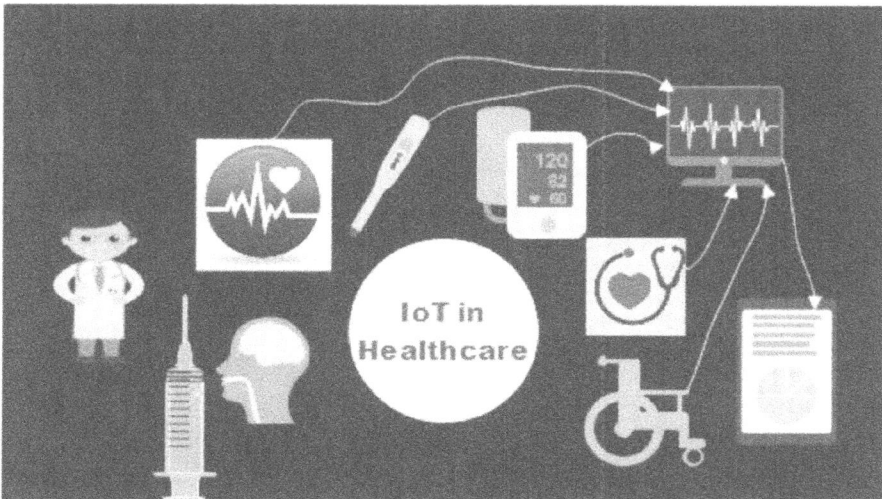

FIGURE 1.13   IoT in healthcare.

and usage. Uses in this field include telemedicine services, patient health alerts, and the capacity to track patient adherence to medicines. RFID, Bluetooth, and Wi-Fi are further IoT components included in this capacity. They greatly enhance the method for measurement and monitor the crucial parameters including temperature, blood pressure, heart rate, cholesterol, and blood sugar levels. These techniques have proven very effective in COVID times while monitoring the patients in remote areas.

c. *Smart farming*: To increase the quantity and quality of fruits and vegetables, control the microclimate such as monitoring temperature, soil moisture, and nutrient levels to enhance crop yields. Compost: Regulates water and suitable temperature conditions in lucerne, hay, and straw to get rid of fungus and other microbes (Figure 1.14). To guarantee the survival and health of the offspring of animal farms, it is important to research air quality and its circulation, identify toxic gases in excretions, and care for the offspring in open farms and huge stalls. Control over fields and growing conditions through monitoring: Greater control over agricultural management, including improved control over fertilizer, electricity, and irrigation, accurate continuous data gathering, and better monitoring of rot and crop waste reduction [22].

d. *Smart environment*:

Detection of air pollution: Excessive carbon dioxide emissions from industries, automotive pollution, farm toxic gas control.

Fire detection in forests: Combustion gas and preventive fire status monitoring and warning zone definitions.

Weather forecasting: To monitor the weather conditions such as humidity and temperature, wind speed, and rain, early detection of earthquakes, etc.

FIGURE 1.14 Smart farming.

Water quality: Investigation of river and seawater suitability for drinking water compatibility.

River flooding: Monitor rivers and dams on rainy days, reservoir water level change monitoring.

Protected species: Track wildlife using GPS/GSM modules to find, track, and send coordinates via messages from tracking collars [5].

e. *Smart living*:

Remote control device: To avoid mishaps and save energy, you may remotely switch the gadget on and off.

Weather: Shows outside weather conditions such pressure, wind speed, humidity, temperature, and rainfall. It also has the capability to broadcast and display long-distance data [17].

Send smart appliances: Refrigerator with LCD screen to display the contents, soon-expiring groceries, products to buy, all information is available in the smartphone app, a washing machine with remote laundry-monitoring capabilities, a cooking range with a smartphone app interface that allows for remote temperature control and oven self-cleaning function monitoring, and security monitoring. By using CCTV cameras and home alarm systems, people may feel secure conducting their daily activities at home. Monitor your energy and water use to get tips on how to save money and use fewer resources, etc. [5] (Figure 1.15).

f. *Intelligent industry*: In an industrial environment, hazardous gases can leak into a chemical plant or mine and be detected, and a chemical plant's toxicity and oxygen concentration can be monitored to ensure worker and

FIGURE 1.15 Smart homes.

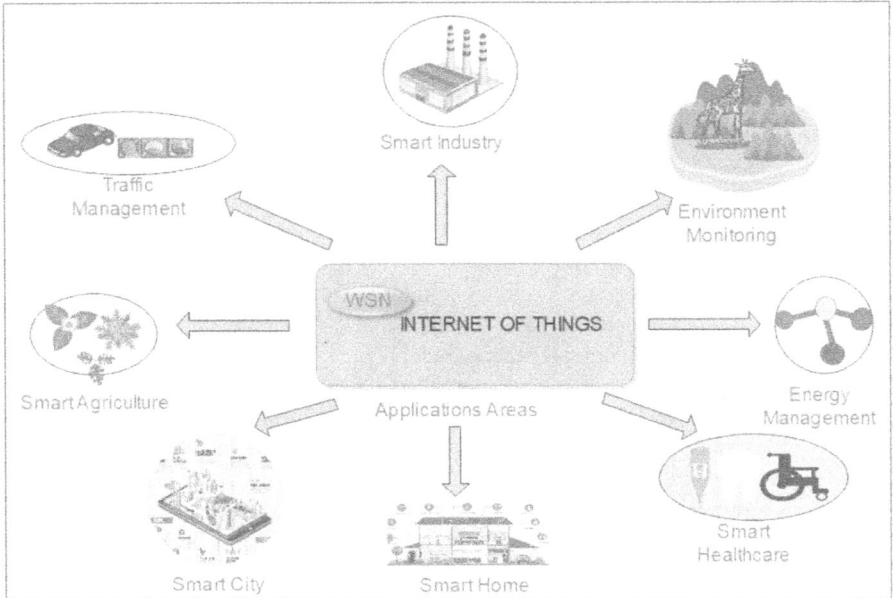

FIGURE 1.16   Applications of IoT and WSN.

product safety. You can connect sensors to your equipment to monitor and send reports, transforming equipment failure early prediction and service maintenance into real component failures. It can be scheduled automatically in advance (Figure 1.16) [22].

1.1.5 CURRENT TRENDS AND CHALLENGES IN IoT AND WSN

1.1.5.1 Current Trends in IoT and WSN

IoT and WSN are both rapidly evolving technologies that are driving significant changes in the way we interact with and utilize technology [5]. Here are some current trends in IoT and WSN:

1. *Edge computing*: Edge computing is becoming increasingly popular in IoT and WSN applications. Edge computing can reduce latency, improve response time, improve reliability, and reduce bandwidth requirements by processing data locally on IoT devices or edge servers. It is a distributed computing paradigm that enables computation and data storage at the edge of the network, closer to the data origin [23]. It enables faster data processing, reduces latency, and addresses bandwidth limitations by processing data locally at the edge instead of sending it to a cloud or data center (Figure 1.17).

   Some commonly used tools and technologies in edge computing are the following [16, 23, 24]:

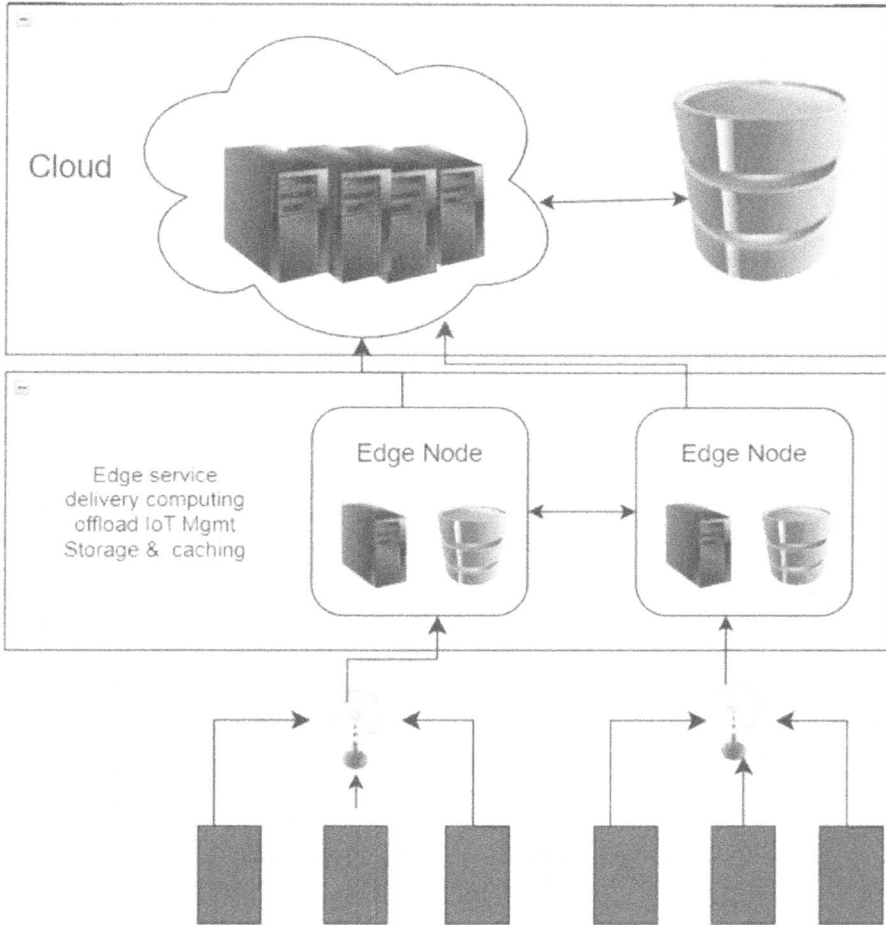

FIGURE 1.17   Edge computing.

- *Edge devices*: Edge devices are the hardware components installed at the edge of the network, such as edge servers, gateways, routers, and IoT devices. These devices have computing power, storage capacity, and connectivity capabilities to process and manage data locally.
- *Edge computing platforms*: Edge computing platforms provide the infrastructure and software framework for managing and orchestrating edge devices and applications. They offer features like edge node management, data ingestion, analytics, security, and integration with cloud services. These platforms include Microsoft Azure IoT Edge, AWS IoT Greengrass, and Google Cloud IoT Edge [24].
- *Containerization*: Containerization technologies like Docker and Kubernetes play a crucial role in edge computing. Containers enable

the packaging and isolation of applications and their dependencies, making it easier to deploy, manage, and scale edge applications across different edge devices and locations.

- *Edge analytics and machine learning*: Edge computing allows for real-time analytics and machine learning inference at the edge. Tools like Apache NiFi, Apache Flink, TensorFlow Lite, and ONNX Runtime enable running analytics and machine learning models on edge devices with limited resources. These tools optimize the models and leverage hardware acceleration to achieve efficient edge inference.
- *Edge security*: Edge computing introduces unique security challenges. Tools like Istio, Envoy, and Nginx help secure edge environments by providing traffic management, authentication, authorization, and encryption capabilities. Additionally, edge security solutions like firewalls, intrusion detection systems (IDS), and data encryption protocols are used to protect data and devices at the network edge.
- *Edge data management*: Edge computing involves dealing with massive amounts of data generated at the edge. Tools such as Apache Kafka, Apache Pulsar, and RabbitMQ facilitate reliable messaging and data streaming between edge devices and backend systems. These tools ensure efficient data transfer, event-driven processing, and data synchronization.
- *Edge monitoring and management*: To monitor and manage edge devices and applications, tools like Prometheus, Grafana, and Nagios are used. These tools provide monitoring, logging, and alerting capabilities, allowing administrators to track the performance, health, and status of edge devices and applications in real-time.
- *Edge connectivity*: Edge computing relies on connectivity technologies to establish communication between edge devices and the central infrastructure. This includes wired connections like Ethernet, wireless protocols such as Bluetooth, Zigbee, Wi-Fi, and cellular networks like 4G LTE and 5G.
- *Edge development frameworks*: Frameworks like Apache Edgent, Eclipse ioFog, and AWS IoT SDK provide software development kits and APIs for building edge applications. These frameworks simplify the development process and offer abstractions for interacting with edge devices, sensors, and data streams.
- *Edge database systems*: Edge computing often requires local storage and databases to store and process data closer to the edge. Tools like SQLite, Apache Cassandra, and Amazon DynamoDB provide lightweight and scalable database solutions suitable for edge environments.

Table 1.2 shows technologies, platforms, and tools used in edge computing which can vary depending on the requirements, use cases, and the underlying infrastructure of the edge computing deployment.

2. *AI and ML*: As the large amount of data generated by IoT devices and WSNs grows, there is a growing need for AI and ML algorithms to analyze and understand this data. AI and ML techniques are being integrated into IoT and WSN systems to enable intelligent decision-making and advanced

TABLE 1.2

Edge Computing in IoT and WSN

	Technologies, Platforms, and Tools	Purpose
Edge Computing	Google Cloud IoT Edge, AWS IoT Greengrass, Microsoft Azure IoT Edge	Computing platforms provide the infrastructure and software framework for managing and orchestrating edge devices and applications.
	Containerization technologies like Docker and Kubernetes	Containers enable the packaging and isolation of applications and their dependencies, making it easier to deploy, manage, and scale edge applications across different edge devices and locations.
	Apache NiFi, Apache Flink, TensorFlow Lite, and ONNX Runtime	Enable running analytics and machine learning models on edge devices with limited resources.
	Istio, Envoy, and Nginx	Help secure edge environments by providing traffic management, authentication, authorization, and encryption capabilities.
	Apache Kafka, Apache Pulsar, and RabbitMQ	Ensure efficient data transfer, event-driven processing, and data synchronization.
	Bluetooth, Zigbee, Wi-Fi, and cellular networks like 4G LTE and 5G	Establish communication between edge devices and the central infrastructure.
	SQLite, Apache Cassandra, and Amazon DynamoDB	Provide lightweight and scalable database solutions suitable for edge environments.

analytics. These technologies help in extracting valuable insights from large-scale sensor data and are being used to optimize resource allocation and processes, detect anomalies, and provide predictive maintenance (Figure 1.18). Here are some key considerations of AI and ML in IoT and WSN, along with related tools and technologies: [12, 16]

- *Predictive analytics*: AI and ML algorithms can analyze sensor data collected from IoT devices and WSN nodes to make predictions and forecasts. This can help in detecting anomalies, predicting failures, optimizing resource allocation, and improving overall system performance. Tools such as TensorFlow, PyTorch, and scikit-learn can be used for developing and deploying ML models for predictive analytics in IoT and WSN [25]. Similarly, Microsoft Azure IoT Edge, and NVIDIA Jetson platforms provide capabilities for running ML models at the edge.

- *Anomaly detection*: ML algorithms can learn normal patterns and behaviors from sensor data and identify anomalies or outliers. This is valuable in detecting potential security breaches, equipment malfunctions, or environmental abnormalities in IoT and WSN deployments.

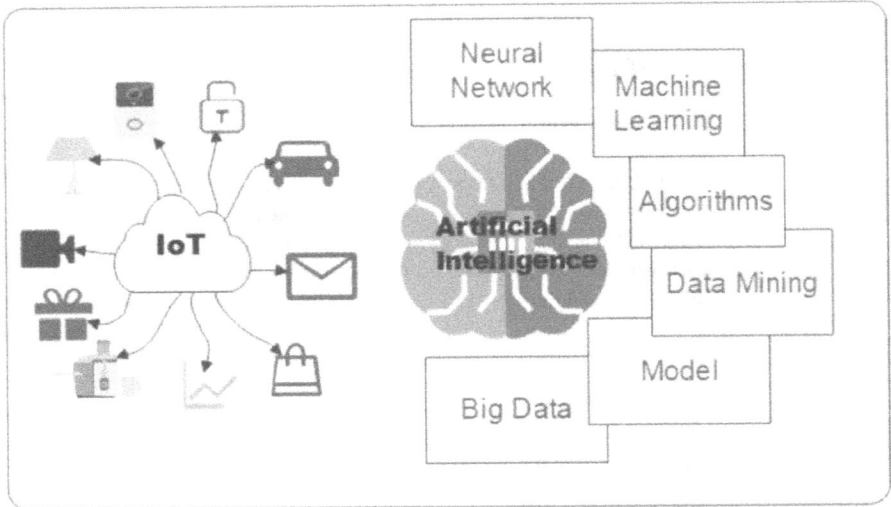

FIGURE 1.18 AI and ML.

Techniques like unsupervised learning, clustering, and statistical analysis can be employed for anomaly detection. Tools like K-means, Isolation Forest, and One-Class SVM are commonly used for anomaly detection [25].

- *Energy optimization*: Energy efficiency is a major concern in WSN and IoT deployments, as sensor nodes often have limited power resources. ML algorithms can be utilized to optimize energy consumption by intelligently scheduling sensing activities, data transmission, and sleep modes. Reinforcement learning and optimization algorithms can be applied to dynamically manage the energy usage of sensor nodes. Tools such as TinyML, Energy-Aware Computing Frameworks (EACOF), and reinforcement learning (RL) libraries can aid in energy optimization [25].
- *Data fusion and integration*: IoT and WSN generate massive amounts of data from multiple sources and sensors. ML techniques like data fusion and integration can combine and analyze data from different sensors or IoT devices to extract meaningful insights and provide a comprehensive view of the system. Tools such as Apache Kafka, Apache Spark, and Apache Flink can assist in real-time data processing, stream analytics, and data fusion.
- *Security and privacy*: AI and ML can enhance security and privacy in IoT and WSN by detecting anomalies, identifying potential threats, and protecting data. ML algorithms can learn patterns of normal behavior and detect deviations or attacks in real-time. Tools such as intrusion detection systems (IDS), anomaly detection algorithms, and encryption techniques can be employed for security and privacy in IoT and WSN.

These examples above show how AI and ML can be applied in the context of IoT and WSN. The choice of tools and technologies depends on
the specific requirements, resources, and constraints of the IoT and WSN
system being developed.

3. *5G Networks*: The introduction of 5G networks is expected to significantly
   improve the performance and reliability of IoT and WSN applications. 5G
   networks offer faster data transfer rates, lower latency, and better connectivity, which will enable more complex and advanced IoT and WSN use
   cases. 5G enables real-time communication, supports massive device connectivity, and facilitates high-bandwidth data transfer, allowing for more
   efficient and reliable IoT and WSN deployments. The ongoing development
   of 6G networks is also exploring further advancements in wireless connectivity for IoT and WSN.

   When it comes to the implementation of 5G in IoT and WSN, there are
   several tools and technologies involved [16, 26, 27]. Here are some key
   elements:

   • *Narrowband IoT (NB-IoT) and long-term evolution for machines (LTE-
     M)*: NB-IoT and LTE-M are low-power, wide-area network technolo-
     gies that operate on existing 4G and 5G networks. They are designed
     specifically for IoT devices that have low data rate and power consump-
     tion requirements, making them suitable for applications such as smart
     metering, asset tracking, and agricultural monitoring.
   • *Network slicing*: Network slicing is a technique in 5G that enables the
     creation of virtual, independent networks on a shared physical infra-
     structure. Each network slice is optimized for specific IoT and WSN
     applications, enabling the allocation of dedicated resources and cus-
     tomized services based on their unique requirements.
   • *Software-defined networking (SDN) and network function virtualiza-
     tion (NFV)*: SDN and NFV technologies are used to optimize and man-
     age the 5G network infrastructure. SDN separates the control plane
     from the data plane, allowing for more flexible and programmable net-
     work management. NFV virtualizes network functions, enabling the
     deployment and scaling of network services more efficiently.
   • *Massive machine-type communications (mMTC)*: 5G provides dedi-
     cated features for mMTC, addressing the connectivity requirements of
     a large number of low-power IoT devices. It enables efficient handling
     of sporadic and infrequent data transmission from IoT sensors, optimiz-
     ing network resources and reducing congestion.
   • *IoT platforms*: IoT platforms provide the middleware and tools neces-
     sary to develop, deploy, and manage IoT and WSN applications, which
     offer functions such as device management, data analytics, and applica-
     tion development interfaces, facilitating the integration and interaction
     between IoT devices and 5G networks.
   • *Security solutions*: With the increased connectivity and data exchange
     in 5G-enabled IoT and WSN systems, robust security solutions are cru-
     cial. This includes authentication mechanisms, encryption protocols,

access control, and intrusion detection systems to protect the devices, data, and communications from potential threats and vulnerabilities.

- *Data analytics and artificial intelligence (AI)*: Data analytics and AI technologies play a crucial role in extracting meaningful insights from the huge amount of data generated by IoT and WSN devices. With 5G, real-time data processing and analysis at the edge of the network, combined with cloud-based analytics, enable enhanced decision-making and predictive capabilities.
- *Protocols and standards*: Various protocols and standards are utilized to enable the interoperability and compatibility of IoT and WSN devices with 5G networks. Protocols like MQTT (Message Queuing Telemetry Transport) and CoAP (Constrained Application Protocol) are commonly used for lightweight and efficient data transmission, while standards such as 3GPP (Third Generation Partnership Project) define the technical specifications for 5G networks.
- *Testing and simulation tools*: Testing and simulation tools specific to IoT and WSN are crucial for verifying the performance, reliability, and compatibility of devices and applications in 5G networks. These tools simulate different network conditions, evaluate the scalability of the IoT ecosystem, and assess the overall quality of service.

These tools and technologies, combined with the capabilities of 5G networks, form a comprehensive ecosystem that enables the seamless integration and efficient operation of IoT and WSN applications, unlocking their full potential in various industries and domains.

4. *Blockchain*: Blockchain technology is being explored as a means of providing secure and decentralized communication and data storage for IoT and WSN applications. Blockchain can provide a decentralized and tamper-proof ledger for recording and verifying transactions between IoT devices and WSN nodes. It ensures the integrity and authenticity of data exchanged, makes it tough for malicious attackers to tamper with or modify the data. It can be used to observe and confirm the source and history of data in IoT and WSN systems. This enables data consumers to verify the integrity and reliability of the data collected by sensors, ensuring its accuracy and trustworthiness. This technology has the potential to enhance security and privacy in these systems and improve data integrity. Smart contracts are self-regulating agreements that are recorded on the blockchain. They can facilitate automated and secure interactions between IoT devices and WSN nodes, enabling autonomous transactions and processes without the need for intermediaries. It enables real-time tracking of goods, ensures the authenticity of products, and enhances trust among participants in the supply chain [13, 16].

Some of the key technologies and tools associated with the integration of blockchain in IoT and WSN are as follows: [28]

- *Distributed ledger technology (DLT)*: Blockchain is a type of DLT that allows for decentralized and transparent record-keeping. It enables multiple participants in an IoT or WSN network to maintain a shared and synchronized ledger of transactions, events, or sensor data. Each block in the chain contains a timestamped and cryptographically secured record, ensuring data immutability.

- *Smart contracts*: These are self-regulating agreements with pre-defined guidelines and conditions encoded into the blockchain. In IoT and WSN, smart contracts can automate interactions and transactions between devices and entities. For instance, a smart contract can automatically trigger a payment when certain conditions are met or enable direct device-to-device communication without intermediaries.
- *Consensus mechanisms*: This mechanism guarantees that all blockchain network participants concur on transaction validity and ledger status. Some commonly used mechanisms include Proof of Work (PoW), Practical Byzantine Fault Tolerance (PBFT), and Proof of Stake. These mechanisms establish trust and prevent malicious activities in a decentralized manner.
- *Identity and access management*: Blockchain can provide decentralized and secure identity management for IoT and WSN devices. It allows devices to have unique, tamper-proof identities on the blockchain, ensuring secure authentication, access control, and data privacy. Decentralized identity solutions like Self-Sovereign Identity (SSI) enable users and devices to have full control over their identities and data.
- *Data integrity and provenance*: Blockchain can guarantee the integrity and provenance of data generated by IoT and WSN devices. By recording data transactions on the blockchain, it becomes tamperproof and verifiable. This helps in validating the authenticity and origin of data, ensuring trustworthiness in applications such as supply chain management, environmental monitoring, and asset tracking.
- *Interoperability and standards*: Interoperability frameworks and standards play an important role in integrating blockchain with IoT and WSN systems. These include protocols such as InterPlanetary File System (IPFS), Hyperledger Fabric, and Ethereum that provide tools and frameworks for developing blockchain-based IoT applications. Initiatives like the Trusted IoT Alliance and the Industrial Internet Consortium work towards establishing interoperability standards and best practices.
- *Decentralized data marketplaces*: Blockchain-based data marketplaces enable the secure and efficient exchange of IoT and WSN data. These marketplaces allow data owners to monetize their data while maintaining control and privacy. Blockchain ensures transparent transactions, data ownership, and fair compensation mechanisms among participants in the marketplace.
- *Blockchain-as-a-service (BaaS)*: BaaS platforms provide pre-configured blockchain infrastructure and services, simplifying the development and deployment of blockchain applications in IoT and WSN. These platforms offer features such as data storage, smart contract deployment, and integration with other cloud services, enabling organizations to leverage blockchain technology without extensive knowledge or resources.

By integrating blockchain technology into IoT and WSN, organizations can enhance security, privacy, and trust in their deployments. The combination of DLT, smart contracts, consensus mechanisms, identity management, and data integrity ensures a reliable and decentralized framework for managing and exchanging data in the interconnected world of IoT and WSN.

There are several tools and platforms available to facilitate the integration. Here are some tools commonly used [28]. Table 1.3 shows tools used in blockchain and their functionalities.

- *Ethereum*: This is a popular blockchain platform that supports the development of decentralized applications (DApps) and smart contracts. It provides a robust and mature ecosystem for building blockchain-based IoT and WSN solutions. Ethereum offers tools like Solidity

TABLE 1.3
Tools for Blockchain

	Tools	Functions
	Ethereum	Popular lockchain platform that supports the development of decentralized applications (DApps) and smart contracts. It provides a robust and mature ecosystem for building blockchain-based IoT and WSN solutions. It offers tools like Solidity web3.js, and Truffle.
	Hyperledger Fabric	Offers tools and libraries for developing and deploying blockchain applications, such as Hyperledger Composer, Hyperledger Caliper, and Hyperledger Explorer.
	IOTA	It utilizes a Directed Acyclic Graph (DAG) structure called the Tangle instead of a traditional blockchain. IOTA focuses on enabling secure, scalable, and feeless micro-transactions between IoT devices. The IOTA Foundation provides libraries, APIs, and tools to develop applications and manage IOTA-based transactions in IoT and WSN.
Blockchain	R3 Corda	Offers tools and libraries for developing distributed applications, managing identities, and building smart contracts.
	MultiChain	Designed for enterprise applications and offers features such as asset issuance, data streams, and enhanced privacy controls. It provides APIs and command-line tools for building custom blockchain solutions for IoT and WSN deployments.
	IBM Blockchain	Provides development frameworks, deployment automation, and integration with other IBM services for building end-to-end blockchain solutions.
	Microsoft Azure Blockchain	Blockchain provides tools, templates, and preconfigured networks for building IoT and WSN solutions on various blockchain platforms, including Ethereum, Hyperledger Fabric, and Corda.

(a smart contract programming language), web3.js (JavaScript library for interacting with Ethereum), and Truffle (a development framework for Ethereum-based projects).

- *Hyperledger Fabric*: An open-source platform hosted by the Linux Foundation. It is designed for enterprise applications, including IoT and WSN. Hyperledger Fabric provides a permissioned blockchain network, allowing participants to have controlled access and privacy. It offers tools and libraries for developing and deploying blockchain applications, such as Hyperledger Composer, Hyperledger Caliper, and Hyperledger Explorer.
- *IOTA*: This is a distributed ledger technology specifically designed for the IoT ecosystem. It utilizes a Directed Acyclic Graph (DAG) structure called the Tangle instead of a traditional blockchain. IOTA focuses on enabling secure, scalable, and feeless micro-transactions between IoT devices. The IOTA Foundation provides libraries, APIs, and tools to develop applications and manage IOTA-based transactions in IoT and WSN.
- *R3 Corda*: This is a blockchain platform designed for building enterprise solutions, including IoT and WSN applications. It emphasizes privacy and scalability and allows for interoperability with existing systems. R3 Corda offers tools and libraries for developing distributed applications, managing identities, and building smart contracts.
- *MultiChain*: This is an open-source blockchain environment that enables the generation and deployment of private, permissioned blockchains. It is designed for enterprise applications and offers features such as asset issuance, data streams, and enhanced privacy controls. MultiChain provides APIs and command-line tools for building custom blockchain solutions for IoT and WSN deployments.
- *IBM Blockchain*: This platform is an enterprise-grade solution for developing, deploying, and managing blockchain networks. It offers tools and services for creating permissioned blockchain networks, including IoT and WSN applications. IBM Blockchain Platform provides development frameworks, deployment automation, and integration with other IBM services for building end-to-end blockchain solutions.
- *Microsoft Azure*: This offers a suite of blockchain services that facilitate the development and deployment of blockchain applications. Azure Blockchain provides tools, templates, and preconfigured networks for building IoT and WSN solutions on various blockchain platforms, including Ethereum, Hyperledger Fabric, and Corda.

These tools and platforms provide developers with the necessary frameworks, libraries, and resources to leverage blockchain technology in IoT and WSN applications. They streamline the development process and offer the flexibility to choose the most suitable blockchain solution based on the specific requirements of the project.

5. *Sustainability and energy efficiency*: Energy efficiency is a critical aspect of IoT and WSN deployments, especially for battery-powered devices. There

is a growing focus on developing energy-efficient sensors, communication protocols, and power management techniques to prolong the lifespan of IoT devices and reduce environmental impact.

6. *Energy harvesting*: Energy harvesting technologies are being developed to power IoT and WSN devices using ambient energy sources such as solar, wind, and vibration. This will enable the deployment of low-power devices that can operate for extended periods without the need for battery replacements.

7. *Interoperability and standardization*: Interoperability among different IoT devices and WSN platforms remains a challenge. Standardization efforts are underway to establish common protocols and frameworks, such as MQTT, CoAP, and OMA LwM2M, to ensure seamless communication and interoperability between devices from different vendors and across diverse IoT ecosystems.

These trends reflect the continuous evolution and advancement of IoT and WSN are continuing to advance and offer new and innovative solutions to a wide range of industries and use cases.

1.1.5.2   Current Challenges in IoT and WSN

Despite the many benefits and opportunities that IoT and WSN offer, there are several challenges that must be mentioned so their full potential can be understood [5, 29, 30]. Here are some current challenges in IoT and WSN:

1. *Security*: One of the biggest challenges facing IoT and WSN is security. With an increasing count of devices are connected to the Internet, the risk of cyberattacks increases. The challenge is to ensure that these devices and networks are secure against various types of cyber threats, such as hacking, malware, and unauthorized access [31].

2. *Interoperability*: Another challenge in IoT and WSN is interoperability. The devices and sensors have their own protocols and communication standards. This can make it difficult to integrate different devices and systems and can lead to inefficiencies and compatibility issues.

3. *Power management*: Many of the IoT and WSN devices are powered by batteries, that have limited life span and need to be replaced frequently. The challenge is to develop more efficient power management systems that can extend the lifespan of batteries, reduce power consumption, and enable devices to operate in remote and harsh environments.

4. *Data management*: IoT and WSN generate vast amounts of data, which can be difficult to manage, process, and analyze. The challenge is to develop efficient and scalable data management systems that can store, process, and analyze data in real-time, and provide insights that can be used to optimize processes and improve decision-making.

5. *Privacy*: As more data is collected and analyzed in IoT and WSN, privacy concerns become more prominent. The challenge is to ensure that personal data is collected and used in an ethical manner and that appropriate safeguards are in place to protect individual privacy and prevent unauthorized access.

1.1.6   SECURITY AND PRIVACY REGARDING IoT AND WSN

Security and privacy are critical issues in IoT and WSN due to the massive amounts of data generated and transmitted by these systems. Security is the biggest concern in IoT and WSN, as these systems are more vulnerable to different kinds of cyberattacks [32, 33]. Here are some key security and privacy challenges in IoT and WSN, and some measures that can be taken to address them:

1. *Device security*: One of the primary concerns in IoT and WSN is securing the devices themselves against unauthorized access and hacking. Measures to address device security can include the use of strong authentication and encryption protocols, regular software updates, and secure boot processes. Authentication protocols can be used to ensure that devices and users are who they claim to be. Authentication measures can include the use of strong passwords, two-factor authentication, and biometric authentication. Keeping your software and firmware up to date is critical to fixing vulnerabilities and reducing the risk of attacks. IoT and WSN devices and systems should be designed to allow for secure and automatic updates.

2. *Data security*: Another critical aspect of IoT and WSN security is ensuring the confidentiality, integrity, and availability of the data generated and transmitted by these systems. This can be achieved through encryption, access controls, and data backup and recovery processes.

   Data encryption is critical to protecting the confidentiality of data transmitted over IoT and WSN. Encryption can be used to protect data transmitted over IoT and WSN by encoding it in such a manner so that can only be read by authorized recipients. Encryption can include both symmetric and asymmetric encryption algorithms and can be applied at different levels of the network stack.

   Controlling access to IoT and WSN devices and systems is essential to preventing unauthorized access and misuse. The access control protocols can be used to restrict access to IoT and WSN devices and systems to authorized users only. These controls can include role-based access control, attribute-based access control, and access control lists.

3. *Network security*: IoT and WSN devices and systems are vulnerable to attacks at the network level, including man-in-the-middle attacks, denial-of-service attacks, and other network-based attacks. Measures must be taken to improve or boost the network security, that can include firewalls, intrusion detection and prevention systems, and regular network scans and vulnerability assessments.

4. *Threat detection and response*: Monitoring for threats and responding to them in a timely manner is important in securing IoT and WSN. This may include the deployment of security information and event management (SIEM) systems, intrusion detection and prevention systems, and incident response plans.

5. *Physical security*: Physical security measures, such as tamper-evident packaging, environmental controls, and secure storage, can help to protect IoT and WSN devices and systems from physical attacks.

6. *Privacy*: As IoT and WSN generate vast amounts of data, protecting the privacy of individuals and organizations becomes increasingly important. This can be achieved through the use of anonymization techniques, data minimization, and strong data protection laws and regulations. *Data minimization*: Collecting only the necessary data is essential to protecting privacy in IoT and WSN. Collecting too much data can increase the risk of data breaches and misuse. Therefore, data minimization techniques such as data filtering and aggregation can be used to reduce the amount of personal data collected. *Anonymization*: Anonymizing data can protect privacy by removing or obscuring personal identifiers from data. Anonymization can be achieved through techniques such as data masking, pseudonymization, and differential privacy.

7. *Privacy policies*: Clear and concise privacy policies should be in place that describe what data is collected, how it is used, and who has access to it. The policies should be easily understandable by users and transparent about data practices.

8. *User consent*: Obtaining explicit user consent before personal data is collected, used, or shared is critical to protecting privacy in IoT and WSN. Consent should be freely given, specific, informed, and unambiguous.

9. *Data breach response*: Developing and testing data breach response plans is important to minimize the impact of any data breaches that may occur. A response plan should include clear communication protocols, a crisis management team, and procedures for containing and investigating data breaches.

10. *Governance*: Effective governance and management of IoT and WSN security and privacy is essential to ensure that all stakeholders are aware of their roles and responsibilities in securing these systems. This can include the establishment of security policies and procedures, regular training and awareness programs, and continuous monitoring and evaluation of security and privacy risks.

Therefore, securing and protecting privacy in IoT and WSN requires a multilayered considerable approach that includes technical, organizational, and regulatory measures. By implementing these measures, it is possible to ensure that IoT and WSN technology is used safely and securely, and that privacy rights are protected for all stakeholders.

1.1.7 CONCLUSION

Both the IoT and WSNs are two related technologies that are transforming our lifestyle, work and communicate. IoT is a grid of networks of various objects, vehicles, buildings, and other objects with sensors, software, and connectivity that enable the collection and sharing of data [34]. WSN is a type of IoT that is specifically focused on the use of wireless sensors for data collection. Both IoT and WSN have the potential to drive innovation, increase efficiency, and improve quality of life in various industries, including healthcare, transportation, manufacturing, and energy. However, they also face several challenges and concerns, including security, privacy, interoperability, and scalability. To address these challenges, various measures can be taken, such as implementing strong security and privacy protocols, ensuring interoperability between devices and systems, and using standardized protocols and data formats and the future of IoT and WSN is promising, and as these technologies continue to evolve and mature, they are likely to play an increasingly important role in our daily lives. Sustainable IoT enables smart cities, energy-efficient buildings, intelligent transportation, and other applications that contribute to a more sustainable and resilient future.

REFERENCES

1. Sfar, A. R., Chtourou, Z., & Challal, Y. (2017, February). "A systemic and cognitive vision for IoT security: A case study of military live simulation and security challenges." In *2017 International Conference on Smart, Monitored and Controlled Cities (SM2C)* (pp. 101–105). IEEE.
2. https://www.statista.com/statistics/471264/iot-number-of-connected-devices-worldwide/#:~:text=By%202025%2C%20forecasts%20suggest%20that%20there%20will%20be,increase%20from%20the%20IoT%20installed%20base%20in%202019.
3. Internet of Things (IoT) Characteristics (linkedin.com).
4. Asir, T. R. G., et al. (2016). "IoT as a service." *International Conference on Innovations in information, Embedded and Communication Systems (ICIIECS)*. Vol. 3 (pp. 1093–1096).
5. Patel, K. K., et al. (2016). "Internet of things-IoT: Definition, characteristics, architecture, enabling technologies, application & future challenges." *International Journal of Engineering Science and Computing*, 6(5), 6122–6131.
6. https://www.elprocus.com/introduction-to-wireless-sensor-networks-types-and-applications/
7. Zhu, Q., et al. (2010). "IoT Gateway: Bridging Wireless Sensor Networks into Internet of Things." *IEEE/IFIP International Conference on Embedded and Ubiquitous Computing*, 347–352.
8. https://www.intechopen.com/chapters/76818
9. https://www.electronicshub.org/wireless-sensor-networks-wsn/
10. https://microcontrollerslab.com/wireless-sensor-networks-wsn-applications/
11. Samad, F., Abbasi, A., Memon, Z.A., Aziz, A., & Rahman, A. (2018). "The future of internet: IPv6 fulfilling the routing needs in internet of things." *International Journal of Future Generation Communication and Networking*, 11(1), 13–22.
12. Ruiz, M., Alvarez, E., Serrano, A., & Garcia, E (2016). "The convergence between Wireless Sensor Networks and the internet of things; Challenges and perspectives: A survey." *IEEE Latin America Transactions*, 14(10), 4249–4254.
13. Poncha, L. J., Abdelhamid, S., Alturjman, S., Ever, E., & Al-Turjman, F. (2018, May). "5G in a convergent Internet of Things era: An overview." In *2018 IEEE International Conference on Communications Workshops (ICC Workshops)* (pp. 1–6). IEEE.
14. Bajaj, K., Sharma, B., & Singh, R. (2020). "Integration of WSN with IoT applications: a vision, architecture, and future challenges." Integration of WSN and IoT for Smart Cities, 79–102.
15. https://www.forbes.com/sites/chrissamcfarlane/2019/10/18/are-smart-cities-the-pathway-to-blockchain-and-cryptocurrency-adoption/?sh=1d990f204609
16. Singh, S., Sharma, P. K., Yoon, B., Shojafar, M., Cho, G. H., & Ra, I. H. (2020). "Convergence of blockchain and artificial intelligence in IoT network for the sustainable smart city." *Sustainable Cities and Society*, 63, 102364..

17. Jain, R. (2018). "A congestion control system based on VANET for small length roads." *Annals of Emerging Technologies in Computing (AETiC)*, 2(1), 17–21, DOI: 10.33166/AETiC.2018.01.003.

18. Soomro, S., Miraz, M. H., Prasanth, A., & Abdullah, M., (2018). "Artificial intelligence enabled IoT: Traffic congestion reduction in smart cities." *IET 2018 Smart Cities Symposium (SCS '18)*, 22–23 April 2018, pp. 81–86, Published by IET Digital Library, Available at https://digital-library.theiet.org/content/conferences//10.1049/cp.2018.1381.

19. Mahmud, S. H., Assan, L., & Islam, R. (2018). "Potentials of internet of things (IoT) in Malaysian construction industry." *Annals of Emerging Technologies in Computing (AETiC)*, 2(1), 44–52. DOI: 10.33166/AETiC.2018.04.004.

20. Zanjal, S. V., & Talmale, G. R. (2016). "Medicine reminder and monitoring system for secure health using IoT." *Procedia Computer Science*, 78, 471–476..

21. Mano, Y., Faical, B. S., Nakamura, L., Gomes, P., Libralon, G, Meneguete, R., Filho, G., Giancristofaro, G., Pessin, G., Krishnamachari, B., & Ueyama, J. (2015). Exploiting IoT technologies for enhancing Health Smart Homes through patient identification and emotion recognition. *Computer Communications*, 89–90, 178–190. DOI: 10.1016/j.comcom.2016.03.010.

22. Sundareswaran, V., & Null, M. S. (20). "Survey on smart agriculture using IoT." *International Journal of Innovative Research in Engineering & Management (IJIREM)*, 5(2), 63–66.

23. https://www.techrepublic.com/article/top-edge-computing-platforms/

24. https://github.com/qijianpeng/awesome-edge-computing

25. https://www.simplilearn.com/best-machine-learning-tools-article

26. https://www.rfpage.com/what-are-the-components-of-5g-technology/

27. https://github.com/calee0219/awesome-5g

28. https://www.techtarget.com/searchcio/feature/Top-9-blockchain-platforms-to-consider

29. Saleh, M. M. (2020). "WSNs and IoT their challenges and applications for healthcare and agriculture: A survey." *Iraqi Journal for Electrical & Electronic Engineering*. The 3rd Scientific Conference of Electrical and Electronic Engineering Researches (SCEEER) | (15–16) June 2020 | BASRAH / IRAQ, (pp. 37–43), DOI: 10.37917/ijeee.sceeer.3rd.6

30. Majid, M., Habib, S., Javed, A. R., Rizwan, M., Srivastava, G., Gadekallu, T. R., & Lin, J. C. W. (2022). "Applications of Wireless Sensor Networks and Internet of Things frameworks in the industry revolution 4.0: A systematic literature review." *Sensors*, 22(6), 2087.

31. Gulati, K., Boddu, R. S. K., Kapila, D., Bangare, S. L., Chandnani, N., & Saravanan, G. (2022). "A review paper on Wireless Sensor Network techniques in internet of things (IoT)." *Materials Today: Proceedings*, 51, 161–165.

32. Sasirekha, S. P., et al. (2020). "Data processing and management in IoT and Wireless Sensor Network." *Journal of Physics: Conference Series*, 1712(1), 1–8.

33. Maphats'oe, T., & Masinde, M. (2016). "A security algorithm for Wireless Sensor Networks in the Internet of Things paradigm." *2016 IST-Africa Week Conference* (pp. 1–10). IEEE. DOI: 10.1109/ISTAFRICA.2016.7530646.

34. Dave Evans. (2011). "The internet of things: How the next evolution of the internet is changing everything." Cisco Internet Business Solutions Group (IBSG), Cisco Systems, Inc., San Jose, CA, USA, White Paper.

2 The Role of Sustainable IoT in Present-Day Life

Parul Saini, Lipika Gupta,
Shivani Malhotra, and Imali Dias

2.1 INTRODUCTION

The Internet of Things (IoT) has caused an important paradigm shift. IoT has made it possible for linked devices to simultaneously communicate with one another and the operator about their current state. The deployment of IoT across various sectors and domains is driven by the improved analytical functionalities, fast response times, and real-time monitoring. COVID-19 has fuelled the adoption of IoT in every sector, especially healthcare [1]. This chapter focuses on what IoT is, why it is important, applications of IoT in different fields for sustainable development of society, how IoT devices are being made sustainable, challenges in the deployment of IoT, and what the future of IoT looks like. This chapter contains data from research papers, articles, and reports from authenticated online sources.

There are plenty of definitions out there as it depends on an individual's perspective. The most basic definition of IoT is as follows: It describes the interconnection of various physical things communicating through networks without any human mediation. These objects can range from simple sensors to complex machines, and they can be found in a wide variety of settings. IoT makes devices such as TVs, refrigerators, washing machines, doorbells, etc. smarter by allowing them to communicate with people and other IoT-enabled things. IoT as a commercial field is thriving with an estimate of 30 billion IoT connections by the end of 2025 [2].

A smart home is the simplest example of an IoT system. It involves integrating the household appliances, such as a smoke detector, water heater, temperature sensor, refrigerator, and AC, to share data with the end user over any smartphone app or voice assistant. For example, a smart home may have a thermostat that can control the room temperature according to the weather conditions and occupancy of the room, a smart lighting system, and smart security cameras that give real-time monitoring and alert the user in case of any risk. In the past, the Internet served as a medium to connect people but now the same Internet has made it possible for the objects and things to communicate with each other. For example, in the morning, when the alarm goes off, an IoT system can effortlessly do the daily tasks like opening the window blinds, starting the coffee machine, turning on the water heater, powering on the music system, etc. While all of this is impressive there is a great deal of behind-the-scenes activity that takes place to ensure the smooth operation of these tasks. A lot of components are involved to carry out the various operations for effective communication between different objects. These devices are:

- **General devices** – Within the context of information exchange these devices are the core components of the data hub utilizing either wired or wireless interfaces to establish connections. Classic examples of such devices include home appliances.
- **Sensing devices** – Equipped with sensors, these devices gather and share data with other devices within the IoT network. They capture data related to environmental or physical factors like temperature, moisture, gas, light, etc.

2.1.1 BASIC BUILDING BLOCKS OF AN IoT SYSTEM

An IoT system consists of various building blocks that perform essential functions like sensing, analysis, and management. Each block has its own unique features adding to the capabilities of the overall system. Figure 2.1 shows the various components used for designing an IoT system.

2.1.1.1 Sensor

The sensor is a key element of an IoT system and detects a certain physical amount and produces an output signal. Its primary function is to collect real-time data from the environment and transmit it to various other units for processing and analysis. Sensors are identified by their distinctive IP addresses since they serve as the primary interface for the enormous network of connected devices. Therefore, the sensors convert signals from one energy domain into an electrical signal. Examples of sensors are a water quality sensor, a gas sensor, a temperature sensor, and a moisture sensor.

2.1.1.2 Gateway

An IoT gateway sets the communication between various IoT devices. It can be quite different from application to application. Apart from bridging the communication gap, it facilitates the users to securely gather, process, and analyze the data. Examples of gateways are BLE, Wi-Fi, LoRaWAN, ZigBee, Bluetooth, and Ethernet.

2.1.1.3 Cloud Server

The vast amount of data generated by the connected devices is effectively managed by a cloud server. This provides a secure platform for all the connected devices to communicate with each other and with the cloud itself. It can quickly adapt to various changes ensuring that the IoT ecosystem remains stable and reliable.

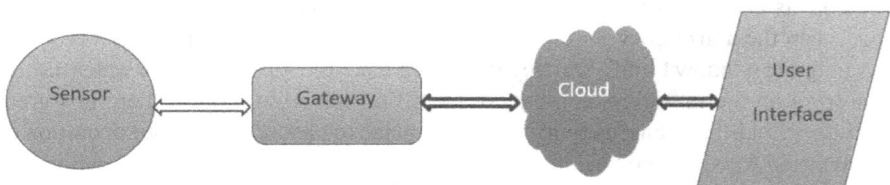

FIGURE 2.1 Basic architecture of an IoT system.

2.1.1.4   User Interface

It provides the means for users to interact with the IoT system and control the connected devices. It can be implemented in various forms such as mobile applications, web-based applications, and dedicated hardware interfaces. It can also enable users to control the devices, set thresholds for alerts, and configure various settings.

To summarize how IoT blocks work, the sensor collects real-time data, sends it to the processor unit, where it is analyzed and useful information is retrieved, and then the data is sent via the gateway to the cloud server. Finally, the user can access the data through a mobile phone or online application for practical use.

2.1.2   Why IoT Is Important

IoT enhances the way people live and work by enabling smart [3]. In recent years, the IoT has successfully caught the attention of those individuals and organizations who are not technologically inclined. Everyone is now recognizing and embracing its comfort, convenience, and the valuable insights it provides. IoT is emerging as a fundamental technology that is revolutionizing a wide range of industries and enhancing our quality of life.

With such an abundance of connected home devices, smart security cameras, remote door locks, and a wide range of appliances connected through applications, this is ample proof of the significant impact that IoT has on human beings.

Apart from its role in providing smart homes, IoT is immensely significant in businesses also. It gives real-time visibility into business operations, offering valuable insights such as the performance of machines and the efficiency of the supply chain and logistics operations. Between 2019 and 2026, the market size of IoT is expected to grow by 25.68%, ultimately reaching the market size of USD 1,319.08 billion by 2026 [4]. Thus, IoT holds immense importance as one of our key technologies and its relevance will continue to rise as businesses increasingly realize the competitive potential offered by IoT devices. The positive influence of IoT is very evident and it is highly likely that it will continue to contribute toward the well-being and progress in all aspects of life in the coming decades [5, 6]. Here are some key points to further understand it's importance:

a. *Better life quality*: IoT devices can help improve the quality of life by improving air quality and energy efficiency, creating better waste management, traffic control, and wearable smart devices, and many other examples.

b. *Automation of processes*: Automation IoT is the future of many businesses. It refers to implementing processes without human intervention. It is done to automate simple or repetitive tasks. It helps to increase efficiency, improve safety, reduce costs, reduce waste, improve building function, and to free people up for other complicated tasks.

c. *Data driven decisions*: The huge amount of data collected from the nearby surroundings through IoT sensors and devices can be analyzed to gain a comprehensive and accurate understanding of the current situation. Based on this data, actions can be taken which can lead to increased productivity in various sectors. Further, IoT devices can monitor and detect potential hazards such as

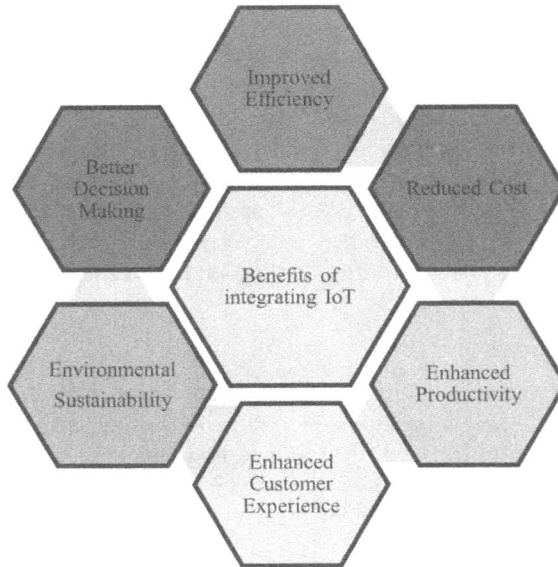

FIGURE 2.2   Benefits of IoT in every field.

gas leaks and triggered fire alarms and alert the relevant parties to take action in good time. With the inclusion of artificial intelligence in IoT systems, data driven decisions can further improve the corresponding actions.

Figure 2.2 shows the benefits of IoT in various fields like healthcare, smart city, agriculture, manufacturing industry, smart homes, and many more.

2.2   SUSTAINABLE IoT

Sustainability of IoT means manufacturing and operating the IoT devices in a way that minimizes the consumption of energy, lowers the carbon footprint, and uses more biodegradable materials [7]. It can also be considered as the long-term deployment of IoT devices without environmental impact. Basically, sustainable IoT embodies energy efficient procedures embraced by IoT devices to foster a sustainable and safer world. Figure 2.3 is an example of creating sustainable development in various sectors using natural solar energy as the source of power for different IoT nodes.

2.2.1   Significance of Sustainable IoT in Present-Day Life

Sustainable IoT has an important role in present-day life as it provides opportunities to address various issues related to the environment, hence promoting sustainable development. It involves the integration of IoT technologies and other principles with an emphasis on reducing environmental harm, conserving resources, improving people's lifestyles, and overall effectiveness. Sustainable IoT offers solutions to

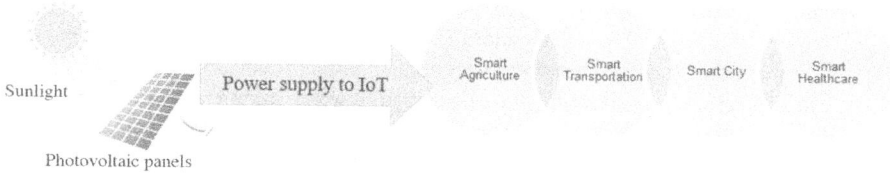

FIGURE 2.3   An illustration of sustainable IoT using solar energy to power IoT nodes.

various sectors such as smart agriculture, smart cities, transportation, waste management, smart healthcare, and smart homes, contributing to enhanced sustainability in our lives. By implementing these solutions, every individual or industry can make informed decisions and work toward reducing the carbon footprint for a sustainable future. But manufacturing, deployment, and proper disposal of IoT devices requires careful planning which will lead us toward a holistic approach to sustainability.

2.2.2   ROLE OF SUSTAINABILITY IN THE NEXT GENERATION

Sustainability plays a pivotal role in shaping the lives of next generation. As the world is facing numerous environmental and social challenges, switching to sustainable practices becomes essential to secure a better future for the coming generations. Poor human practices pose a great threat to nature, which is evident in various forms such as carbon emissions, floods, droughts, etc. The solution to all these problems is sustainable development of the whole of society. This means acquiring knowledge about green technology and adopting practices that prioritize the preservation and development of nature. Achieving sustainability for the coming generations is intertwined with economic gain, while simultaneously protecting our ecosystem. Here are some of the key roles of sustainability in the next generation:

- *Environmental stewardship*: Sustainability focuses on conserving and protecting the natural environment. The next generation needs to understand the importance of environmental stewardship and take actions to mitigate climate change, preserve natural resources, and protect biodiversity. By accepting sustainable practices, such as reducing greenhouse gas emissions, promoting renewable energy, and implementing responsible land and water management, the next generation can contribute to a better planet.
- *Education and awareness*: Sustainability education is essential for the next generation to understand the interconnection of environmental, social, and economic issues. By incorporating sustainability into educational curricula, young people can develop the knowledge, skills, and mindset necessary to tackle sustainability challenges efficiently. Building awareness about sustainable practices and promoting sustainable lifestyles from an early age can empower the future generation to make well-versed decisions and become responsible global citizens.

- *Innovation and technology*: The power of innovation and technology gives the opportunity to the next generation to harness the energy to drive sustainable solutions. Developments in areas such as renewable energy, clean transportation, sustainable agriculture, and waste management can play a substantial role in transitioning toward a more sustainable world. By encouraging research, development, and entrepreneurship in these areas, future generations can contribute to solving persistent sustainability issues.
- *Economic flexibility*: Sustainability stresses the need for an economic system that cares about long-term feasibility without compromising the environment or future generations' welfare. The future generations should adopt sustainable business practices, encourage circular economy models, and participate in renewable energy and clean technologies. In doing so, they can make a more robust and wealthy economy that benefits both people and the planet.
- *Social equity*: Sustainability recognizes the interconnection between environmental, economic, and social issues. The next generation should strive for a more equitable society where everyone has access to basic needs, equal opportunities, and social justice. The next generation can contribute toward a reasonable world by addressing challenges like poverty, social exclusion, and inequality.

2.2.2.1 What is Sustainable IoT Needed?

The need for sustainable IoT devices arises from the increasing use of IoT devices in our daily lives and their impact on the environment. IoT devices are being used in homes, offices, industries, and public places. However, it is important to consider the environmental impacts of using IoT devices and to ensure that they are used in a way that is sustainable and responsible [8]. Figure 2.4 illustrates some of the ways

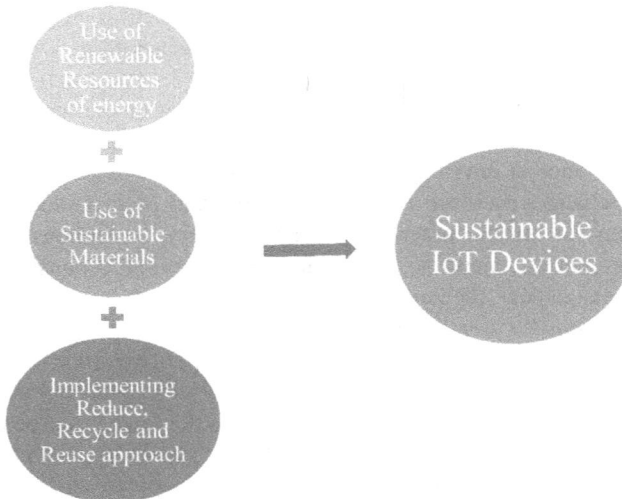

FIGURE 2.4   Major contributing factors of sustainable IoT devices.

which can lead to the development of sustainable IoT devices and systems, which in turn can contribute toward a sustainable society.

The major contributing factors for manufacturing sustainable IoT devices are explained in brief:

- *Use of renewable energy resources*: By harnessing energy from natural sources like solar, wind, thermal, and hydro, the emission of harmful gases into the environment could be reduced significantly and the burden on fossil fuel reserves could be lessened. The IoT systems and devices can be powered using these resources. Thus, relying on these resources, IoT devices can operate without creating toxicity and this contributes to sustainable IoT devices.
- *Use of sustainable materials*: Using sustainable materials for manufacturing IoT devices is one of the key aspects of a larger sustainability framework. A few examples of sustainable materials are natural fibers, recycled plastics, and bio-based plastics. These materials play a crucial role in making IoT devices sustainable through e-waste reduction, extended product life span, and reduced environmental impact.
- *Implementing three Rs approach*: The three Rs approach, i.e., Reduce, Reuse, and Recycle, plays an important role in achieving sustainability. This approach facilitates a circular economy where devices and resources are kept in use for a longer duration, materials are recovered and reused, and e-waste is minimized. Implementing this approach will create a more sustainable future by conserving resources, minimizing waste, and reducing energy consumption.

There are several reasons why sustainability is important in IoT devices:

Environmental impact: IoT devices are responsible for creating an adverse impact on the environment due to the resources used at various stages in manufacturing, disposal, and delivery. One of the worst environmental impacts is the generation of waste. Extraction of raw materials for the production of IoT devices has negative impacts on the environment in terms of pollution and habitat destruction.

Energy use: The second most important impact of IoT devices on the environment is the use of energy. Most IoT devices demand a substantial amount of energy for operation and charging [9]. Unfortunately, this energy comes from non-renewable resources of energy leading to greenhouse gas emissions and adverse climate change. So, in order to reduce the greenhouse effect, save energy, and reduce emissions, various approaches are needed to make IoT devices sustainable [10].

E-waste: IoT devices have a short life span and are required to be replaced frequently, which ultimately gives rise to huge amounts of electronic waste. To avoid this, efforts should be made to design IoT devices with long-life spans, thus reducing the need for frequent replacements and, hence, reducing e-waste. E-waste contains toxic materials like lead and mercury, which can pose a great risk to human health and the environment. So, for the sake of human well-being and to reduce their impact on environment, durable materials that can be upgraded or repaired should be used for manufacturing IoT devices [11].

The main problem in the deployment of IoT devices is the higher consumption of power and energy and excess emission of greenhouse gases. Hence, the need to adopt energy-efficient procedures to implement IoT in different areas for the purposes of reducing power consumption and greenhouse effects. The focus should be on developing energy-efficient devices that can operate for longer periods on a single battery charge. This can involve using low-power wireless communication protocols, optimizing software and hardware for energy efficiency, and using energy harvesting technologies that can generate power from sources such as solar, wind, or thermal energy. The following factors should be considered when designing an IoT device to make it more sustainable:

1. *Power optimization*: Power consumption is a major obstacle in deploying IoT devices for several reasons. While designing an IoT device, low-power consumption should taken into account so that the battery lasts longer, which in turn increases the life of an IoT device [12, 13]. Below are some points to consider when optimizing power consumption:
    a. *Implementing sleep modes*: Implementing sleep mode in IoT devices could save a huge amount of energy when they are not in use for longer periods of time. That's a great step toward power optimization. There are basically three sleep modes that an IoT device's processor supports – active, light sleep, and deep sleep. We can build IoT devices with these three modes, but while implementing this feature, we should bear in mind that the device should be able to return quickly to active mode.
    b. *Decreasing the number of transistors*: As most of the chips manufactured today incorporate a larger number of transistors than is actually required for specific tasks, there exists ample room for power optimization by optimizing transistor count. With the use of electronic design automation tools, we can reduce the number of transistors in a circuit by transforming the logical equations of a system. Fewer transistors means less power consumption. Another way of reducing power consumption is sizing transistors through the use of automatic sizing tools, also called cell selection [14].
    c. *Discarding various features to save battery life*: At the initial design stage, developers should consider all the features of an IoT product. They must learn how particular features impact power consumption, for example a colored display screen will consume more power compared to a black and white screen. Similarly, avoiding excessive and unwanted push notifications will increase the battery life. Push notifications should be generated for only for situations when urgent action is needed, or the end user should be given the flexibility to choose when to use this feature.

d. *Selecting the appropriate wireless protocol*: When selecting wireless protocols for sustainable IoT, it is essential to consider factors like energy efficiency, range, and reliability to ensure that the chosen protocol meets sustainability goals [15]. These protocols must be selected by keeping in mind where the application is being used; for example, the BLE protocol should be selected for short- to medium-range data transmission, which in turn will save power. Low power wide area network (LPWAN) can be used for a range up to 15 km, which can make battery-powered IoT devices last longer. Cellular and satellite-based wireless protocols can offer long-range connectivity, making them a good option for IoT devices deployed in remote locations.

e. *Wake-up radio (WUR) technology*: This technology from IEEE 802.11ba will significantly extend the battery life of devices and sensors used in IoT. This technology will place a small low-power radio in IoT devices that will tell the main radio to wake up and use power only when needed. This will significantly extend the battery life of the devices.

f. *Use of energy harvesting*: Energy harvesting techniques such as solar cells, piezoelectricity, or thermoelectricity can help generate power for IoT devices, reducing the dependence on batteries [16].

g. *Optimizing code*: The code can be optimized by using the right data types, using a smaller number of delays, writing simple code, and using a smaller number of instructions. All these features could help optimize power usage.

h. *Flexibility and durability*: The devices must be manufactured, installed, and shipped only once. After that, these devices should be able to adapt according to requirements or to be used flexibly for longer periods of time, which will automatically reduce the power consumption.

2. *Reducing the carbon footprint*: IoT devices can have a significant impact on carbon emissions, both during their manufacturing process and their use. The carbon footprint of IoT devices is primarily determined by their energy consumption and the materials used in their production. Manufacturers can focus on using more sustainable materials, improving production processes to minimize waste and energy use, and designing devices that are more energy-efficient. Major contributors to the carbon footprint are fuel combustion, water pollution, and emissions of harmful gases [17].

Figure 2.5 shows the contributing factors to the carbon footprint and how it can be reduced using sustainable approaches. The carbon footprint can be reduced to some extent using the following methods:

Switching to renewable resources of energy for electricity generation: Substitution of non-renewable resources of energy (coal, natural gas, fuel) with sustainable resources of energy such as wind, solar, and hydropower to generate electricity. This change will enable the production of electricity with minimal greenhouse gas emissions, thus contributing to mitigating the effects of climate change [18–20].

Changing lifestyle and habits: By making small lifestyle changes to help combat climate change, every human can contribute toward a healthier

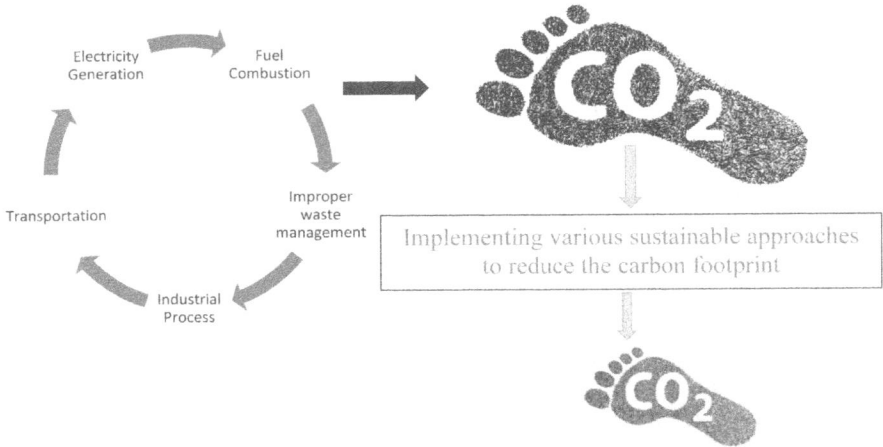

FIGURE 2.5 Reducing the carbon footprint using sustainable approaches.

planet by significant reduction in the carbon footprint. For example, by using reusable bags instead of single-use plastic bags, reducing meat consumption, using energy-efficient appliances, planting more trees, and using more public transport, we can contribute toward lowering the carbon footprint.

Reducing e-waste: Most electronics contain traces of silver, copper, and even gold that can be recycled to make something new instead of using valuable fossil fuels to mine brand-new metals. Furthermore, the plastic that houses the software and hardware can carry a serious carbon impact because its production requires an extensive amount of energy in the form of fossil fuels. Almost all electronics can be recycled so long as they have a battery. Though some have criticized recycling because of its own environmental impact, but in reality, this cost is outweighed by the energy and resources saved by not starting from scratch. By collaborating with the recycling companies, the e-waste can be recycled and reused, thus reducing the carbon footprint to a great extent. Waste to energy system (WTE) is another way to reduce the amount of waste that ends up in landfills, thus reducing the carbon footprint. This approach generates energy in a more sustainable and environmentally friendly way.

Reducing the size of RFID tags: This will decrease the amount of excess and toxic waste as these tags are hardly reusable as they are made up of non-biodegradable materials. Or we can do further research and find a biodegradable material that can be reused [21].

Use of sustainable materials for the manufacturing of IoT devices: Using sustainable materials and manufacturing processes will reduce the carbon footprint of IoT devices. As the demand for IoT devices is increasing day by day, the need to switch to green materials is also increasing: for example, PLA (Poly Lactic Acid) has unique mechanical and optical properties making it a good sensing material, but this material can't withstand high temperatures and is not suitable for electronic devices. This halts the process and restricts the use of this material for manufacturing the circuits. As a result, research trends focus on the use of biodegradable and eco-friendly materials for the manufacturing of IoT devices [22].

Energy efficient approaches: Several energy efficient approaches could be implemented in various fields to reduce the carbon footprint.

Industrial processes can improve their energy efficiency through the use of low-power components, optimized software, and efficient power management, hence reducing their carbon footprint [23]. Some of the key points for increasing energy efficiency are mentioned in Figure 2.6.

1. *Carbon capture utilization and storage (CCUS) technology*: The adoption of low-carbon technologies such as CCUS can play a crucial role in reducing carbon emissions generated by industrial processes. CCUS technologies capture carbon dioxide before it is released into the atmosphere and either store it underground or use it in other processes [24].

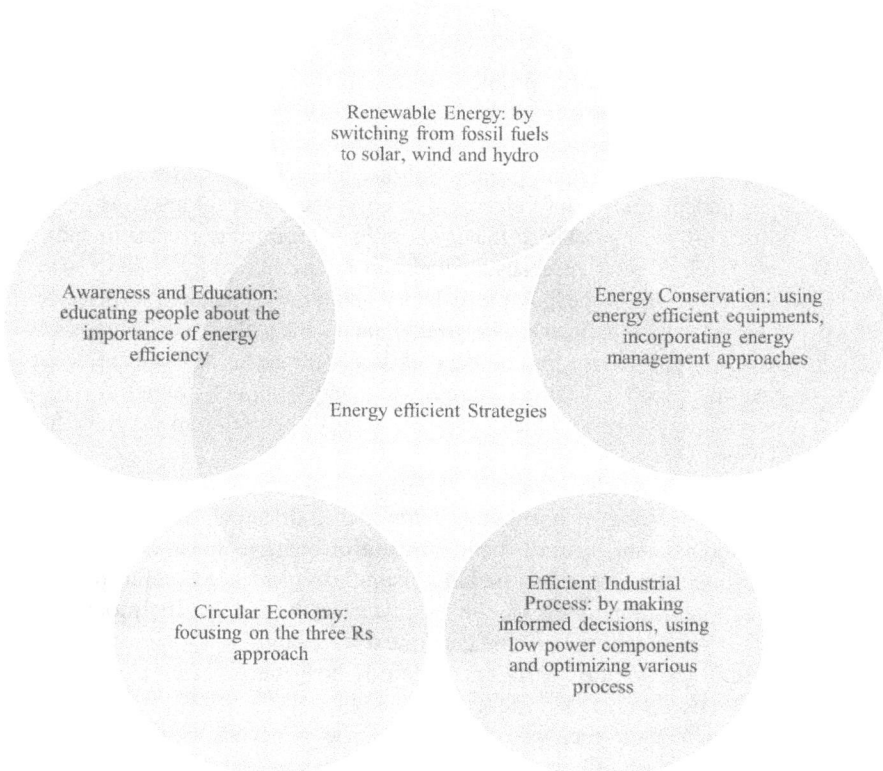

Renewable Energy: by switching from fossil fuels to solar, wind and hydro

Awareness and Education: educating people about the importance of energy efficiency

Energy Conservation: using energy efficient equipments, incorporating energy management approaches

Energy efficient Strategies

Circular Economy: focusing on the three Rs approach

Efficient Industrial Process: by making informed decisions, using low power components and optimizing various process

FIGURE 2.6 Key strategies for increasing energy efficiency.

2. *Transportation*: The transportation sector stands as one of the primary contributors to greenhouse gas emissions. Implementing active transportation such as cycling, walking, public transportation, and smart transportation systems like smart parking and smart traffic management, can help optimize transportation efficiency and reduce carbon emissions. In addition, more advanced technologies in transport should be used, such as hybrid or electric engines with lower greenhouse gas emissions, and lastly, we can buy locally produced foods because they are not transported over long distances.

3. *5G technology*: In the coming years, 5G-technology-enabled IoT sensors will reduce the carbon footprint to a larger extent, as predicted by some telecom consultants, because of the three core capabilities of this technology: low latency, faster data transfer speed, and massive device connectivity. As 5G technology is being deployed, there is an increased focus on its environmental footprint. This technology has the potential to contribute toward lowering the carbon footprint in several ways. Figure 2.7 shows various ways of implementing 5G technology to reduce the carbon footprint.

3. *Switching to battery- less power sources*: As we know, batteries are indeed the most common solution for powering the IoT devices, but this has a lot of disadvantages in terms of cost, production, toxicity, and e-waste. To overcome this problem, researchers are finding ways to move toward battery-less power sources by proposing different energy harvesting techniques[25].

   a. *Piezoelectric*: The main property of piezoelectric material is converting mechanical energy into electrical energy which can be used to power small IoT sensors. These materials offer a promising avenue for powering IoT devices in a sustainable and environmentally friendly way. The numerous types of piezoelectric materials include Piezoelectric Ceramics, Piezoelectric Polymers, and Piezoelectric Composites. Recently, Singapore researchers have developed a new piezoelectric material which is 40 times more flexible than competing materials. This new achievement could lead to better energy harvesting in wearable IoT devices.

   b. *Thermoelectric*: These materials offer a promising avenue for powering IoT devices, by harnessing temperature differences that are often overlooked and turning them into useful energy. Some examples of thermoelectric materials include bismuth telluride, lead telluride, and silicon-germanium alloys. These materials have high thermoelectric efficiency, meaning they can generate a large amount of electricity from a small temperature difference. Here is how thermoelectric materials can be used to harness energy in different ways:

      • *Waste heat recovery*: Thermoelectric materials can be used to recover waste heat from industrial processes, engines, or electronic devices and convert it into electricity to power IoT sensors and devices.

Optimizing energy
consumption by
leveraging real time
data

Optimized resource
utilization by allowing
a large number of IoT
devices to be
connected
simultaneously

How 5G technology is helpful in
lowering carbon footprint

Reducing the need for
physical hardware by
virtualizing network
functions and moving
them to cloud

Reducing distance
between devices
and base station as
this technology
requires denser
infrastructure with
smaller cells.

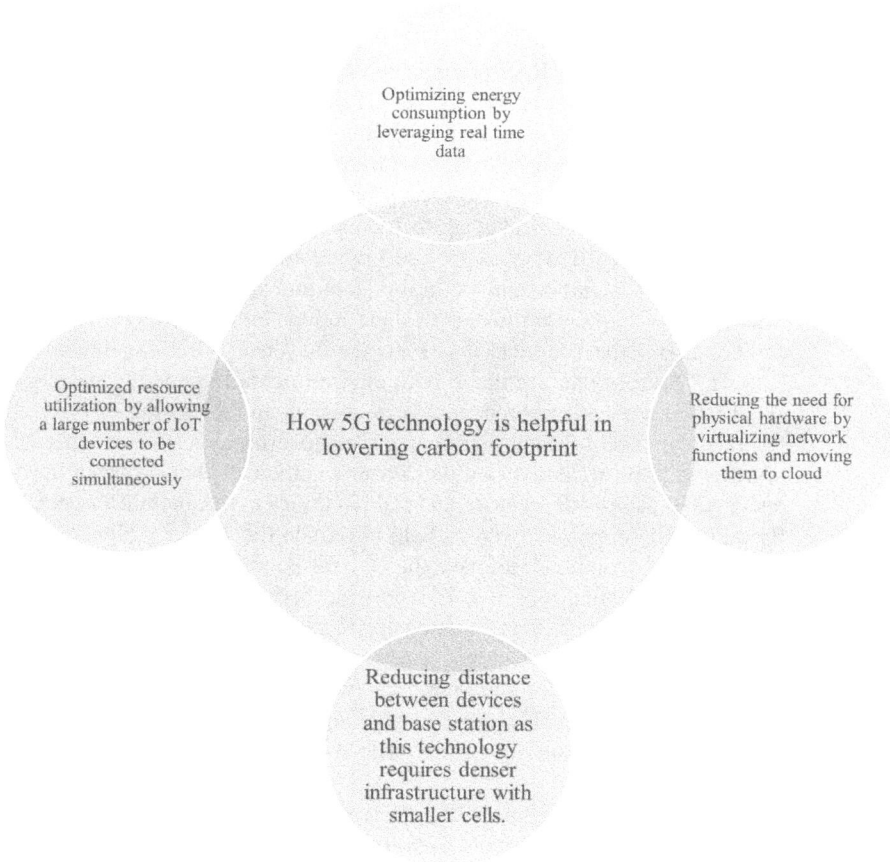

FIGURE 2.7   Contribution of 5G technology to lowering the carbon footprint.

- *Temperature sensors*: Thermoelectric materials can be used to create temperature sensors that generate an electrical signal when exposed to temperature changes. This signal can be used to power or communicate with other IoT devices.
- *Self-powered sensors*: By integrating thermoelectric materials with low-power IoT sensors, it is possible to create self-powered sensors that require no external power source.

c. *Photovoltaic*: Photovoltaic (PV) technology, also known as solar power, is a popular option for powering IoT devices. PV technology has the capability to convert sunlight into electrical energy, providing a power source for IoT devices. Some factors to consider when using PV technology in IoT devices include the efficiency of the solar panel, the amount of sunlight available, and the size and weight of the device. Advances in PV technology, such as the development of perovskite solar cells and tandem solar cells, are also making solar power more efficient and affordable.

d. *RF energy harvesting*: Radio frequency (RF) energy harvesting technology can be used to power the various IoT nodes, thus reducing the need for batteries and extending their operational lifetime. This technology involves harvesting energy from radio frequency signals that are already present in the environment. RF energy harvesting can be used to power wearable devices in the healthcare sector and in military applications [25].

To ensure that IoT devices are used in a way that is sustainable and responsible, it is important for individuals and organizations to consider the environmental impacts of their actions and to make informed choices about the products and services they use. This can include choosing devices that are made with environmentally friendly materials and that are designed for reuse or recycling, and disposing of waste responsibly. In addition, it is important for governments and industries to develop and implement regulations and standards that promote the responsible use of IoT devices and reduce their environmental impact. By taking these actions, we can help to reduce the adverse impacts of IoT on the environment and pave the way for a more sustainable future [26]. Table 2.1 mentions the different methods/technology/hardware modifications/material modifications proposed by different authors for making sustainable IoT systems. The performance factor and the major advantages of the highlighted methods, which contribute toward making the system sustainable, are also given in the table.

4. **High-tech companies and organizations working on manufacturing sustainable IoT devices**

To make IoT devices sustainable, different organizations and high-tech companies are designing products with the sustainable development goals (SDGs) of the United Nations in mind [39, 40]. They are following the three Rs approach, i.e., Reduce, Reuse, and Recycle. Below is a list of some of these companies and organizations:

a. **LG** is an electronics and appliance maker whose aim is to achieve some of the SDGs by 2030 as follows: reducing greenhouse gas emissions from product manufacturing by 50%; incorporating 60,000 tons of recycled plastic into its manufacturing process; and increasing the recovery of electronic waste by up to 8 million tons by 2030.

b. **Schreder** is a Belgium headquartered company that is dedicated to providing innovative and sustainable lighting solutions for public spaces, while also minimizing its own environmental impact. The company has started a range of initiatives to minimize the consumption of energy and reduce waste in its operations, such as using renewable energy sources and promoting recycling and waste reduction programs.

c. **Johnson Controls** is a multinational conglomerate that specializes in designing, manufacturing, and selling building automation systems, HVAC equipment, fire and security systems, and energy storage solutions. It is also committed to sustainability and reducing its

TABLE 2.1
Different Factors and Technologies Used for Making Sustainable IoT Systems

Ref. No.	Sustainability Factor	Technology/Model/ Hardware	Performance	Advantages
[27]	Power reduction	Semi-transparent organic photovoltaic cells (PVC) to power IoT nodes in low-light interior conditions.	Efficiency is good in low-light indoor areas.	Flexible and easy to process and manufacture PVC; lightweight tailorable design; strong UV-visible wavelength absorption; ideal for indoor applications.
[28]	Power reduction	IoT devices powered by renewable RF energy.	The model indicated that supply voltage and payload can have significant impact on the energy consumed for various tasks.	A model designed to explain the energy harvesting procedure and power consumption of IoT devices in practical situations.
[29]	Reduced energy consumption	The laser-induced graphene (LIG) technology, comprising a multi-operating mode triboelectric nanogenerator (MTENG) and a micro-supercapacitor (MSC).	Incorporation of a lotus-leaf-inspired bionic structure on the device surface enables waterproofing and self-cleaning capabilities ensuring stable electrical output.	Highly integrated and flexible device with remarkable capabilities like self-cleaning, self-charging, and energy harvesting.
[30]	Toxic waste reduction	Green RFID tag antenna with incorporated humidity sensor wireless sensor network (WSN) modules.	An eco-friendly multi-module RF industrial solution having exceptional performance in the operational range.	Wider range, multidimensional tags with sensors, standalone.
[31]	Power reduction and carbon footprint reduction	Lagrange multiplier method for increasing the utility lifetime while reducing the carbon footprint using wake-up radio (WUR), wireless energy harvesting (WEH), and error control coding (ECC).	Improved network utility, prolong the lifetime, transmission power control, efficient routing.	Improved reliability, energy harvesting, maintenance-free operation.
[32]	Energy consumption reduction and time reduction	Utilization of inexpensive automated irrigation systems via Arduino.	When specific thresholds are reached, the irrigation system is initiated, which results in enhanced convenience, reduced energy consumption, improved irrigation efficiency.	By improving the management of water resources in regions with limited water availability, it leads to enhanced production outcomes.

(Continued)

TABLE 2.1 *(Continued)*
Different Factors and Technologies Used for Making Sustainable IoT Systems

Ref. No.	Sustainability Factor	Technology/Model/ Hardware	Performance	Advantages
[33]	Power reduction	Hardware used: Arduino Uno and Raspberry Pi. The design of the smart irrigation system incorporates photovoltaic panels and a variety of control devices	Minimized water wastage that would otherwise occur.	The key benefit of the proposed irrigation system lies in its capability to accurately determine the water requirements of plants and provide precise supply accordingly. This designed system exhibits qualities of sustainability, efficiency, reliability, and ease of accessibility.
[34]	Reduced energy consumption	Uses new energy-efficient routing algorithms.	By implementing sleep nodes, the communication between nodes was effectively reduced.	The proposed scheme surpasses traditional WSN schemes in terms of energy efficiency and flexibility.
[35]	Less energy consumption and power reduction	The utilization of energy-efficient hardware and software solutions that minimize resource consumption.	Computing performance was maintained without degradation and reduced the consumption of energy.	Developed hardware solutions with the objective of designing devices that exhibit reduced energy consumption.
[36]	Reduced energy consumption	Adoption of hardware and software like power-saving virtual machine techniques.	The implementation of virtual energy-efficient resource allocation mechanisms resulted in notable enhancements in power and energy factors.	Robust and precise model and evaluation approaches were employed to effectively address energy-saving policies.
[37]	Power Reduction	Gate clustering technique is used.	On average, there was an 8% reduction in the number of transistors and a 27% decrease in the number of connections.	The suggested design successfully achieved a reduction in power consumption, thereby contributing to the sustainability of IoT.
[38]	Energy consumption reduction	Edge computing technology and a Deep Reinforcement Learning based energy scheduling scheme was applied.	Managed to achieve a reduced energy cost while also causing less delay when compared to conventional schemes.	Exhibiting greater efficiency in comparison to traditional schemes.

environmental impact. The company has set several environmental goals, which includes two main factors – the first one is to reduce the emission of greenhouse gases, and the second is to increase the use of renewable resources of energy (wind, solar, hydropower). Their IoT platform collects data from different sources and uses this data to optimize building operations and improve energy management.

d. **New Sun Road** is a California-based public benefit corporation that manufactures a suite of IoT products for energy management in developing countries. The company's software platform, called SunRise, is designed to manage energy production, storage, and consumption in real-time. The platform also incorporates machine learning algorithms to predict energy demand and optimize system performance. The company's primary focus is on developing microgrids. New Sun Road's microgrid solutions combine advanced software and hardware technologies to enable remote monitoring, control, and optimization of energy systems.

e. **HDR** is a Nebraska-based company that develops buildings with a focus on renewable energy. The company focuses on projects based on solar, hydro, and wind energy to reduce the emission of greenhouse gases which are produced in abundance when using non-renewable resources of energy. It has received several awards and certifications, including LEED (Leadership in Energy and Environmental Design) certifications for many of its buildings.

f. **Watershed** is a California-based sustainability consulting firm that works with businesses, governments, and non-profit organizations to develop and implement sustainable strategies. It offers services which include ecosystem restoration and sustainable development. It helps its clients to identify sustainability opportunities, set sustainability goals, and develop action plans to achieve those goals. It has developed several sustainability tools and resources that are available to the public. These resources include the "Sustainability Compass", a web-based tool that helps businesses assess their sustainability performance and identify opportunities for improvement.

g. **Mill** is a California-based company that specializes in manufacturing and selling kitchen trash bins to reduce the harmful waste. These specially made bins dry and grind down food scraps, turning them into compact grounds. Once the bin fills up, customers box the grounds up and mail them back to Mill, where they're taken to farms instead of landfills.

h. **Green Electronics Council (GEC)** is an organization that works to promote sustainable electronics through its EPEAT (Electronic Product Environmental Assessment Tool) certification program, which assesses the environmental impact of electronic products, including IoT devices [41].

i. **Sustainable Electronics Initiative (SEI)** is a research and education centre at the University of Illinois at Urbana-Champaign that focuses on promoting sustainable electronics design, manufacture, and disposal. They also offer training and resources for professionals in the electronics industry [42].

These are just a few examples of organizations working in the field of sustainable IoT devices. There are many more out there, and their work is critical to ensuring that IoT technology is developed and deployed in a way that is environmentally responsible and sustainable [43].

2.3 USING SUSTAINABLE IoT TOWARD THE DEVELOPMENT OF A SUSTAINABLE SOCIETY

Sustainability in its broadest sense means satisfying the current generation's needs while safeguarding the ability of future generations to fulfil their own requirements. It involves achieving an equilibrium between economic development, environmental preservation, and social welfare. These three factors are the pillars of sustainability. We are moving toward Industry Revolution 4.0 which combines the physical, biological, and digital systems to create new opportunities to change the way we live our lives [44]. Industry Revolution 4.0 is expected to lead to significant improvements in productivity, innovation, and competitiveness across many sectors, including manufacturing, healthcare, transportation, agriculture, and energy [45].

Figure 2.8 shows two major factors for the sustainable development of society. Sustainability by IoT involves IoT as an enabler to create a sustainable environment. It involves the integration of IoT technologies into sustainable practices. For example, in the context of smart cities, IoT can be used to monitor traffic to reduce congestion and hence emissions, regulate the energy consumption in homes and office buildings, and manage water and waste management efficiently. IoT enables the collection and analysis of a vast amount of data, allowing real-time monitoring and management of resources. Sustainability into IoT focuses on designing and manufacturing energy efficient IoT hardware and software with the aim of creating

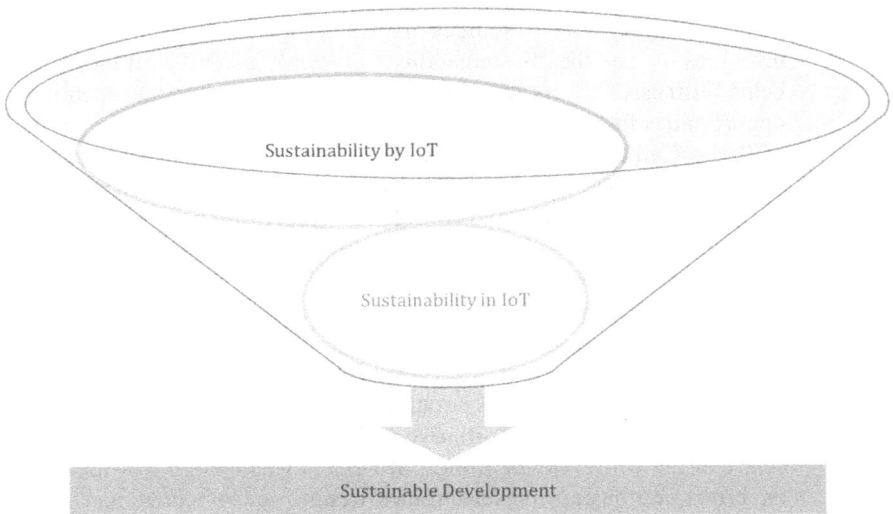

FIGURE 2.8 Two contributing factors for sustainable development.

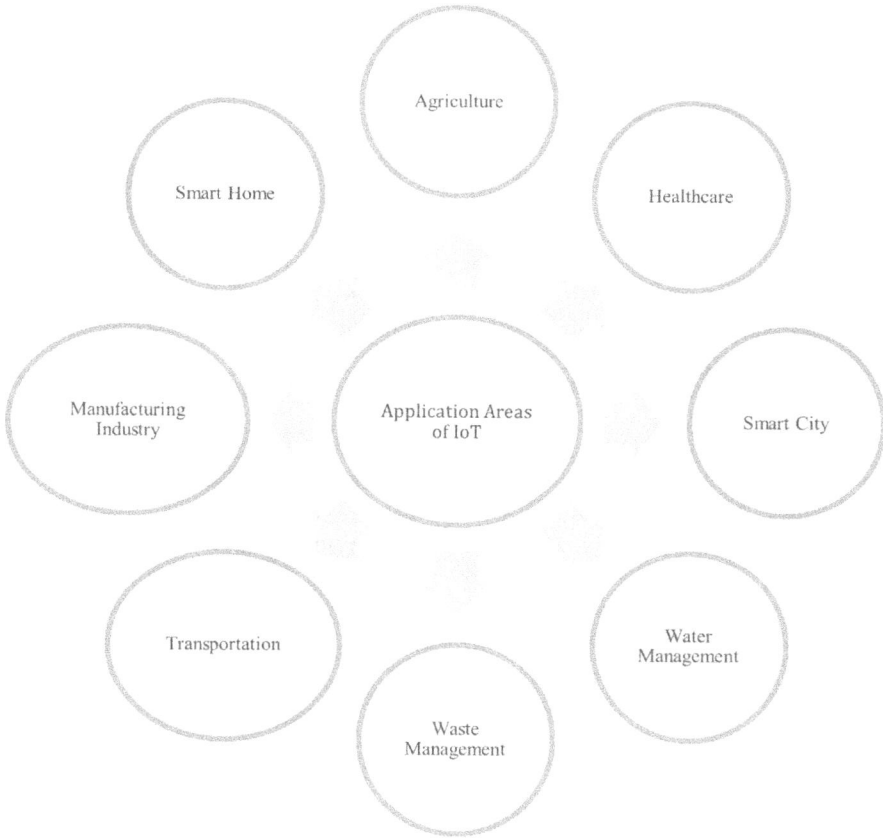

FIGURE 2.9 Application areas of IoT.

a sustainable future. Sustainability into IoT encompasses several key aspects such as environmental impact, resource efficiency, circular economy, social equity, and economic viability. Figure 2.9 shows the various sectors like agriculture, healthcare, manufacturing industries, smart cities, transportation, waste management, and water management where IoT is being implemented [46, 47].

2.3.1 SDGs of the United Nations

Before explaining the various application areas of IoT, a brief discussion about the Sustainable Development Goals (SDGs) of the United Nations is necessary to understand the mapping of IoT in different fields using SDGs. This set of 17 goals was established by the United Nations in 2015 for the sustainable development of society. The aim of these goals is to address the social, economic, and environmental challenges in order to make this world a more peaceful and better place. Each of these goals addresses particular areas of sustainable development. Achieving SDGs requires combined efforts from individuals, industries, and government bodies to implement these goals in every sector. The 17 SDGs are as follows:

- End poverty in all its forms.
- End hunger by promoting sustainable agriculture.
- Promote well-being and ensure healthy lives for all.
- Ensure quality education for all.
- Achieve gender equality and promote women's and girls' empowerment.
- Ensure universal access to clean water and sustainable sanitation systems.
- Ensure universal availability of clean and affordable energy to all.
- Encourage economic progress for sustainable economic development and decent work for everyone.
- Develop robust infrastructure, promote sustainable industrialization and innovation.
- Reduce disparities within and among countries.
- Create safe, secure, resilient, and sustainable cities for all.
- Ensure sustainable consumption and production practices.
- Adopt sustainable approaches to lower the harmful impact on the climate.
- Conserve and responsibly use the seas, oceans, and marine resources to promote sustainable development.
- Support the sustainable use of terrestrial ecosystems, reverse land degradation, and effectively manage forests.
- Promote harmonious and peaceful societies, ensure transparency in justice to all.
- Enhance the means of implementation for sustainable development.

It becomes clear from these goals how IoT can be applied, and outcomes mapped, in different sectors.

2.3.1.1   IoT in Agriculture

There are so many challenges that are faced by the agriculture sector, such as climatic changes, environmental impacts, pests and diseases, extreme weather events, labour shortage, and food security, and IoT has been instrumental in addressing these challenges. Smart farming represents a technologically advanced and innovative approach that uses IoT to enhance overall efficiency of agriculture [48]. Smart farming encompasses a range of activities, from preparing soil to harvesting crops. The integration of these activities with cloud-based platforms allows farmers to monitor and manage various aspects of farming operations like soil preparation, irrigation, use of fertilizers, weather conditions, and pest infestations [49, 50]. For example, by leveraging IoT sensors and weather monitoring devices, farmers can make informed decisions regarding irrigation, crop protection, and overall farm management. In addition to weather monitoring, IoT applications are extensively used in various agricultural aspects like livestock monitoring, field observation, vehicle tracking, etc. [51]. Implementation of IoT in the agriculture sector leads to more sustainable and efficient farming practices [52].

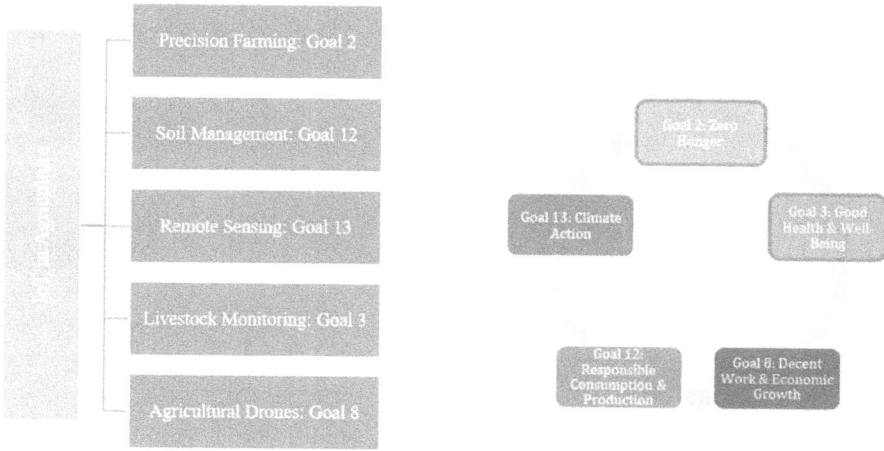

FIGURE 2.10   Benefits of IoT in agriculture with SDGs.

Eekifoods, an agritech startup, has developed an IoT mechanism which auto-mates farm tasks like irrigation, nutrient management, and climate control with high precision and accuracy and eases the work of labor. **Wiseconn Engineering** is working on a smart irrigation system which helps farmers produce better crops more efficiently. Farmers can use sensors to monitor the weather conditions and analyze the soil quality through the data so they know when to irrigate, which crop to grow, etc. Some other Indian Agritech Startups in precision farming and qual-ity control are Aarav Unmanned Systems, Aiobono, Aquaconnect, BharatAgri, CropIn, Fasal, GramworkX, Intello, Original4Sure, Agdhi. Figure 2.10 shows how IoT in agriculture is satisfying some of the sustainable development goals of the United Nations.

2.3.1.2   IoT in Healthcare

IoT in healthcare refers to the incorporation of IoT technologies into various health-care processes and systems. The emergence of IoT has opened up possibilities for life-saving applications in the healthcare sector. By collecting the real-time data from bedside devices, enabling real-time monitoring of patient information and pro-moting prompt diagnosis, the complete system of patient care could be enhanced and improved [53]. By the year 2090, it is estimated that 87% of healthcare orga-nizations will have successfully implemented IoT technology. More than one-third of healthcare organizations fail to utilize real-time data from connected devices which leads to risk of errors in diagnosis and treatment, potential for data loss and other insufficiencies. Efficient and accurate healthcare relies on timely and correct information and the advent of IoT has promoted the integration of a vast array of medical devices [54]. As implementation of IoT technology continues to gain trac-tion, we are witnessing an increase in the connectivity of various medical devices, which enhances the potential for better healthcare outcomes. The capability to

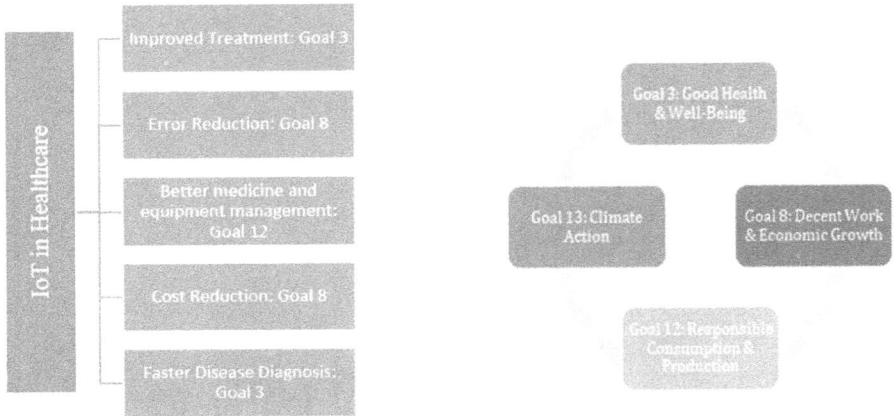

FIGURE 2.11   Benefits of IoT in healthcare with SDGs.

conveniently monitor and manage patient health can save the invaluable time of caregivers on a daily basis. High-quality care is possible with the aid of IoT as healthcare professionals can offer remote diagnosis to track the medical assets efficiently without visiting the patients physically [55, 56]. Locating the relevant department for patients, as well as for caregivers, has become easy by utilizing sensors and Wi-Fi. This reduces confusion and enhances efficiency while retrieving the necessary information needed for optimal patient care. Real-time data from patient heart monitors to temperature gauges already exists, and now it can be leveraged to create a safer and more effective healthcare environment [57]. Now with the help of a mobile application, IoT data can be securely and effectively managed by patients, as well as staff. Figure 2.11 shows the mapping of IoT in healthcare using the SDGs of the United Nations.

The future of IoT in healthcare is bright, as these days people are more interested in collecting their personal health data and gaining insights about their bodies (Figure 2.12). Some of the tech companies implementing IoT in healthcare are Intel, Google, Amazon Web Services, Samsung, Microsoft, COMCAST, Apple, CISCO, Tencent, and Meta.

2.3.1.3   IoT in the Smart City

A smart city is an urban development that uses the benefits of IoT to gather and utilize valuable information for effective management of resources and assets optimization [58]. This encompasses the collection of data from both citizens and mechanical devices to further enable the smart monitoring and management of different systems like transport, waste management, water disposal, electricity generation, etc. A smart city aims to enhance the quality of life of the citizens by promoting a healthy and sustainable environment using these smart solutions [59, 60].

IoT is revolutionizing what is possible across civic sectors [61]. It is keeping citizens and property safe with smart cameras featuring video and audio analytics that recognize criminal behaviour, identify faces, track occupancy, and detect the sound

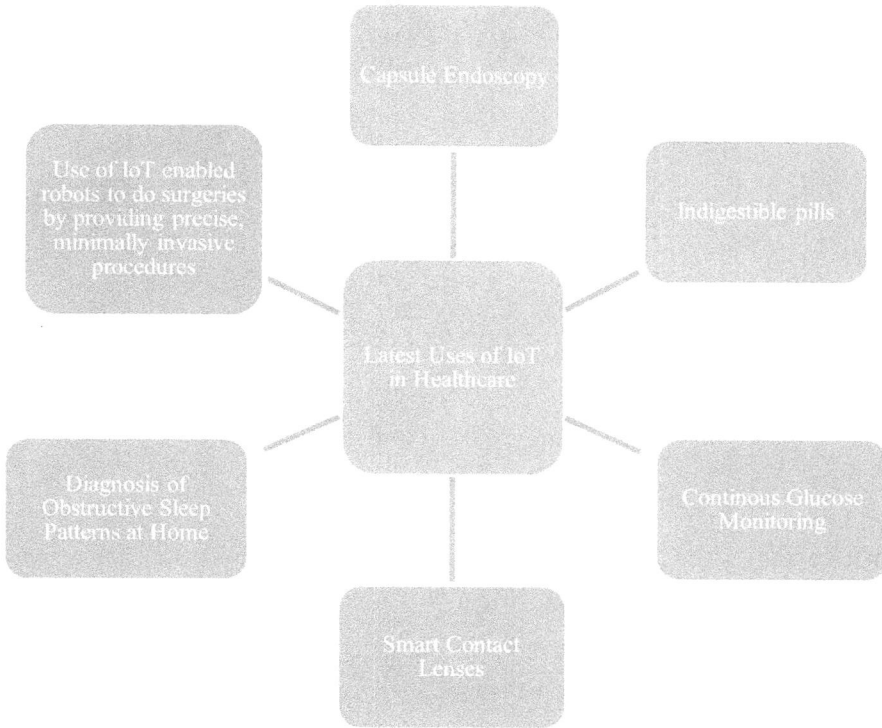

FIGURE 2.12   Some of the latest developments of IoT-enabled healthcare products.

of gunshots and breaking glass. It is keeping us healthy with the use of smart air-quality sensors that can monitor pollutants outside and maintain optimal ventilation levels inside to reduce the risk of virus transmission. It is optimizing energy usage with smart meters that analyze and automate power management to utilize resources more effectively and cut down on operating costs [62]. It is better-informing citizens with smart digital signage that provides them with real-time information and serves targeted ads determined by AI, and so much more. IoT and smart cities are unleashing a more adaptable, sustainable, and enjoyable environment. Given the wide-ranging potential of this technology it is no wonder that global IoT and smart cities market share is expected to grow very sharply in the next few years. Figure 2.13 shows how the deployment of IoT in alignment with UN's SDGs in cities could lead toward the sustainable development of society.

Ingram Micro Technology Solutions has developed scalable, turnkey solutions along with training and workshops to help simplify the complexities of IoT and smart cities and successfully go to market. These solutions include hardware and software products that enable cities to collect, manage, and analyze data from a range of IoT devices. This data can then be used to optimize city services, such as traffic management, waste management, and energy consumption. **Energy Efficiency Services Limited (EESL)** is an India-based company that provides IoT-based smart meters to

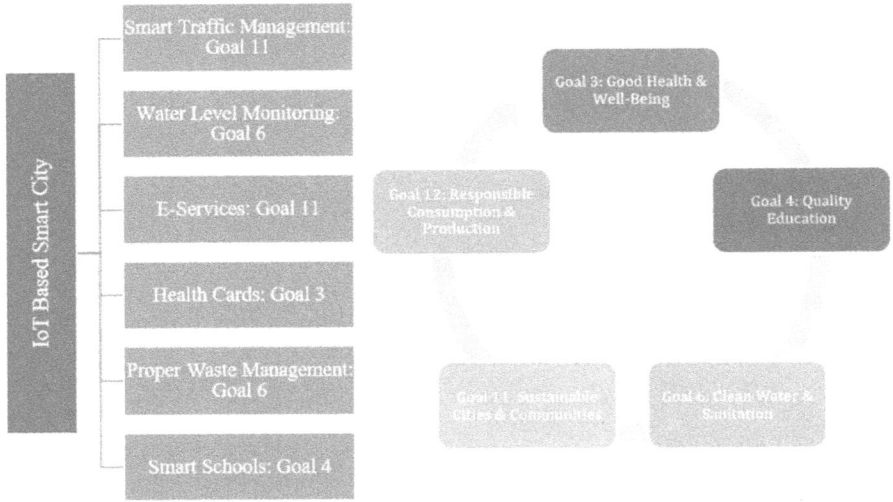

FIGURE 2.13 IoT in the smart city with the sustainable development goals (SDGs) of the United Nations.

facilitate the deployment of energy-efficient technologies, promote energy conservation, and reduce greenhouse gas emissions.

2.3.1.4 IoT in Manufacturing Industry

IoT is transforming manufacturing for suppliers as well as customers. Manufacturing industries are embracing the potential of IoT [63]. Manufacturers rely on the integration of physical things like machines, tools, and skilled persons to drive the production process and deliver the end products utilized by the customers/users on a daily basis. By connecting the physical assets and the surrounding environment to a network, IoT enables the smooth sharing of real-time data and precise information. Analysis of this data enables the manufacturers to speed-up product development and delivery and plan more effectively. IoT data plays an important role in optimizing the flow of information and goods across the supply chain. For instance, sensors installed in delivery trucks help companies to plan for better routes to save time and fuel, and the accurate delivery of the products from the factory to the consumers. Adopting IoT technology enables customer participation in product design for a better outcome and enhanced customer experience. It is estimated that IoT has the potential to elevate global productivity by up to 25% by the year 2025, which could generate an economic value of approximately USD 11 trillion.

The evolution of IoT is changing people's perspectives. The integration of disruptive trends in our daily lives has triggered a surge in demand for customization, elevated customer expectations, and heightened the complexity of supply chain management. All these transformative changes have a direct impact on manufacturing industry as people increasingly demand a huge variety of customizable

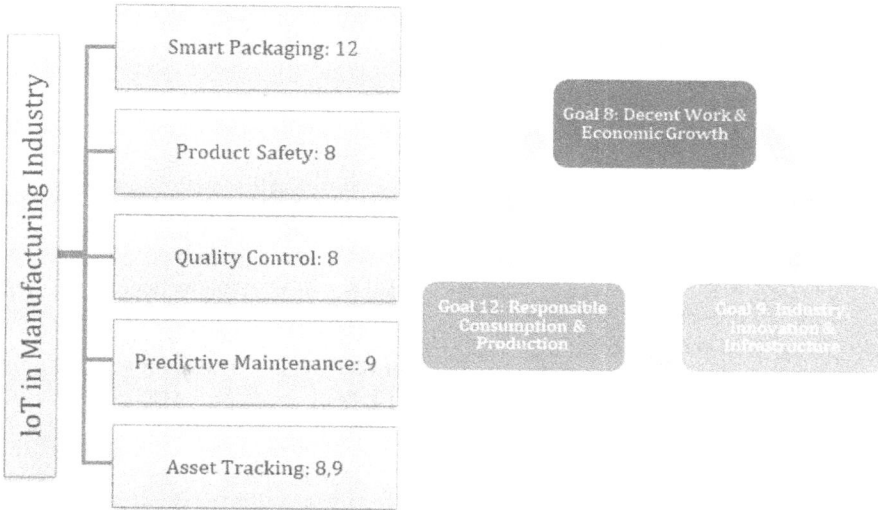

FIGURE 2.14   IoT-based manufacturing industries with SDGs of the United Nations.

products that cater to their preferences and needs [64]. This industry is also enjoying the flavor of disruption through connected devices [65, 66]. Figure 2.14 gives a clear description of how IoT in manufacturing industry is satisfying the SDGs of United Nations.

2.4   CONCLUSION AND FUTURE OF IoT

In this chapter, the basic concept of IoT has been explained, followed by an analysis of the need for sustainable IoT. Various methods and strategies that are being researched and adopted to make sustainable IoT devices and systems have been discussed. It is evident that multiple sectors, including the essential services, commercial, and customer care sectors are recognizing the potential of IoT and leveraging its benefits more than ever before. As the adoption of IoT continues to surge, businesses that embrace the IoT revolution early have a tremendous opportunity for growth and success. The companies and institutions that successfully harness the benefits of IoT to empower themselves have the potential to establish competitive advantages that are not easy to ignore. Therefore, it is important to adapt the practices used for making sustainable IoT devices and systems in order to implement the applications needed to create a sustainable society. This field is steadily growing and is expected to continue its rapid expansion and transformation. The future of IoT holds tremendous promise and is set to reshape various aspects of our lives. However, just like any new technology, there are security concerns with IoT. As more devices become interconnected, it is critical that they are protected from potential cyberattacks. This can be accomplished by putting in place strong security measures like encryption, firewalls and regular software updates. In conclusion, the IoT is already reshaping industries and enhancing people's lives. It has the potential to make our homes smarter, healthcare more efficient, agriculture more productive, and cities more sustainable. With continuing innovation and investment, the potential for IoT is virtually endless.

REFERENCES

1. M. Umair, M. A. Cheema, O. Cheema, H. Li, and H. Lu, "Impact of COVID-19 on IoT Adoption in Healthcare, Smart Homes, Smart Buildings, Smart Cities, Transportation and Industrial IoT," *Sensors*, vol. 21, no. 11, p. 3838, 2021, doi: 10.3390/s21113838
2. Y. Song, F. R. Yu, L. Zhou, X. Yang, and Z. He, "Applications of the Internet of Things (IoT) in Smart Logistics: A Comprehensive Survey," *IEEE Internet Things J.*, vol. 8, no. 6, pp. 4250–4274, 2020.
3. N. Alam, P. Vats, and N. Kashyap, "Internet of Things: A Literature Review," 2017 Recent Developments in Control, Automation & Power Engineering (RDCAPE), 2018, vol. 3, no. 05, pp. 192–197, doi: 10.1109/RDCAPE.2017.8358265
4. Cision, "Internet of Things (IoT) Market Worth $1319.08 Billion, Globally, by 2026 at 25.68% CAGR: Verified Market Research," *Cision*. https://www.prnewswire.com/ (accessed May 11, 2023).
5. J. Donaldson, "Internet of Things: The Importance of IoT in our Everyday Lives," *IoT, Technology*, 2019. https://www.mojix.com/internet-of-things-everyday-lives/%0Ahttps:// mojix.com/internet-of-things-everyday-lives/ (accessed May 12, 2023).
6. "What is IoT and How It Works." https://info.hummingbirdnetworks.com/blog/what-is- iot-and-how-it-works (accessed May 12, 2023).
7. V. Gupta, S. Tripathi, and S. De, "Green Sensing and Communication: A Step Towards Sustainable IoT Systems," *J. Indian Inst. Sci.*, vol. 100, no. 2, pp. 383–398, 2020, doi: 10.1007/s41745-020-00163-8
8. S. U. Rehman, P. Singh, S. Manickam, and S. Praptodiyono, "Towards sustainable IoT ecosystem," in *2020 2nd International Conference on Industrial Electrical and Electronics (ICIEE)*, 2020, pp. 135–138.
9. X. Liu, and N. Ansari, "Toward Green IoT: Energy Solutions and Key Challenges," *IEEE Commun. Mag.*, vol. 57, no. 3, pp. 104–110, 2019.
10. A. Schneider, and F. T. Council, "Green Technologies For A Sustainable Future," 2021. https://www.forbes.com/sites/forbestechcouncil/2021/01/12/green-technologies-for-a- sustainable-future/?sh=5df8741d6c23 (accessed May 12, 2023).
11. "Green IoT: Sustainable Design and Technologies - Speranza." https://www.speranzainc. com/green-iot-sustainable-design-and-technologies/ (accessed May 12, 2023).
12. N. Hossein Motlagh, M. Mohammadrezaei, J. Hunt, and B. Zakeri, "Internet of Things (IoT) and the Energy Sector," *Energies*, vol. 13, no. 2, p. 494, 2020.
13. Z. Qin, F. Y. Li, G. Y. Li, J. A. McCann, and Q. Ni, "Low-Power Wide-Area Networks for Sustainable IoT," *IEEE Wirel. Commun.*, vol. 26, no. 3, pp. 140–145, 2019.
14. R. Reis, "Strategies for Reducing Power Consumption and Increasing Reliability in IoT," in *Internet of Things. Information Processing in an Increasingly Connected World: First IFIP International Cross-Domain Conference, IFIPIoT 2018, Held at the 24th IFIP World Computer Congress, WCC 2018, Poznan, Poland, September 18-19, 2018, Revised Selected*, 2019, pp. 76–88.
15. N. N. Srinidhi, S. M. D. Kumar, and K. R. Venugopal, "Network Optimizations in the Internet of Things: A Review," *Eng. Sci. Technol. an Int. J.*, vol. 22, no. 1, pp. 1–21, 2019.
16. L. Liu, X. Guo, W. Liu, and C. Lee, "Recent Progress in the Energy Harvesting Technology—from Self-Powered Sensors to Self-Sustained IoT, and New Applications," *Nanomaterials*, vol. 11, no. 11, p. 2975, 2021.

17. "Carbon FootPrint - GeeksforGeeks." https://www.geeksforgeeks.org/carbon-footprint/ (accessed May 12, 2023).
18. J. C. R. Kumar, and M. A. Majid, "Renewable Energy for Sustainable Development in India: Current Status, Future Prospects, Challenges, Employment, and Investment Opportunities," *Energy. Sustain. Soc.*, vol. 10, no. 1, pp. 1–36, 2020, doi: 10.1186/s13705-019-0232-1
19. A. Qazi *et al.*, "Towards Sustainable Energy: A Systematic Review of Renewable Energy Sources, Technologies, and Public Opinions," *IEEE Access*, vol. 7, pp. 63837–63851, 2019.
20. V. Pecunia, L. G. Occhipinti, and R. L. Z. Hoye, "Emerging Indoor Photovoltaic Technologies for Sustainable Internet of Things," *Adv. Energy Mater*, vol. 11, no. 29, p. 2100698, 2021, doi: 10.1002/aenm.202100698
21. S. H. Alsamhi, O. Ma, M. S. Ansari, and Q. Meng, "Greening Internet of Things for Smart Everythings with a Green-Environment Life: A Survey and Future Prospects," *arXiv Prepr. arXiv1805.00844*, 2018.
22. M. Patil, S. Boraste, and P. Minde, "A Comprehensive Review on Emerging Trends in Smart Green Building Technologies and Sustainable Materials," *Mater. Today Proc.*, vol. 65, pp. 1813–1822, 2022, doi: 10.1016/j.matpr.2022.04.866
23. V. Tahiliani, and M. Dizalwar, "Green IoT Systems: An Energy Efficient Perspective," in 2018 Eleventh International Conference on Contemporary Computing (IC3), 2018, pp. 1–6.
24. Sara Budinis; Mathilde Fajardy; Carl Greenfield; Rachael Moore, "Carbon Capture, Utilisation and Storage – Analysis - IEA," 2022. https://www.iea.org/reports/carbon-capture-utilisation-and-storage-2 (accessed May 12, 2023).
25. T. Sanislav, G. D. Mois, S. Zeadally, and S. C. Folea, "Energy Harvesting Techniques for Internet of Things (IoT)," *IEEE Access*, vol. 9, pp. 39530–39549, 2021.
26. H. Rahmani *et al.*, "Next-Generation IoT Devices: Sustainable Eco-Friendly Manufacturing, Energy Harvesting, and Wireless Connectivity," *IEEE J. Microwaves*, vol. 3, no. 1, pp. 237–255, 2023.
27. S. Kim, M. Jahandar, J. H. Jeong, and D. C. Lim, "Recent Progress in Solar Cell Technology for Low-Light Indoor Applications," *Curr. Altern. Energy*, vol. 3, no. 1, pp. 3–17, 2019, doi: 10.2174/1570180816666190112141857
28. Y. Luo, and L. Pu, "Practical Issues of Energy Harvesting and Data Transmissions in Sustainable IoT," *arXiv Prepr. arXiv2001.00087*, 2019.
29. X. Li, C. Jiang, F. Zhao, Y. Shao, Y. Ying, and J. Ping, "A Self-Charging Device with Bionic Self-Cleaning Interface for Energy Harvesting," *Nano Energy*, vol. 73, p. 104738, 2020.
30. Y. Amin, "Printable Green RFID Antennas for Embedded Sensors," KTH Royal Institute of Technology, 2013.
31. C. Mahapatra, Z. Sheng, P. Kamalinejad, V. C. M. Leung, and S. Mirabbasi, "Optimal Power Control in Green Wireless Sensor Networks with Wireless Energy Harvesting, Wake-up Radio and Transmission Control," *IEEE Access*, vol. 5, pp. 501–518, 2016.
32. Y. A. Rivas-Sánchez, M. F. Moreno-Pérez, and J. Roldán-Cañas, "Environment Control with Low-Cost Microcontrollers and Microprocessors: Application for Green Walls," *Sustainability*, vol. 11, no. 3, p. 782, 2019.
33. S. Ali, H. Saif, H. Rashed, H. AlSharqi, and A. Natsheh, "Photovoltaic Energy Conversion Smart Irrigation System-Dubai Case Study (Goodbye Overwatering & Waste Energy, Hello Water & Energy Saving)," in *2018 IEEE 7th World Conference on Photovoltaic Energy Conversion (WCPEC)(A Joint Conference of 45th IEEE PVSC, 28th PVSEC & 34th EU PVSEC)*, 2018, pp. 2395–2398.
34. S. Rani, R. Talwar, J. Malhotra, S. H. Ahmed, M. Sarkar, and H. Song, "A Novel Scheme for an Energy Efficient Internet of Things Based on Wireless Sensor Networks," *Sensors*, vol. 15, no. 11, pp. 28603–28626, 2015.

35. T. Shree, R. Kumar, and N. Kumar, "Green Computing in Cloud Computing," in *2020 2nd International Conference on Advances in Computing, Communication Control and Networking (ICACCCN)*, 2020, pp. 903–905.
36. C. Zhu, V. C. M. Leung, L. Shu, and E. C.-H. Ngai, "Green Internet of Things for Smart World," *IEEE Access*, vol. 3, pp. 2151–2162, 2015.
37. C. Conceição, G. Moura, F. Pisoni, and R. Reis, "A Cell Clustering Technique to Reduce Transistor Count," in *2017 24th IEEE International Conference on Electronics, Circuits and Systems (ICECS)*, 2017, pp. 186–189.
38. Y. Liu, C. Yang, L. Jiang, S. Xie, and Y. Zhang, "Intelligent Edge Computing for IoT-Based Energy Management in Smart Cities," *IEEE Netw.*, vol. 33, no. 2, pp. 111–117, 2019.
39. "32 Environmental Companies Building a Sustainable World | Built In." https://builtin.com/greentech/environmental-companies (accessed May 12, 2023).
40. "IoT for ESG: Advancing Environmental Goals | Techfunnel." https://www.techfunnel.com/information-technology/iot-for-esg/ (accessed May 12, 2023).
41. "Global Electronics Council | Home." https://globalelectronicscouncil.org/ (accessed May 12, 2023).
42. "Sustainable Electronics Initiative – Promoting a more sustainable system for designing, producing, using, and managing electronic devices." https://sustainable-electronics.istc.illinois.edu/ (accessed May 12, 2023).
43. "Sustainable Development with IoT - Challenges and Solutions | Order Group." https://ordergroup.co/blog/sustainable-development-with-iot-challenges-and-solutions/ (accessed May 12, 2023).
44. N. S. Valeyeva, R. V. Kupriyanov, E. Valeeva, and N. V. Kraysman, "Influence of the Fourth Industrial Revolution (Industry 4.0) on the System of the Engineering Education," in *The Impact of the 4th Industrial Revolution on Engineering Education: Proceedings of the 22nd International Conference on Interactive Collaborative Learning (ICL2019)– Volume 2 22*, 2020, pp. 316–325.
45. Á Verdejo Espinosa, J. L. López, F. Mata, and M. E. Estevez, "Application of IoT in Healthcare: Keys to Implementation of the Sustainable Development Goals," *Sensors*, vol. 21, no. 7, p. 2330, 2021.
46. J. Terra, "Real-World IoT Applications in 2023," *Simplilearn*, 2022. https://www.simplilearn.com/iot-applications-article (accessed May 12, 2023).
47. Rinu Gour, "Top 10 Applications of IoT," *DZone*, 2018. https://dzone.com/articles/top-10-uses-of-the-internet-of-things (accessed May 12, 2023).
48. R. Dagar, S. Som, and S. K. Khatri, "Smart Farming–IoT in Agriculture," in *2018 International Conference on Inventive Research in Computing Applications (ICIRCA)*, 2018, pp. 1052–1056.
49. A. A. R. Madushanki, M. N. Halgamuge, W. A. H. S. Wirasagoda, and A. Syed, "Adoption of the Internet of Things (IoT) in Agriculture and Smart Farming Towards Urban Greening: A Review," *Int. J. Adv. Comput. Sci. Appl.*, vol. 10, no. 4, pp. 11–28, 2019.
50. L. Gupta, S. Malhotra, and A. Kumar, "Study of applications of Internet of Things and Machine Learning for Smart Agriculture," in *2022 IEEE International Conference on Current Development in Engineering and Technology (CCET)*, 2022, pp. 1–5.
51. I. Mohanraj, K. Ashokumar, and J. Naren, "Field Monitoring and Automation Using IoT in Agriculture Domain," *Procedia Comput. Sci.*, vol. 93, pp. 931–939, 2016, doi: 10.1016/j.procs.2016.07.275
52. V. P. Kour, and S. Arora, "Recent Developments of the Internet of Things in Agriculture: A Survey," *IEEE Access*, vol. 8, pp. 129924–129957, 2020, doi: 10.1109/ACCESS.2020.3009298

53. M. N. Bhuiyan, M. M. Rahman, M. M. Billah, and D. Saha, "Internet of Things (IoT): A Review of Its Enabling Technologies in Healthcare Applications, Standards Protocols, Security, and Market Opportunities," *IEEE Internet Things J.*, vol. 8, no. 13, pp. 10474–10498, 2021.

54. M. Parmar, and H. Jit Kaur, "Impact of IoT in Biomedical Applications: Part I," in *Electronic Devices, Circuits, and Systems for Biomedical Applications: Challenges and Intelligent Approach*, Elsevier, 2021, pp. 423–439. doi: 10.1016/B978-0-323-85172-5.00002-2

55. Z. N. Aghdam, A. M. Rahmani, and M. Hosseinzadeh, "The Role of the Internet of Things in Healthcare: Future Trends and Challenges," *Comput. Methods Programs Biomed.*, vol. 199, p. 105903, 2021.

56. H. N. Saha, D. Paul, S. Chaudhury, S. Haldar, and R. Mukherjee, "Internet of Thing Based Healthcare Monitoring System," in *2017 8th IEEE Annual Information Technology, Electronics and Mobile Communication Conference (IEMCON)*, 2017, pp. 531–535.

57. D. Tiwari, D. Prasad, K. Guleria, and P. Ghosh, "IoT Based Smart Healthcare Monitoring Systems: A Review," in 2021 6th International Conference on Signal Processing, Computing and Control (ISPCC), 2021, pp. 465–469.

58. S. I. Foundation, "Smart Cities: Solutions to Achieve Sustainable Urbanization," *Historic flight*, 2021. https://solarimpulse.com/smart-cities-solutions?utm_term=smartcity definition&utm_campaign=Solutions&utm_source=adwords&utm_medium= ppc&hsa_acc=1409680977&hsa_cam=11451944566&hsa_grp=117528790928&hsa_ad= 475011813029&hsa_src=g&hsa_tgt=kwd-48725135771&hsa_k (accessed May 12, 2023).

59. P. Bellini, P. Nesi, and G. Pantaleo, "IoT-Enabled Smart Cities: A Review of Concepts, Frameworks and Key Technologies," *Appl. Sci.*, vol. 12, no. 3, p. 1607, 2022, doi: 10.3390/app12031607

60. Y. Qian, D. Wu, W. Bao, and P. Lorenz, "The Internet of Things for Smart Cities: Technologies and Applications," *IEEE Netw.*, vol. 33, no. 2, pp. 4–5, 2019, doi: 10.1109/ MNET.2019.8675165

61. A. S. Syed, D. Sierra-Sosa, A. Kumar, and A. Elmaghraby, "IoT in Smart Cities: A Survey of Technologies, Practices and Challenges," *Smart Cities*, vol. 4, no. 2, pp. 429–475, 2021, doi: 10.3390/smartcities4020024

62. Z. Chen, C. B. Sivaparthipan, and B. A. Muthu, "IoT Based Smart and Intelligent Smart City Energy Optimization," *Sustain. Energy Technol. Assessments*, vol. 49, p. 101724, 2022, doi: 10.1016/j.seta.2021.101724

63. T. Kalsoom *et al.*, "Impact of IoT on Manufacturing Industry 4.0: A New Triangular Systematic Review," *Sustainability*, vol. 13, no. 22, p. 12506, 2021.

64. E. Manavalan, and K. Jayakrishna, "A Review of Internet of Things (IoT) Embedded Sustainable Supply Chain for Industry 4.0 Requirements," *Comput.& Ind. Eng.*, vol. 127, pp. 925–953, 2019.

65. A. Kumar, S. Sharma, N. Goyal, A. Singh, X. Cheng, and P. Singh, "Secure and Energy-Efficient Smart Building Architecture with Emerging Technology IoT," *Comput. Commun.*, vol. 176, pp. 207–217, 2021, doi: 10.1016/j.comcom.2021.06.003

66. S. Rani, S. H. Ahmed, and R. Rastogi, "Dynamic Clustering Approach Based on Wireless Sensor Networks Genetic Algorithm for IoT Applications," *Wirel. Networks*, vol. 26, no. 4, pp. 2307–2316, 2020, doi: 10.1007/s11276-019-02083-7

3 Designing an Integrated IoT-WSN Framework for Smart City Applications

Sanjivani Hemant Kulkarni, Namrata Gawande, Ashwin S. Chatpalliwar, Namita Parati, Kishor S. Wagh, Deepak T. Mane, and Basant Tiwari

3.1 INTRODUCTION

Due to the rising need for effective management of urban services, the idea of a "smart city" has attracted a lot of interest recently. Cities face many difficulties in providing basic services, managing resources, and improving the quality of life for their citizens as urban populations continue to grow [1]. A "smart city" can monitor and control all its numerous moving elements in real time using cutting-edge tools like the Internet of Things (IoT) and Wireless Sensor Networks (WSN). Since it makes use of both technologies' advantages, the IoT-WSN architecture is essential to the effective operation of smart cities.

The integrated IoT-WSN design collects and transmits data from diverse urban assets using sensors, controllers, gateways, and other devices over a wireless network. The IoT and WSN are two examples of cutting-edge technologies that smart cities use to address these issues and transform conventional urban infrastructure into a linked, intelligent ecosystem. The ecosystem creates a productive system for controlling urban services by fusing the strength of IoT and WSN technologies [2]. These resources can include anything from parking spots and lighting to waste disposal facilities and air quality monitoring stations. Advanced data analytics tools and techniques are then used to process and analyze the obtained data. In order to optimize resource allocation, increase operational efficiency, and enhance the overall quality of services provided to residents, city administrators, and stakeholders can use this information to identify patterns and trends and make data-driven decisions.

A wide range of smart city sectors find applications for the combined IoT-WSN platform. One such application is traffic management, where real-time information from sensors installed throughout road networks may be used to track traffic flow, spot congestion, improve signal timings, and provide drivers with dynamic routing instructions [3]. This decreases travel time, fuel use, and carbon emissions in addition to increasing traffic efficiency.

Waste management is another important application, where sensors built into trash cans or waste collection trucks can track waste levels, improve collection routes, and facilitate prompt disposal. This assists cities in lowering operational expenses, reducing their negative environmental effects, and maintaining hygienic conditions.

DOI: 10.1201/9781003437079-3

Another important area where the integrated IoT-WSN framework can have a big influence is energy management. Smart cities can track patterns of energy consumption, optimize energy distribution, and encourage residential and commercial energy saving by implementing smart meters, sensors, and actuators. This results in less wasted energy, lower utility costs, and a more environmentally friendly energy infrastructure.

Another area where the integrated IoT-WSN framework might be extremely helpful is water management. Water distribution networks can use sensors to check water quality, find leaks or other irregularities, and enable preventive maintenance. As a result, water resources are used more effectively, water loss is decreased, and the reliability of the water supply for people is improved [4].

Smart cities must also prioritize public safety, and the integrated IoT-WSN framework is crucial in this regard. Smart cities can improve public safety, identify emergencies, and act quickly to protect their citizens by implementing surveillance cameras, smart street lighting systems, and environmental sensors.

Designing an integrated IoT-WSN framework for applications in smart cities necessitates careful consideration of several variables. Determining the suitable hardware and software components, establishing the network architecture, and putting in place strong security measures to safeguard privacy are a few of these. Furthermore, in order to support the expansion and integration of new technologies and services in the future, scalability, interoperability, and standards compliance are crucial factors [5]. The combined IoT-WSN framework has enormous promise for creating cities that are smart, effective, and sustainable. Smart cities can gather real-time data [6], analyze it intelligently, and make educated decisions to enhance the quality of life for their citizens by utilizing the power of IoT and WSN technologies.

3.1.1   OVERVIEW OF THE IoT-WSN FRAMEWORK

The landscape of developing smart cities has lately undergone a radical change thanks to a new and more understandable architecture that combines IoT with WSN. The IoT concept is centered on connecting various devices so that they may easily share data online. To collect and share environmental data, a WSN is built up of sensor nodes that are wirelessly connected to one another [7]. These two technologies work well together to provide a powerful and efficient way to manage municipal operations and build the ecosystems of interconnected, smart cities.

The architecture of an IoT-WSN framework depends on a number of crucial components that perform well together. Its core is made up of sensors, which are necessary for acquiring vital data about the environment. These sensors come in several sizes and shapes, and they may be used to monitor many different parameters, such as temperature, humidity, air quality, noise levels, and much more. By gathering data in real-time, these sensors provide invaluable insights into the performance and operation of municipal services.

In addition to the sensors, the controllers assist efficient system administration. For instance, controllers in traffic management systems in the context of smart cities can be used to optimize signal timings based on real-time vehicle movement data, thereby lowering congestion and enhancing traffic flow. The integration of controllers into weather monitoring systems enables prompt responses to rapidly changing weather conditions [8]. These controllers make sure that numerous urban services are effectively controlled and coordinated, improving their overall effectiveness and responsiveness.

The communication link between the sensors and the main computer network is provided by gateways which oversee sending the information gathered by the sensors to the central processing unit for additional examination and decision-making. The gateways guarantee seamless network access and effective data transmission, enabling the easy exchange of data between devices and the central processing unit.

The enormous volumes of data that the sensors have collected must be processed and analyzed, and this is where data analytics tools come into play. These tools make it possible to spot trends, patterns, and anomalies in the data, which offers insightful information for improving municipal services. Smart cities can make data-driven decisions, improve resource allocation, and increase overall operational efficiency by utilizing advanced analytics approaches. Additionally, predictive modelling is made possible by data analytics, which enables cities to foresee future patterns and take preventative action to improve service delivery [9].

The IoT-WSN framework uses a smart network architecture to provide safe and effective data transfer between devices. Depending on the individual needs of the smart city application and the resources available, different communication protocols might be used. Popular protocols like Bluetooth, Wi-Fi, ZigBee, and Lora WAN all have their own benefits in terms of range, power usage, data transmission rate, and network scalability. In order to build dependable and smooth connectivity and guarantee that data is sent accurately and effectively throughout the network, the choice of communication protocol is essential. In this way, heating, ventilating, and air conditioning (HVAC) systems can be optimized using data on energy use gathered from building sensors, resulting in energy savings and improved sustainability. As a result, the quality of life for local citizens is increased, sustainability is improved, and smarter, more connected urban settings are developed.

The IoT-WSN framework's adaptability enables customization based on the needs of each smart city application. The architecture can change to accept new technologies and breakthroughs as cities develop, guaranteeing scalability and future-proofing. Urban management could be revolutionized by the integrated IoT-WSN architecture, making cities more sustainable, efficient, and resilient. The fusion of IoT and WSN technologies will be crucial in forming the cities of the future as we move into the era of smart cities [10]. Figure 3.1 shows how the IoT-WSN technologies work together to administer urban services effectively and communicate real-time data.

IoT-WSN architecture offers a powerful method for controlling urban services since it enables real-time monitoring and administration of a wide range of city components. The platform's versatility enables a variety of smart city applications, from public services and transport to public safety and environmental protection [11]. If a successfully integrated IoT-WSN architecture can be built, the IoT and WSN have the potential to make cities smarter, more connected, and more environmentally friendly [12].

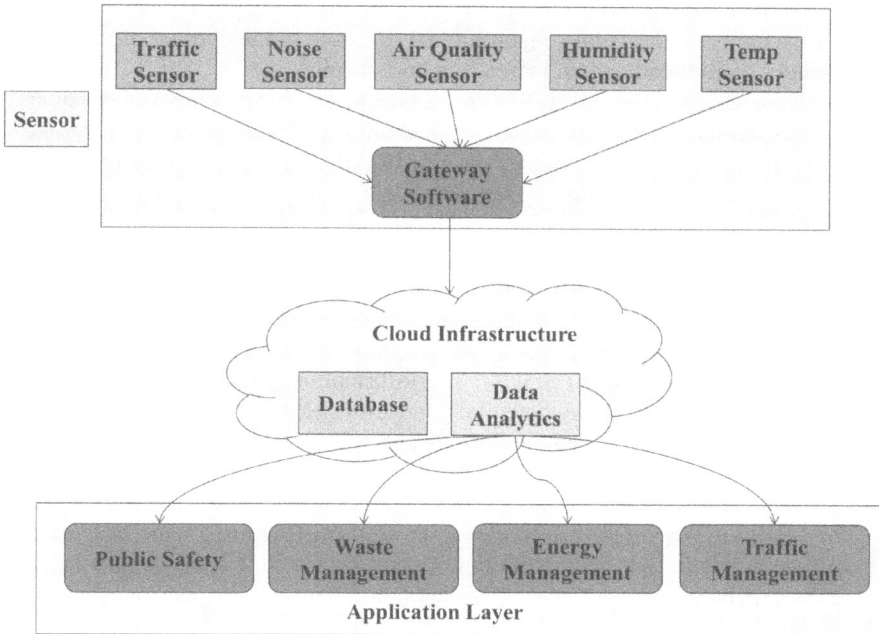

FIGURE 3.1   IoT-WSN integrated framework.

3.1.2 SMART CITY APPLICATIONS

The integrated IoT-WSN framework can be useful in a smart city for the following applications [13–18]:

i. *Management of traffic flows*: The IoT-WSN framework can be used to track traffic volumes in real time and change signal times accordingly. In order to predict where congestion will occur, sensors can be installed at strategic points to monitor traffic flow and velocity. Once the framework is in place, the timing of the lights can be modified to ease traffic flow and decrease congestion. As a result, the city will be more efficient and air pollution will be reduced [19, 20].

ii. *Energy management*: The IoT-WSN framework can be used for energy management by keeping tabs on a building's energy consumption and making necessary adjustments to the heating, ventilation, and air conditioning systems. Putting sensors in buildings to monitor things like temperature, humidity, and occupancy can help pinpoint energy leaks. Once the structure is in place, HVAC systems may be fine-tuned to cut back on energy use and costs.

iii. *Waste management*: Garbage pickup might take advantage of the IoT-WSN architecture by monitoring the quantity of trash cans and modifying collection schedules as necessary. It is feasible to identify the locations where garbage production is concentrated by putting sensors in trash cans and analyzing the collected data. The framework may be applied to enhance the trash collection schedule, resulting in a decrease in the number of garbage trucks on the road and an increase in cost savings.

iv. *Water management*: The IoT-WSN architecture offers a number of possible applications in water management, including finding plumbing leaks and monitoring water usage. Sensors can be installed in structures and throughout the water distribution system to collect data on water usage and pressure. This information can be used to identify problem areas. Once a leak has been found and fixed, the structure can be used to reduce water loss.

v. *Public safety*: The IoT and wireless sensor networks' infrastructure may be utilized to keep an eye on objects and be prepared for any emergencies that might occur in public spaces. In public places, sensors may be deployed to look out for suspicious behavior. The information gathered would then be analyzed to highlight any potentially harmful situations. After that, the system may alert the relevant authorities and provide them access to the most recent information.

The fusion of WSN with IoT creates a strong infrastructure for the management of multiple municipal services. Real-time data collection and analysis may help cities lower expenses, increase living standards, and enhance services to inhabitants. The framework is adaptable and may be customized to match the unique needs of each municipality. It can also expand to include new projects as they are developed.

3.2 DEFINING DATA REQUIREMENTS

It is crucial to establish the data needs of the system in order to create an efficient IoT-WSN framework for smart city applications. Determining the data types, data collection intervals, and data granularity needed for analysis is a part of this process [21–24].

3.2.1 TYPES OF DATA

The types of data that need to be collected will depend on the specific application of the IoT-WSN framework. Some common types of data that may be collected in smart city applications include:

1. *Environmental data*: This may include data on temperature, humidity, air quality, noise levels, and weather conditions.
2. *Traffic data*: This may include data on vehicle counts, speed, and congestion levels.
3. Waste management data: This may include data on waste generation rates, recycling rates, and landfill capacities.
4. *Energy consumption data*: This may include data on energy consumption patterns in buildings, streetlights, and other infrastructure.
5. *Health data*: This may include data about a person's health, such as their pulse, blood pressure, and amount of exercise undertaken.

3.2.2 FREQUENCY OF DATA COLLECTION

How frequently data is gathered will depend on how the IoT-WSN architecture is implemented, as well as how the data is going to be used. To enable real-time traffic management, some data, such as traffic statistics, may need to be collected in real-time. Some information, for example, from garbage pickup, might only need to be gathered once daily or once weekly.

3.2.3 DATA GRANULARITY

Depending on the IoT-WSN framework application, different levels of data granularity may be necessary. Aggregate data, such as regional garbage collection rates, may be sufficient for some purposes, in contrast to the detail necessary for others (such as vehicle counts and speeds). The capabilities of the sensor nodes and the network design will also have an impact on the level of detail in the data. For the collection of extremely detailed data, more advanced sensors and network infrastructure may be needed, thereby increasing the system cost. A crucial initial step in creating a successful IoT-WSN architecture for smart city applications is determining the data needs. To enable efficient analysis and decision-making, it is essential to thoroughly assess the many data types that must be acquired, the frequency with which data should be gathered, and the level of granularity necessary.

3.3 SELECTING SENSORS

The choice of sensors is an important step in creating an IoT-WSN architecture for smart city applications. Low-cost sensors are required for this study's focus on smart city applications since they can accurately and dependably gather data for that use. In this section, we will look at some of the most significant sensors that could be included in the infrastructure of a smart city [10, 25–30].

  i. *Temperature sensors*: These sensors are employed to determine the ambient temperature. They are frequently used in smart city applications including building automation and energy management. Temperature sensors can be used to monitor the surrounding temperature indoors, outside, or in manufacturing.
 ii. *Humidity sensors*: These sensors can gauge how much moisture is present in the air. A few examples of the heating, ventilation, and air conditioning (HVAC) systems used in smart cities are agricultural and building automation systems. The relative wetness of interior rooms, outdoor surroundings, and manufacturing facilities may be evaluated using humidity sensors.
iii. *Air quality sensors*: These sensors can track pollutant concentrations. They are extensively utilized in smart city applications, for example in traffic management, public health improvement, and environmental monitoring. Ozone, particulate matter, nitrogen oxides, and carbon monoxide are just a few of the pollutants that air quality monitors can track.
 iv. *Noise sensors*: These sensors measure the level of background noise in a certain region. They are widely used in a range of smart city applications,

TABLE 3.1

Comparison of Sensor Types Utilized in Smart City Applications

Sensor Type	Measurement	Application	Advantages	Disadvantages
Temperature	Temperature	Building automation, energy management	Low cost, easy to install	Limited to temperature measurement only
Humidity	Moisture	Agriculture, HVAC systems, building automation	Helps prevent mold and moisture damage	Limited to moisture measurement only
Air Quality	Pollutants	Environmental monitoring, traffic management, public health	Can detect various pollutants	Can be expensive, requires regular calibration
Noise	Sound level	Traffic management, public safety, noise pollution control	Can measure noise levels accurately	May be affected by other sounds or environmental factors
Vibration	Vibration	Structural health monitoring, industrial automation, transportation systems	Can detect vibration in structures or machinery	Can be expensive, may require complex installation

including managing traffic, ensuring public safety, and controlling noise pollution. Noise sensors can monitor decibel levels in a range of settings, such as congested streets, open spaces, and noisy factories.

v. *Vibration sensors*: These sensors gauge the intensity of environmental vibrations. They are essential to many smart city applications, including public transport, factory automation, and monitoring health infrastructure. Using vibration sensors, the level of vibration in a building, bridge, or other structure may be determined. Installing these sensors might be advantageous for both the infrastructure of public transport and vehicles.

Table 3.1 summarizes the many sensor types utilized in smart city applications, along with the benefits and drawbacks of each. The table demonstrates that there are trade-offs to consider, such as cost and accuracy, when deciding on a sensor. To make sure the IoT-WSN framework efficiently supports the smart city application, the right sensors must be chosen after thorough consideration of the data requirements and application demands.

3.4 SELECTING COMMUNICATION PROTOCOLS

Once sensors have been chosen, the right communication protocol must be specified so that data can be sent from the sensors to the central monitoring system. Considerations including data rate, range, power consumption, and cost will inform the protocol selection. In smart city applications, the following four types of communication protocols are typically used [31–33]:

3.4.1   Wɪ-Fɪ

A popular communication protocol called Wi-Fi (Wireless Fidelity) offers wireless connectivity for a range of gadgets and applications. It is appropriate for a variety of smart city applications due to its high data transfer rates and operation in the 2.4 GHz and 5 GHz frequency ranges. We will go deeper into Wi-Fi in this section, looking at its main benefits, features, and uses in the context of the integrated IoT-WSN framework for smart cities.

With the help of Wi-Fi technology, devices can wirelessly join a local area network (LAN), which enables them to communicate and share data with one another and the internet. It is dependent on the IEEE 802.11 standard family, which establishes the requirements for wireless networks. Access points (APs) that send and receive data signals make up Wi-Fi networks, together with client devices that connect to these APs to access the network.

The high data transmission rates of Wi-Fi are one of its key benefits. Wi-Fi networks can establish fast and dependable connections, which makes them appropriate for use with high-bandwidth applications like video streaming, real-time monitoring, and cloud-based services. As a result, Wi-Fi is a great option for many smart city use cases, including surveillance systems, open Wi-Fi hotspots, and intelligent transportation systems.

The ubiquitous use and accessibility of Wi-Fi is another important characteristic. Worldwide, homes, businesses, public areas, and cities all have Wi-Fi networks. This large infrastructure facilitates the deployment of Wi-Fi-enabled devices and offers extensive coverage for applications related to smart cities. To connect sensors, devices, and infrastructure and to provide smooth communication and data exchange, municipalities can make use of existing Wi-Fi networks.

Wi-Fi networks can also be readily scaled and expanded to support an increase in the number of users and devices. To boost network capacity and increase coverage, several access points can be set up. Roaming is a feature of Wi-Fi networks that enables seamless switching between access points for devices. This scalability is essential in smart cities since there are increasingly more connected devices and people that need wireless access.

Wi-Fi can be used as a communication protocol for many IoT devices and sensors within the integrated IoT-WSN architecture. It makes it possible for these gadgets to connect to the internet and send information to centralized systems for additional processing and analysis. Wi-Fi-enabled sensors, for instance, can gather information on energy use, garbage management, traffic movement, and air quality in smart city applications, offering crucial insights for urban planning and management.

Strong security protections are provided by Wi-Fi to safeguard data being transmitted across the network. Wi-Fi supports several security protocols, including WPA2 and WPA3, which offer encryption and authentication tools to guarantee secure communication. This is essential for protecting sensitive data in applications for smart cities, such as transaction data, personal information, and surveillance footage.

Additionally, Wi-Fi technology provides a variety of deployment possibilities. It supports a variety of network topologies, including mesh networks, where several APs come together to create a self-healing network that can quickly adjust to environmental changes. Mesh networks improve network coverage and stability, which qualifies them for extensive smart city deployments.

As a result of its widespread adoption, Wi-Fi is a communication technology that is essential to the integrated IoT-WSN framework for smart city applications. It is the best option for establishing connections between devices and facilitating seamless communication in smart city environments due to its fast data transfer rates, wide availability, scalability, strong security features, and flexibility. Wi-Fi networks can enable a variety of applications, including smart transportation systems, public Wi-Fi hotspots, and surveillance. Cities can create effective, dependable, and secure wireless connectivity by utilizing Wi-Fi technology, establishing the groundwork for a smarter and more interconnected urban ecology.

3.4.2 BLUETOOTH

A wireless system called Bluetooth makes it possible for devices to interact and exchange data across short distances. It is now a commonplace technology that can be found in many gadgets, including smartphones, computers, wearables, and IoT devices. The key characteristics, benefits, and uses of Bluetooth will be covered in more detail in this section where we analyze the integrated IoT-WSN framework for smart cities.

Bluetooth employs low-power radio waves to link devices and operates in the 2.4 GHz frequency band. It adheres to the Bluetooth Core Specification, which establishes the communication standards and protocols. In the master-slave architecture used by Bluetooth devices, one device serves as the master and manages communication with one or more slave devices. The low power consumption of Bluetooth is one of its main benefits. The energy-efficient nature of Bluetooth technology makes it appropriate for battery-operated devices and applications that need a long battery life. As many IoT devices and sensors rely on finite power sources in smart city applications, this trait is very advantageous. Both traditional Bluetooth and Bluetooth Low Energy (BLE) are available as Bluetooth communication options. For applications that need more bandwidth, such as audio streaming and file transfer, classic Bluetooth offers faster data transmission rates. BLE, on the other hand, is perfect for IoT devices and sensors that need recurring data updates because it is optimized for low power consumption and supports intermittent data transfer.

Bluetooth's simplicity of use and smooth communication are further benefits. No complicated setups are required to pair and connect Bluetooth devices. Devices that have been linked can instantly reconnect when they come into range, giving users a convenient and trouble-free experience. This functionality is crucial in settings like smart cities where a lot of devices need to connect and communicate with one another easily.

Additionally, Bluetooth enables several profiles and services that make certain features and applications possible. For instance, the Bluetooth Health Device Profile (HDP) enables the transmission of health-related data from wearable devices to healthcare systems, while the Bluetooth Hands-Free Profile (HFP) permits hands-free communication in vehicles. These profiles increase Bluetooth's functionalities and enable a variety of use cases for smart cities.

Furthermore, Bluetooth offers strong security safeguards to safeguard data while it is being transmitted. To provide safe communication between devices, it supports encryption and authentication techniques. This is crucial in applications for smart cities because devices may communicate sensitive data like personal data, payment information, and surveillance footage.

Over time, Bluetooth technology has developed new and improved versions. For instance, Bluetooth 5.0 brought about advancements in data transmission rates, range, and battery economy. With the help of Bluetooth Mesh, a feature added to Bluetooth 5.0, huge networks may be built where devices can connect with one another in a self-organizing fashion. This is especially helpful in the deployment of smart cities when a lot of devices need to be connected. Bluetooth can be used as a communication protocol for many IoT devices and sensors in the integrated IoT-WSN architecture. It makes it possible for data to be transferred between devices and to link to centralized systems for additional processing and analysis. For managing smart cities, Bluetooth-enabled sensors, for instance, can gather information on environmental conditions, occupancy levels, and asset monitoring.

In conclusion, the integrated IoT-WSN framework for smart city applications relies heavily on Bluetooth, a flexible and widely used communication technology. It is suited for a variety of IoT devices and sensors due to its low power consumption, simplicity of use, smooth communication, security features, and support for numerous profiles. In smart city contexts, Bluetooth technology offers effective and dependable communication between devices, enabling smooth integration and interoperability. Cities can use Bluetooth to develop networked ecosystems that facilitate effective data collecting, analysis, and decision-making for enhanced urban services and a higher standard of living for citizens.

3.4.3 ZIGBEE

Zigbee is a low-power wireless technology for short-range, low-data-rate communication. Healthcare, smart homes, and industrial automation use it extensively. This section discusses Zigbee's primary benefits, features, and usage in the integrated IoT-WSN framework for smart cities.

Zigbee runs at 2.4 GHz and employs IEEE 802.15.4 for its physical and media access control (MAC) layers. It provides reliable and secure connectivity in multi-device environments. Zigbee uses a mesh networking topology to generate a self-organizing and self-healing network. Mesh networks allow devices to send messages via several routes, improving network coverage and resilience. If a device fails or is unavailable, the network automatically reroutes communications. Self-healing is necessary in smart city deployments where network dependability is crucial. Zigbee supports tree, mesh, and star networks. Smart city applications determine the network configuration. For applications with a single hub that connects with numerous end devices, a star network is better than a mesh network.

Zigbee's power efficiency is a benefit. It extends battery life in battery-powered devices. Zigbee's energy efficiency is ideal for smart city applications because devices and sensors are often situated in remote or hard-to-reach regions. Smart city applications can deliver modest amounts of data using Zigbee's 20–250 kbps transmission speeds. Low-data-rate capacity simplifies the communication system and conserves energy. It benefits smart metering, home automation, and environmental monitoring. Zigbee provides reliable, strong communication.

Smart city applications that transfer sensitive data require secure communication protocols. Zigbee secures data with several methods. AES-128 symmetric encryption protects device-to-device data transfer. Zigbee authentication and key establishment techniques restrict network access to permitted devices. Zigbee devices are cheap, small, and simple. These properties make Zigbee suitable for smart city device deployment.

Zigbee networks can grow with a city's devices. In the combined IoT-WSN architecture, Zigbee can communicate with smart city applications. It lets gadgets and sensors collect data on the environment, occupancy, energy usage, and other aspects. After collecting data, a central system can analyze and make judgments about what action is needed. Smart city management employs Zigbee. Smart lighting systems use Zigbee-enabled lightbulbs, switches, and sensors to adjust lighting based on occupancy and ambient illumination. Smart meters use Zigbee to track energy use and communicate data to utilities for invoicing and load management.

To summarize, Zigbee is a low-power wireless communication system that has several smart city applications. Its low power consumption, mesh networking, dependability, security, and scalability make it valuable for large-scale sensor and device networks. Zigbee enables smart city device connectivity for data collection, analysis, and decision-making to improve urban services and quality of life. Zigbee can create networked, sustainable ecosystems that maximize resource utilization and improve urban systems.

3.4.4 LoRaWAN

For long-range, low-power wireless IoT applications, LoRaWAN was developed. This innovative technology allows low-data-rate wide-area networks with long battery lives. In this section, we will discuss LoRaWAN's primary advantages and uses in the integrated IoT-WSN framework for smart cities.

LoRaWAN uses unlicensed ISM channels like 868 MHz in Europe and 915 MHz in North America in the sub-GHz band. Its long-range characteristics allow devices to send data over many kilometers, even in highly populated places with many barriers and interference. Due to its long-range connection, LoRaWAN is ideal for smart city applications that require wide-area coverage.

LoRaWAN saves power. The technology reduces energy usage to extend battery life by many months to years. Because of its low power requirements and minimal maintenance and operational costs, LoRaWAN is ideal for remote or inaccessible device deployments.

LoRaWAN uses a star-of-stars network to connect end devices to a gateway, which connects to the network server. The gateway sends endpoint data to the network server for examination. This star-of-stars architecture simplifies network construction and data handling.

The protocol can support modest data rates of a few hundred to several kilobits per second, depending on settings. Many IoT applications, such as environmental monitoring, asset tracking, and smart agriculture, require periodic data transfer. LoRaWAN uses chirp spread spectrum modulation for robust communication. This modulation approach helps LoRaWAN overcome multipath fading and interference to carry data reliably in challenging radio frequency settings. LoRaWAN uses sophisticated signal processing algorithms to recover weak signals for long-distance communication. LoRaWAN needs security since IoT data is sensitive. Security safeguards protect data confidentiality and integrity in the protocol. End-to-end encryption secures device-to-server communication. LoRaWAN's authentication and authorization techniques restrict network access to permitted devices. LoRaWAN networks are scalable and handle multiple devices. The protocol streamlines device integration and dynamic bandwidth allocation based on application needs. Because of its scalability, LoRaWAN is suitable for smart city installations where many networked devices and sensors supply urban services. The combined IoT-WSN architecture can employ LoRaWAN for smart city applications. It allows devices to communicate long-distance, making it easier to monitor the environment, infrastructure, assets, and more. The central system can assess LoRaWAN data and make judgements. LoRaWAN has several smart city administrations uses. In smart parking systems, LoRaWAN-capable sensors can detect parking space occupancy in real time to help drivers and municipal planners optimum parking use. Cities can use LoRaWAN sensors to monitor air quality, temperature, humidity, and noise to control environmental conditions. LoRaWAN lets smart metering programs remotely monitor and regulate energy consumption by connecting utility meters to the network. This helps cities save energy, discover abnormalities, and distribute energy better. LoRaWAN sensors can detect trash in bins, enhance collection routes, and reduce operational costs for smart waste management.

Lastly, LoRaWAN is a powerful and versatile IoT protocol for smart cities. Due to its extended range, low power consumption, scalability, and security, it is ideal for wide-area deployments with many devices and sensors. Using LoRaWAN in the integrated IoT-WSN architecture can improve infrastructure management, resource allocation, and city services.

The pros and cons of the most popular communication protocols used in smart city applications are summarized in Table 3.2. It is clear from the table that there are a variety of factors to consider while deciding on a protocol, including data rate, range, power consumption, and cost. To make sure the IoT-WSN framework efficiently supports the smart city application, it is important to consider data requirements and application needs while choosing a suitable communication protocol.

3.5 ROLE OF 5G AND BEYOND

3.5.1 5G PROTOCOLS USED IN SMART CITY APPLICATIONS FOR COMMUNICATION

The importance of 5G and other protocols in communication applications for smart cities cannot be overstated. 5G and beyond 5G play a very important role – in the communication needs of applications for smart cities, 5G protocols are essential. High-speed data transfer, low latency, widespread device connectivity, network slicing, edge computing, enhanced bandwidth, dependability, energy efficiency, security, and integration with cutting-edge technology are all features they offer. These protocols serve as the building blocks for sophisticated smart city services, boosting citizens' quality of life and promoting sustainable urban growth.

TABLE 3.2

Protocols Used in Smart City Applications for Communication

Communication Protocol	Wi-Fi	Bluetooth	Zigbee	LoRaWAN
Range	Short range (up to 100 m)	Short range (up to 100 m)	Short range (up to 100 m)	Long range (up to several kilometers)
Power Consumption	Moderate to high	Low to moderate	Low to moderate	Low
Data Rate	High	Moderate	Low	Low to moderate
Network Topology	Point-to-point or infrastructure	Point-to-point or infrastructure	Mesh	Star-of-stars
Security	Strong security	Moderate security	Strong security	Strong security
Scalability	Limited scalability	Limited scalability	High scalability	High scalability
Frequency Band	2.4 GHz, 5 GHz	2.4 GHz	2.4 GHz	Sub-GHz (868 MHz, 915 MHz)
Applications	Home/Office networks,	Personal devices	Home automation	Wide-area IoT
	Internet access	Audio streaming Wearable devices	Industrial control Healthcare	Smart city Applications

There are many opportunities to enhance urban living as a result of the adoption of 5G and beyond 5G protocols in smart city applications. These protocols offer the connectivity and capabilities required to promote sustainable urban development, from improving transit systems and public safety to enabling environmental monitoring, energy management, healthcare services, and smart buildings. Cities can use these protocols to build more effective, livable, and connected environments for their residents by using the benefits of high-speed data transfer, low latency, and huge device connectivity. These cutting-edge communication protocols offer the infrastructure required to support the enormous connectivity needs and data exchange in a smart city setting. Let us look more closely at their functions:

3.5.1.1 5G Communication Protocol

i. *High-speed data transmission*: 5G protocols greatly outperform earlier generations in terms of data transmission speeds, providing seamless connectivity and real-time communication for a variety of smart city applications. This is especially crucial for high bandwidth applications like video surveillance, augmented reality, and self-driving cars.

ii. *Low latency*: The ultra-low latency of 5G technologies is achieved through reducing communication delays. This is crucial for applications like remote healthcare, traffic control, and emergency services that call for quick responses. Real-time monitoring, analysis, and decision-making in situations requiring quick responses are made possible by the low latency.

iii. *Massive device connectivity*: 5G technologies allow for the simultaneous connection of a huge number of devices to the network. Applications for smart cities that use a wide range of sensors, IoT devices, and infrastructure components require this feature in order to function. This connectivity makes it possible to gather, aggregate, and analyze data from various sources quickly and effectively, producing in-depth insights and better services.

iv. *Network slicing*: The 5G standard includes the idea of network slicing, which enables the construction of virtual networks specifically suited for applications related to smart cities. Each network slice can be tailored to meet specific needs, including those related to capacity, latency, and security. This adaptability enables effective resource management and guarantees that each application gets the network resources it needs for peak performance.

v. *Edge computing*: By extending computing and storage capabilities closer to the network's edge, 5G protocols support edge computing. As a result, there is less need for a centralized cloud infrastructure, and real-time data processing is made possible. Edge computing can improve the effectiveness of smart city services like intelligent traffic management and smart grid systems and is especially advantageous for latency-sensitive applications.

3.5.1.2 Beyond 5G Communication Protocol

1. *Increased bandwidth*: Protocols designed for use beyond 5G are intended to offer even greater data rates and more bandwidth than 5G. This makes it possible for massive amounts of data, including sophisticated sensor data or high-resolution video feeds, produced by smart city applications to be transmitted seamlessly.

2. Beyond 5G protocols place a strong emphasis on ultra-reliable communications to ensure high levels of dependability, availability, and resilience in crucial applications. This is crucial for smart city services that deal with emergency response, infrastructure management, and public safety.

3. *Improved energy efficiency*: Beyond 5G protocols seek to improve energy efficiency by lowering power consumption across linked devices and network infrastructure. This enables IoT devices to have longer battery lives and lessens the harmful effect on the environment of communication networks, both of which are essential for the deployment of sustainable smart cities.

4. *Enhanced security and privacy*: Beyond 5G protocols take into account the growing security and privacy issues in applications for smart cities. They offer cutting-edge encryption methods, authentication systems, and privacy-enhancing tools to safeguard sensitive information and block hacker attacks. Maintaining the integrity and security of vital infrastructure and citizen information depends on this.

5. *Artificial intelligence (AI)* and the Internet of Things (IoT) are two developing technologies that beyond 5G protocols aspire to interact with invisibly. In order to support intelligent decision-making, predictive analytics, and automation in smart city applications, they offer optimized connectivity and data exchange capabilities.

3.6 CLOUD INFRASTRUCTURE

When it comes to smart city applications, the cloud infrastructure is a crucial part of the overall integrated IoT-WSN framework. It is a cheap and scalable solution for storing and processing all that sensor data. Storage, processing, and cloud services are the three pillars upon which the cloud is built. Figure 3.2 shows the integration of the cloud infrastructure into the IoT-WSN framework.

3.6.1 STORAGE OF DATA

Using cloud storage to keep all the data collected by the sensors is a safe and scalable option. Structured, semi-structured, and unstructured data are all viable options for storing the information. Data redundancy, data replication, and data encryption are only some of the data security features made available by cloud storage. The information can be kept in a variety of locations, including object stores, block stores, and file stores. Object storage works well for archiving vast amounts of unstructured data like media files and is a flexible and inexpensive data storage solution that can be accessed from any location. Block storage is ideal for storing databases and other types of organized data because of the quick access it enables. Documents, spreadsheets, and presentations are just some of the file types that can be stored in a file store [34].

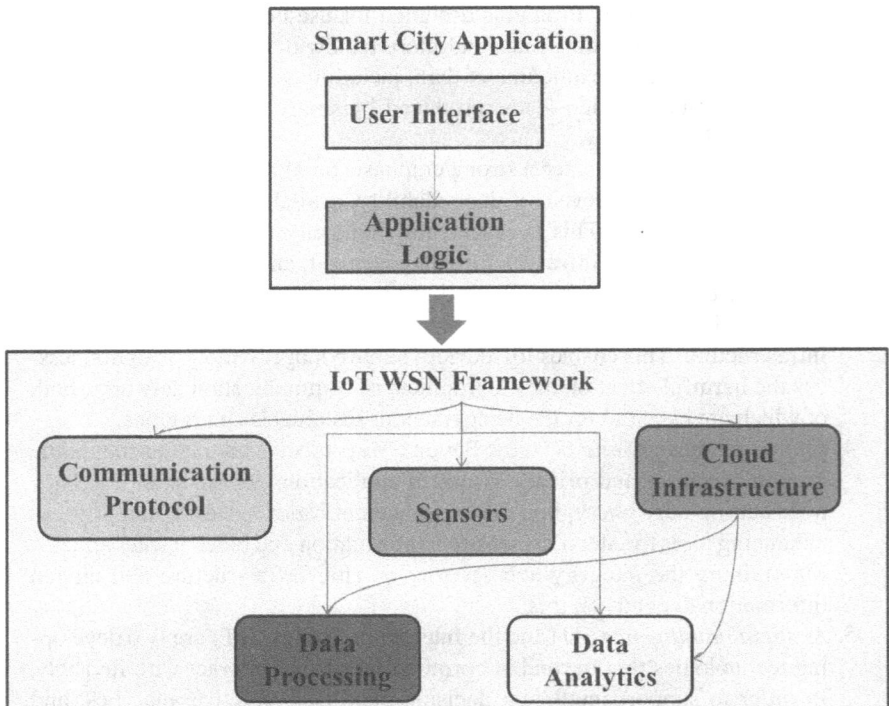

FIGURE 3.2 IoT-WSN framework with cloud infrastructure.

3.6.2 PROCESSING OF DATA

The massive amounts of data created by smart city applications may be efficiently processed on the cloud. It is up to the application to decide whether it needs the processing to happen in real time or in batches. Data warehouses, data analytics, and machine learning (ML) are only few of the technologies that can be used to perform the processing. Depending on the needs of the application, the processing can also be performed locally. Structured data, such as that found in databases, is well suited to the storage and processing capabilities of a data warehouse. Real-time data analysis and trend detection are two of the many uses for data analytics. Model training and data-driven inference are two applications where ML excels. In order to speed up responses and decrease latency, the processing might be done either at the edge or on-premise [35].

3.6.3 CLOUD SERVICES

Smart city applications can be built, deployed, and managed with the help of cloud services. Some of the services that may be found in the cloud are data analytics, machine learning, and IoT platform services. Auto-scaling, load balancing, and security are just some of the additional benefits that smart city applications can gain from cloud services. Data analytics services give you the means to examine data in real time and spot trends and patterns. Tools for training models and producing data-driven predictions are provided by ML services. Connecting devices and aggregating sensor data are made possible by the services offered by IoT platforms. Applications for the smart city will be responsive and able to process massive volumes of data thanks to auto-scaling and load balancing services. Data security and confidentiality are guaranteed by the security services provided. For these reasons, storing and processing the massive volumes of data produced by smart city applications may be done efficiently and affordably on the cloud. By providing a framework for building, deploying, and managing smart city applications, cloud services pave the way for the creation of cutting-edge new smart city software. The IoT-WSN architecture utilizes a cloud-based data storage and processing system to permit continuous monitoring and analysis of smart city systems in real time [36].

3.7 DATA PROCESSING

For smart city applications, the IoT-WSN framework requires data processing. Data created by sensors in a smart city's infrastructure must be gathered, stored, and analyzed. The primary objective of data processing is to derive useful information for city officials so they can make better decisions and enhance the effectiveness of local services. The following are the steps involved in data processing for smart city applications:

    i. Data collection is the starting point for processing information gathered by the smart city's sensors. Environment-related information can include temperature, humidity, air quality, and noise level.

ii. After information has been gathered, it must be archived in a data warehouse or the cloud. Data volume and application requirements will determine the optimal storage system.

iii. Sensor data is often imprecise and requires cleaning before it can be used. Prior to analysis, data must be cleaned to get rid of anomalies and inconsistencies.

iv. The process of transforming data into a readily analyzed format is known as "data transformation." Statistical approaches can be used to normalize the data, which may require aggregating data from several sensors or transforming the data into a time series.

v. The next stage is to analyze the data to draw conclusions. In order to analyze massive amounts of data and find patterns and links, smart city applications frequently employ ML techniques.

vi. After the data has been analyzed, it must be visualized so that it may be comprehended quickly and easily. The information can be displayed graphically using data visualization tools like charts and graphs.

vii. Processing large amounts of data requires a lot of time and energy. In order to process data, it is possible to use cloud computing platforms to access the required computer resources. When data is processed in the cloud, it can be analyzed in real time, allowing local officials to act immediately on the insights gained.

Data processing is an essential part of the IoT-WSN architecture for smart city uses. Data from smart city sensors must be gathered, stored, cleaned, transformed, analyzed, and visualized. Data processing can provide valuable insights that help city officials make better decisions and enhance the effectiveness of city operations [37].

3.8 MACHINE LEARNING ALGORITHMS

Data processing in an IoT-WSN framework for smart city applications can be greatly aided using machine learning (ML) methods. In vast datasets, these algorithms may spot patterns and connections that may be invisible to the human eye. As a result, city officials will be better able to make choices and streamline governmental operations. Data processing in smart city applications often makes use of the following ML algorithms:

i. *Supervised ML*: A supervised learning algorithm is an ML algorithm that is trained using an annotated data set. Input labels are used by the algorithm to learn how to produce the desired output. Data classification tasks, such as determining trash type or traffic congestion, are amenable to this approach.

ii. In an unsupervised learning ML approach, the system is trained using data that has not been labelled in any way. The algorithm is trained to discover connections and patterns in the data without being provided with any labels. Groups of buildings with similar energy consumption trends can be identified using this technique.

iii. Using trial and error to learn from its surroundings, a system employs an ML algorithm called reinforcement learning. The algorithm learns to maximize rewards by receiving rewards or avoiding punishments for its actions. By changing things like traffic lights to minimize congestion, this algorithm can help smart cities run more efficiently.

iv. To learn from data, deep-learning ML algorithms employ simulated neural networks. These neural networks are useful for voice and picture recognition because of their ability to recognize complicated patterns in data. Video camera data can be analyzed with deep-learning algorithms for use in smart city applications like crime prevention and infrastructure health monitoring.

v. Time series analysis is a statistical method for studying information collected over time. Predicting future trends or spotting abnormalities in data is a popular use case for this kind of analysis in smart city applications. Time series analysis has several applications, including the study of traffic patterns and the forecasting of energy needs.

The data processing capabilities of the IoT-WSN framework for smart city applications are greatly enhanced using ML techniques. To aid city administrators in making educated decisions and enhancing the effectiveness of city operations, these algorithms can detect patterns and links in big datasets. As shown in Table 3.3, the smart city industry makes extensive use of a variety of ML algorithms, including supervised learning, unsupervised learning, reinforcement learning, deep learning, and time series analysis.

TABLE 3.3
Comparative Study of Various Machine Learning Algorithms Used in an IoT-WSN Framework for Smart City Applications

Algorithm	Type	Use	Advantages	Disadvantages
Supervised Learning [38]	Classification and regression	Predicting future outcomes based on labeled historical data	Easy to use and interpret, accurate results	Requires large amount of labeled data
Unsupervised Learning [39]	Clustering and dimensionality reduction	Discovering hidden patterns and relationships in data	Does not require labeled data, can discover previously unknown patterns	Results may not be accurate due to the lack of labeled data
Reinforcement Learning [40]	Decision-making	Optimizing actions to maximize rewards in a dynamic environment	Can learn complex strategies, can adapt to changing environments	Requires a lot of training data, can be computationally expensive
Deep Learning [41]	Neural networks	Image and speech recognition, natural language processing	Can identify complex patterns in data, can learn from large datasets	Requires a lot of training data and computational resources
Time Series Analysis [42]	Statistical technique	Predicting future trends or identifying anomalies in time-series data	Can predict future trends with high accuracy, can identify anomalies in data	May not be effective for non-time-series data

3.9   ENSURING SCALABILITY

The system requires a storage and processing architecture which can scale to accommodate massive amounts of data. This is possible with the help of Hadoop, Spark, and Kafka, among other distributed computing frameworks. The importance of ensuring that the IoT-WSN architecture can handle the load and scale effectively grows with the number of connected devices. Scalable communication protocols like MQTT, CoAP, and AMQP can help with this. These protocols ensure that devices can reliably and efficiently connect to the cloud.

Bringing computer and storage resources closer to the devices is the goal of edge computing, a distributed computing paradigm. Particularly useful for smart city applications that depend on real-time data processing, this can assist in lowering the system's latency and bandwidth needs. The system can scale to a larger number of devices by shifting some of the work to the edge.

Containerization is a lightweight virtualization approach that provides a sandboxed environment for software to function in. This can make it easier to deploy and maintain applications within the IoT-WSN infrastructure. Containerization allows the system to readily scale up or down in response to changes in demand, which can boost its efficiency and scalability.

The IoT-WSN framework's success in smart city applications relies heavily on ensuring scalability. The system can handle a huge amount of data and many devices by incorporating mechanisms such as distributed file systems, distributed computing frameworks, scalable communication protocols, edge computing, and containerization.

3.9.1   LARGE AMOUNTS OF DATA

A typical difficulty encountered by IoT-WSN frameworks used in smart city applications is dealing with massive volumes of data. Large-scale use of sensors and other devices results in copious amounts of data that must be gathered, stored, and processed in real time before any useful insights can be drawn. This information can be gathered from a wide range of sources, including surveillance cameras, traffic monitors, and environmental sensors.

Numerous methods for handling massive volumes of data are available for dealing with this problem. Distributed computing is one approach; it includes partitioning the data into smaller chunks and processing them simultaneously across different servers. This facilitates task redistribution and expedites data processing. The information can also be stored in the cloud, making it available from any device, at any time. Data compression is another method, used to shrink files before they are stored or sent. In this class, lossless compression accomplishes this goal without information loss.

Another approach for handling massive amounts of data is data aggregation. In order to accomplish this, it is necessary to combine information from several resources into one comprehensive database. It is also crucial to make sure the data processing infrastructure can deal with massive amounts of data. This requires that data centers with high-bandwidth connections, distributed computing clusters, and high-performance servers can all work together. Data processing can be sped up and performance can be enhanced by employing specialized hardware like graphics processing units (GPUs).

Finally, in IoT-WSN frameworks for smart city applications, data management is a critical difficulty that must be addressed. Distributed computing, cloud-based storage solutions, data compression, data aggregation, and other methods, combined with high-performance infrastructure, can help with this. Further reduction in data amount and improvement in data quality can be achieved using efficient data management practices like filtering, cleansing, and reduction. Data filtering is the process of picking useful data based on criteria you provide. Data cleansing is the process of discovering and fixing data flaws. Data reduction eliminates extraneous or useless information. Data transmission to the cloud for processing can be minimized by adopting edge computing approaches. Instead of sending data to be processed in a centralized server or cloud, it is processed locally on the device or at the edge of the network in edge computing. This lessens the quantity of data that must be communicated, which in turn decreases latency and boosts the overall performance of the system.

In conclusion, IoT-WSN frameworks for smart city applications face a significant issue in managing massive amounts of data. Scalability and effective management of the massive amounts of data generated by such systems can be ensured using methods like distributed computing, cloud-based storage solutions, data compression, and aggregation, as well as efficient data management practices and edge computing.

3.9.2 LARGE NUMBER OF DEVICES

Many sensors can be spread out around the city and its infrastructure in IoT-WSN frameworks for smart city applications. Device discovery, configuration, and maintenance can be especially difficult when dealing with a high number of devices. To overcome this obstacle, a scalable architecture that can support many devices must be implemented. Devices, for instance, can be clustered or grouped according to their location or function using a hierarchical architecture. It is possible for each cluster to have its own coordinator device which would be in charge of administering the other devices in the group. This has the potential to improve network performance by decreasing traffic on the network.

In addition, using low-power wireless technologies like Zigbee, LoRaWAN, or Bluetooth can aid in reducing the power consumption of devices, which in turn can assist in extending the life of the battery and decrease the frequency of maintenance.

Adopting common protocols and interfaces that enable devices to communicate and interact with each other is crucial for easing the device discovery and configuration processes. Network devices can be found and managed through protocols like the Simple Network Management Protocol (SNMP). IoT device software and firmware can be managed and updated with the help of the Lightweight M2M (LwM2M) protocol. Devices can also be monitored and managed from a central place using remote device management services. These systems allow administrators to check on the status of each device, keep tabs on key performance indicators, and adjust the network's software and hardware from a single location.

When it comes to implementing IoT-WSN frameworks for smart city applications, device management presents a significant problem. Effective management of many devices and scalability as the system expands can be achieved using scalable designs, low-power wireless technologies, standard protocols and interfaces, and remote device management platforms. In addition, IoT-WSN frameworks for smart city applications can benefit greatly from edge computing's ability to solve scalability issues. Instead of sending all data to the cloud for processing, data is processed locally on devices or gateways in edge computing. For time-sensitive applications like traffic control, this can lessen network traffic and speed up reaction times.

One method for handling a multitude of gadgets is to make use of cloud computing services. When dealing with massive amounts of data, cloud computing's scalable and adaptable storage, processing, and analytics resources are useful. In addition to making it easier to scale the system as the number of devices and applications increases, cloud computing can also enable seamless integration of many IoT-WSN networks across different geographies.

In conclusion, in IoT-WSN frameworks for smart city applications, the management of a high number of devices is a major difficulty. Addressing this problem and guaranteeing scalability as the system expands can be accomplished using scalable designs, low-power wireless technologies, standard protocols and interfaces, remote device management platforms, edge computing, and cloud computing resources.

3.10 CONCLUSION

It is important to consider several key variables while designing an integrated IoT-WSN framework for smart city applications, such as sensor selection, communication protocols, cloud infrastructure, data processing, and security measures. Knowing the precise data requirements is essential for ensuring efficient data collection, processing, and storage. Additionally, using the right communication protocols and sensors is crucial for operation efficiency. Scalable and flexible resources for data storage, processing, and analytics are provided by cloud infrastructure when combined with ML algorithms. In IoT-WSN frameworks, data security is essential. This can be accomplished by security methods including data encryption, user authentication, and access control. Another crucial issue to overcome is scalability, particularly given the growing number of devices and applications. Edge computing and cloud computing resources can be combined to manage massive volumes of data and devices, and address scalability issues. A thorough grasp of data requirements, sensors, communication protocols, cloud infrastructure, data processing, security, and scalability is necessary to build an integrated IoT-WSN framework for smart city applications. It is feasible to create a dependable and effective framework that can support a variety of smart city applications by doing an in-depth study of these components and choosing the suitable technology.

REFERENCES

1. V. Hosseinnezhad et al., "IoT-based smart cities: A survey", *Proceeding of the 16th IEEE International Conference on Environment and Electrical Engineering (EEEIC)*, pp. 1–6, 2016.

2. V. Khetani, Y. Gandhi, S. Bhattacharya, S. N. Ajani and S. Limkar, "Cross-domain analysis of ML and DL: Evaluating their impact in diverse domains", *International Journal of Intelligent Systems and Applications in Engineering*, vol. 11, no. 7s, pp. 253–262, 2023.
3. N. C. Luong, D. T. Hoang, P. Wang, D. Niyato, I. D. Kim and Z. Han, "Data collection and wireless communication in the Internet of Things (IoT) using economic analysis and pricing models: A survey", *IEEE Communication Surveys Tutorials*, vol. 18, no. 4, pp. 2546–2590, 2016.
4. R. Khatoun, Hammi, S. Zeadally, A. Fayad and L. Khoukhi, "IoT technologies for smart cities", *IET Networks*, vol. 7, no. 1, pp. 1–13, 2018.
5. M. Frincu and R. Draghici, "Towards a scalable cloud enabled smart home automation architecture for demand response", *Proceeding of the IEEE PES Innovative Smart Grid Technologies Conference Europe (ISGT-Europe)*, pp. 1–6, 2016.
6. P. Vlacheas et al., "Enabling smart cities through a cognitive management framework for the internet of things", *IEEE Communications Magazine*, vol. 51, no. 6, pp. 102–111, 2013.
7. L. Sanchez et al., "SmartSantander: IoT experimentation over a smart city testbed", *Computer Networks*, vol. 61, pp. 217–238, 2014.
8. R. Lea and M. Blackstock, "City hub: A cloud-based IoT platform for smart cities", *Proceeding of the 6th IEEE International Conference on Cloud Computing Technology and Science (CloudCom)*, 2014.
9. A. Zanella, N. Bui, A. Castellani, L. Vangelista and M. Zorzi, "Internet of things for smart cities", *IEEE Internet of Things Journal*, vol. 1, no. 1, pp. 22–32, 2014.
10. R. Petrolo, V. Loscrì and N. Mitton, "Towards a smart city based on a cloud of things a survey on the smart city vision and paradigms", *Transaction on Emerging Telecommunication Technologies*, vol. 28, no. 1, pp. 54–58, 2017.
11. W. Shuai, P. Maillé and A. Pelov, "Charging electric vehicles in the smart city: A survey of economy-driven approaches", *IEEE Transactions on Intelligent Transportation Systems*, vol. 17, no. 8, pp. 2089–2106, 2016.
12. S. Djahel, R. Doolan, G.-M. Muntean and J. Murphy, "A communications-oriented perspective on traffic management systems for smart cities: Challenges and innovative approaches", *IEEE Communication Surveys Tutorials*, vol. 17, no. 1, pp. 125–151, 2015.
13. S. N. Ajani, P. V. Potnurwar, V. K. Bongirwar, A. V. Potnurwar, A. Joshi and N. Parati, "Dynamic RRT* algorithm for probabilistic path prediction in dynamic environment", *International Journal of Intelligent Systems and Applications in Engineering*, vol. 11, no. 7s, pp. 263–271, 2023.
14. K. E. Psannis, Stergiou, B.-G. Kim and B. Gupta, "Secure integration of IoT and cloud computing", *Future Generation Computer Systems*, vol. 78, pp. 964–975, 2018.
15. K. Psannis, Kim and H. Bhaskar, "Special section on emerging multimedia technology for smart surveillance system with IoT environment", *The Journal of Supercomputing*, vol. 73, no. 3, pp. 923–925, 2017.
16. A. Matheus, Chaturvedi, S. H. Nguyen and T. H. Kolbe, "Securing spatial data infrastructures for distributed smart city applications and services", *Future Generation Computing Systems*, vol. 101, pp. 723–736, 2019.
17. N. Lafioune and M. St-Jacque, "Towards the creation of a searchable 3D smart city model", *Innovation & Management Review*, vol. 17, no. 3, pp. 285–305, 2020.
18. S. Ajani and M. Wanjari, "An efficient approach for clustering uncertain data mining based on hash indexing and Voronoi clustering", 2013 5th International Conference and Computational Intelligence and Communication Networks, 2013, pp. 486–490.
19. S. Milovanov et al., "Building virtual 3D city model for smart cities applications: A case study on campus area of the University of Novi Sad", *ISPRS International Journal of Geo-Information*, vol. 9, no. 8, pp. 476, 2020, doi: 10.3390/ijgi9080476.
20. W. Lu, Xue, Z. Chen and C. J. Webster, "From LiDAR point cloud towards digital twin city: Clustering city objects based on Gestalt principles", *ISPRS Journal of Photogrammetry and Remote Sensing*, pp. 167, 2020.

21. A. Al-Fuqaha, M. Guizani, M. Mohammadi, M. Aledhari and M. Ayyash, "Internet of Things: A survey on enabling technologies protocols and applications", *IEEE Communications Surveys Tutorials*, vol. 17, no. 4, pp. 2347–2376, 2015.
22. A. Whitmore, A. Agarwal and L. Da Xu, "The Internet of Things a survey of topics and trends", *Information Systems Frontiers*, vol. 17, no. 2, pp. 261–274, 2015.
23. W. Zhou, Y. Jia, A. Peng, Y. Zhang and P. Liu, "The effect of IoT new features on security and privacy: New threats existing solutions and challenges yet to be solved", *IEEE Internet of Things Journal*, vol. 6, no. 2, pp. 1606–1616, 2019.
24. J. Kim, Gilani, J. Song, D. Seed and C. Wang, "Semantic enablement in IoT service layers—Standard progress and challenges", *IEEE Internet Computing*, vol. 22, no. 4, pp. 56–63, 2018.
25. S. C. Mukhopadhyay and N. Suryadevara, "Internet of Things: Challenges and opportunities", Internet of Things, 2014.
26. P. El-Mougy, M. Ibnkahla and L. Hegazy, "Software-defined wireless network architectures for the Internet-of-Things", *IEEE Local Computer Networks Conference Workshops*, pp. 804–811, 2015.
27. R. Vilalta et al., "Improving security in Internet of Things with software defined networking", *IEEE Global Communications Conference (GLOBECOM)*, pp. 1–6, 2016.
28. J. Yick, B. Mukherjee and D. Ghosal, "Wireless sensor network survey", *Computer Networks*, vol. 52, no. 12, pp. 2292–2330, 2008.
29. C. Perera, A. Zaslavsky, P. Christen, M. Compton and D. Georgakopoulos, "Context-aware sensor search, selection and ranking model for internet of things middleware," in *2013 IEEE 14th International Conference on Mobile Data Management*, 2013, vol. 1, pp. 314–322.
30. F. Viani, A. Polo, M. Donelli and E. Giarola, "A relocable and resilient distributed measurement system for electromagnetic exposure assessment", *IEEE Sensors Journal*, vol. 16, no. 11, pp. 4595–4604, 2016.
31. B. Afshar, M. Fathy, M. Asgari, M. Shahverdy and P. Shahverdi, "A machine learning-based approach to detect polluting vehicles in smart cities", *2022 Sixth International Conference on Smart Cities, Internet of Things and Applications (SCIoT)*, Mashhad, Iran, Islamic Republic of, 2022, pp. 1–5, doi: 10.1109/SCIoT56583.2022.9953644
32. M. Aboubakar, M. Kellil, A. Bouabdallah and P. Roux, "Using machine learning to estimate the optimal transmission range for RPL networks", *NOMS 2020 - 2020 IEEE/IFIP Network Operations and Management Symposium*, Budapest, Hungary, 2020, pp. 1–5, doi: 10.1109/NOMS47738.2020.9110297
33. X. Liu and E. Ngai, "Distributed machine learning for Internet-of-Things in smart cities", *2019 IEEE International Conference on Industrial Internet (ICII)*, Orlando, FL, USA, 2019, pp. 368–374, doi: 10.1109/ICII.2019.00069
34. S. U. N. Goparaju, S. S. S. Vaddhiparthy, C. Pradeep, A. Vattem and D. Gangadharan, "Design of an IoT system for machine learning calibrated TDS measurement in smart campus", *2021 IEEE 7th World Forum on Internet of Things (WF-IoT)*, New Orleans, LA, USA, 2021, pp. 877–882, doi: 10.1109/WF-IoT51360.2021.9595057
35. T. Khandelwal, Muskan, M. Khandelwal and P. S. Pandey, "Women safety device designed using IoT and machine learning", *2018 IEEE SmartWorld, Ubiquitous Intelligence & Computing, Advanced & Trusted Computing, Scalable Computing & Communications, Cloud & Big Data Computing, Internet of People and Smart City Innovation (SmartWorld/SCALCOM/UIC/ATC/CBDCom/IOP/SCI)*, Guangzhou, China, 2018, pp. 1204–1210, doi: 10.1109/SmartWorld.2018.00210
36. N. Zenasni, C. Habib and J. Nassar, "A fuzzy logic based offloading system for distributed deep learning in wireless sensor networks", *2022 International Joint Conference on Neural Networks (IJCNN)*, Padua, Italy, 2022, pp. 1–8, doi: 10.1109/IJCNN55064.2022.9892817

37. M. Anjum, M. A. Khan, S. A. Hassan, A. Mahmood, H. K. Qureshi and M. Gidlund, "RSSI fingerprinting-based localization using machine learning in LoRa networks", *IEEE Internet of Things Magazine*, vol. 3, no. 4, pp. 53–59, 2020, doi: 10.1109/IOTM. 0001.2000019

38. M. Benedetti, L. Ioriatti, M. Martinelli and F. Viani, "Wireless sensor network: A pervasive technology for earth observation", *IEEE Journal of Selected Topics in Applied Earth Observations and Remote Sensing*, vol. 3, no. 4, pp. 488–496, 2010.

39. M. Mesiti, S. Valtolina, L. Ferrari, M. S. Dao and K. Zettsu, "An editable live ETL system for Ambient Intelligence environments", in 2015 IEEE 2nd World Forum on Internet of Things (WF-IoT), 2015, pp. 393–394.

40. S. Obermeier, S. Böttcher and D. Kleine, "CLCP #150; A distributed cross-layer commit protocol for mobile ad hoc networks", in *2008 IEEE International Symposium on Parallel and Distributed Processing with Applications*, 2008, pp. 361–370.

41. SO. Flauzac, C. Gonzalez, A. Hachani and F. Nolot, "SDN based architecture for IoT and improvement of the security", *IEEE Information Networking and Applications Workshops*, pp. 688–693, 2015.

42. Mishra and M. Kumar, Role of Technology in Smart Governance: Smart City Safe City, March 2017, [online] Available:

4 Principles of Artificial Intelligence in the Internet of Things

Ankita Sharma and Shalli Rani

4.1 INTRODUCTION

The term "smart" is frequently used to characterize technologies that can do tasks autonomously. Even though cellphones are deemed "smart," they cannot perform many tasks automatically. When the owner is driving, a smartphone, for example, cannot automatically turn off notifications or message alerts. A smarter smartphone might be able to eliminate driving distractions caused by alarms. This might be accomplished by using a wireless link between the user, their smartphone, and the car. In another case, if the user becomes ill, the smartphone should be able to call for emergency assistance or a family member. This might be accomplished by utilizing the smartphone's global positioning system (GPS) to determine the owner's location and then dialing 911 or another emergency number automatically. These are just a few ideas for making smartphones smarter.

As technology advances, we may expect to see many more methods for smartphones to assist us in our daily lives. The physical world is becoming more and more interconnected. With the rise of the Internet of Things (IoT), a network of physical items embedded with sensors and software, we are already seeing this trend. These items can gather and communicate information about their surroundings, which can then be used to improve efficiency, safety, and productivity. Smart thermostats, for example, may learn our patterns and alter the temperature in our houses accordingly. Smart traffic lights can communicate with one another in order to improve traffic flow. Smart parking meters can also notify us when a parking space becomes available. As IoT expands, we will require AI to make these gadgets and objects "smart." AI will enable people to understand their surroundings, learn from their experiences, and make independent judgements [1]. This will relieve us of the need to manually control these devices and allow us to focus on other activities. In line with John McCarthy's viewpoint [2], AI is the field of study and application that aims to imbue machines with intelligence, enabling them to comprehend human language, solve problems, and pursue objectives just like humans. Nevertheless, some authors suggest that AI's ultimate goal is to metaphorically replicate the human mind within a computer. One method used to determine machine intelligence is the Turing Test. AI-powered self-driving cars, for example, will be able to navigate roadways safely and efficiently without human intervention. More accurate than human doctors, AI-powered medical equipment will be able to identify illnesses and suggest therapies. Furthermore, AI-powered robots will be able to perform dangerous

88

DOI: 10.1201/9781003437079-4

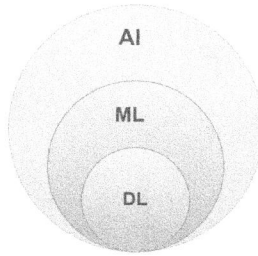

FIGURE 4.1   Relationship between AI, ML, and DL.

or repetitive activities currently performed by people. AI techniques are the tools and procedures used to create AI systems [3], and include machine learning (ML), deep learning (DL), neural networks, natural language processing (NLP), computer vision, robotics, and expert systems.

- ML is a subset of AI, as so it is a form of AI that enables computers to learn without being explicitly programmed, as shown in Figure 4.1. The ML algorithms provide the data to train the machines, and the machines then utilize that data to learn how to make predictions or judgments.
- DL is a subset of ML and AI, and learns from data using artificial neural networks.
- Neural networks, which are inspired by the human brain, can learn complicated patterns from data.
- Natural language processing (NLP), is a field in computer science that explores the interaction between computers and human (natural) languages. This tool is used to extract meaning from text, translate languages, and generate text.
- Computer vision is a branch of computer science that deals with the interpretation of digital images and movies. The algorithms are used to recognize objects, track motion, and produce three-dimensional representations.
- Robotics is an engineering discipline concerned with the design, manufacture, operation, and use of robots. The techniques are used to create robots that can conduct activities on their own.
- Expert systems are computer programs that answer issues by combining knowledge and reasoning. These are frequently employed in medical, financial, and legal domains.

For the advancement in AI technology, we should expect to see more sophisticated AI techniques developed. AI is a rapidly evolving field that is still in its infancy [4]. One of the most difficult aspects of AI is developing machines that are capable of learning and thinking like humans. Another problem is ensuring that AI systems are ethical, fair, and transparent. This means they should not be used to discriminate against or harm people, and their choices should be clear and responsible. A further problem is that AI applications must handle privacy, security, and job displacement concerns. AI systems acquire and utilize a large amount of data,

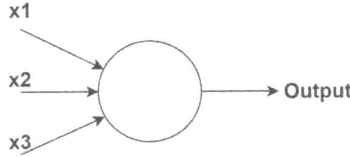

FIGURE 4.2 The working of AI.

raising privacy concerns. They can also be used to compromise systems and steal data, raising security problems. There is concern that AI will result in widespread job displacement as computers become capable of performing more and more tasks currently performed by humans.

The x_1, x_2, x_3 are the inputs to a neural system; the center circle represents a transfer function. It is an analytical response to the inputs. The answer will take the inputs into account and generate just one result. The working of AI is shown in Figure 4.2.

4.2 AI PRINCIPLES

As AI technology develops, it is critical that humans continue revising and improving the principles of AI to guarantee that AI is utilized for beneficial, rather than harmful, effects. The eight most fundamental principles of AI are discussed here and are also shown in Figure 4.3.

- *Human-centered values and fairness*: AI systems should be constructed and refined in such a way that human values are respected, and justice is promoted. This means that AI systems should not be employed to prejudice or harm people, and they should be built to be open and accountable to mankind.

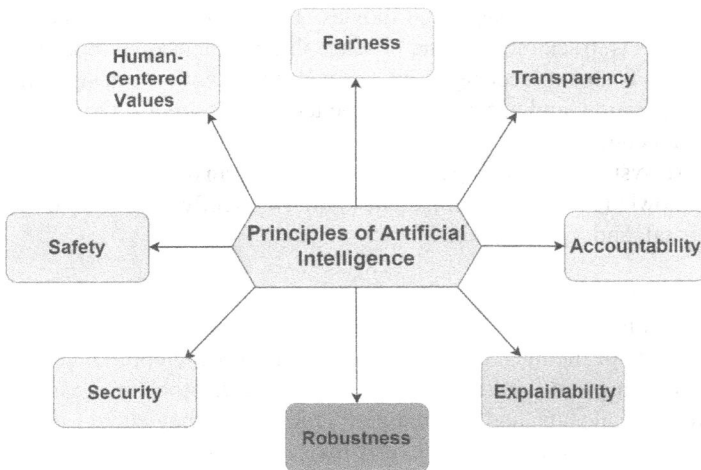

FIGURE 4.3 AI principles.

- *Transparency and explainability*: Humans should be able to understand and explain AI systems. This implies individuals need to comprehend how AI systems operate as well as why they make the judgments they do. This is critical for a variety of reasons, including making sure that AI systems are impartial and equitable and that users can depend on them [5].
- *Robustness, security, and safety*: AI systems must be strong, protected, and trustworthy. This implies they must be resistant to threats and unanticipated inputs, and they must not endanger users or the surroundings.
- *Accountability*: Humans should hold AI systems accountable. This implies that there must be explicit regulations and processes in place for holding AI system developers and users accountable for their conduct. This is critical for ensuring that AI systems are used responsibly and ethically [2].

4.3   ARTIFICIAL INTELLIGENCE AND THE INTERNET OF THINGS

The Internet of Things (IoT) is an interconnected collection of peripheral devices equipped with software, cameras, sensors, and network connectivity in order to gather and transfer data. Artificial intelligence (AI) is an area of computer science concerned with the development of autonomous beings or systems capable of reasoning, learning, and acting intelligently [6]. Many industries, including manufacturing, healthcare, public transit, and energy, have the potential to be transformed by the combination of IoT and AI. For instance, AI can be used to evaluate data generated by IoT devices in order to boost product quality, predict equipment breakdowns, and regulate traffic patterns.

AI research applied to IoT is no longer up to date. There have been numerous proposals and concepts for AI-based IoT applications in the past.

After reading the signals from nature, the sensors translate them into electrical voltages that may be processed and sent to the receiver. After that, an app is used to display the signals. The internet portal also allows users to view the signal details [7]. It is not unexpected that a lot of individuals today utilize speech detection in IoT to operate equipment. A recorder with a training system must be set up before employing voice control on appliances so that the sound can be detected.

Data mining is another way artificial intelligence is used in IoT. Data handling and storage space reduction are accomplished through the use of data mining. This implies that there will be a propensity to take longer to find the needed data as the amount of data in the network increases. Data mining technique is used to shorten the time needed to find the desired data. A diagram of the steps used in data mining is shown in Figure 4.4.

- *Data collection*: The gathering of the data is the initial phase. Many other sources, including records, transactional logs, and questionnaires, can provide this information.
- *Data cleaning*: After data has been gathered, it must be cleaned to eliminate any inaccuracies or discrepancies. This is a crucial stage because even minor mistakes can significantly affect the outcomes of the process [8].

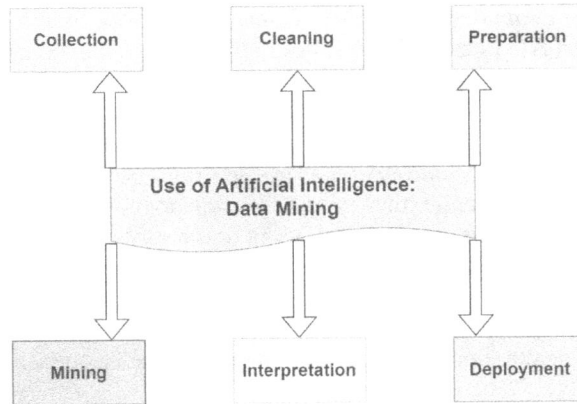

FIGURE 4.4 Use of AI data mining.

- *Data preparation*: The following step is to get the data ready for the process
  of data mining, which could entail converting it into a format that is compatible with data mining techniques or generating new features from what
  is already there.
- *Data mining*: In the following stage, technologies are used to look for developments and patterns in the information being collected. There are numerous data mining methods available, and each one is appropriate for a certain
  kind of data mining task.
- *Data interpretation*: The outcomes of the process need to be interpreted
  after application. This entails comprehending the implications of the findings and how they relate to the commercial queries that drove the data mining operation [9].
- *Data deployment*: The data mining project's outcomes must be released as
  a final step. This can entail developing a model that can be applied to forecasting, or it might entail writing a report that compiles the outcomes of the
  data mining operation.

4.4 AI TECHNIQUES

The main AI techniques are ML, DL, NLP, computer vision, neural networks,
and robotics, and the first four of these are discussed in detail in this chapter.
Underpinning these techniques are reinforcement learning, transfer learning, and
evolutionary algorithms. The authors highlight the significance of DL in [6], an
effective technique for image recognition, audio recognition, and NLP. They emphasize the wide-ranging applications of DL and the significant advancements it has
brought to the field of AI. The report by the authors in [10] underscores the value
of reinforcement learning, a robust method for training agents to make decisions
in complex scenarios. It discusses how reinforcement learning has proven effective
in developing AI systems for games such as Go and chess, showcasing its broad
application across different domains. According to authors in [9], transfer learning

refers to the capability of AI models to leverage knowledge acquired from one task to enhance performance on another task. This approach has demonstrated its effectiveness across various applications, such as computer vision and natural language processing. The study conducted by authors [11] emphasizes the immense potential of transfer learning and its wide-ranging applications across different industries. In the study conducted by authors in [12], evolutionary algorithms are defined as optimization methods that draw inspiration from the principles of natural selection. These techniques have proven to be highly effective in addressing various optimization challenges in AI, including tasks like neural network construction and robot control. The report emphasizes the efficacy of evolutionary algorithms in optimizing complex systems.

The following subsections discuss four of the six techniques used in AI which are ML, DL, computer vision, NLP, robotics, and neural networks, as shown in Figure 4.5.

4.4.1 MACHINE LEARNING

One of the subfields of AI is called machine learning (ML), and it may be one of the most crucial ones [13]. ML relies on the improvement of its own methods [14], and this depends on the evaluation of data, the application of some algorithms for processing data, and the application of other methods for making choices based on the already studied data.

Making intelligent data analysis easier, using learning algorithms, is one of the objectives of ML. Learning algorithms come in a variety of forms and can be categorized based on the results they produce. Different types of algorithms fall into this category, the most important being decision trees (DT), k-nearest neighbors (kNN), support vector machines (SVM), and naive bayes (NB). NB is the parametric

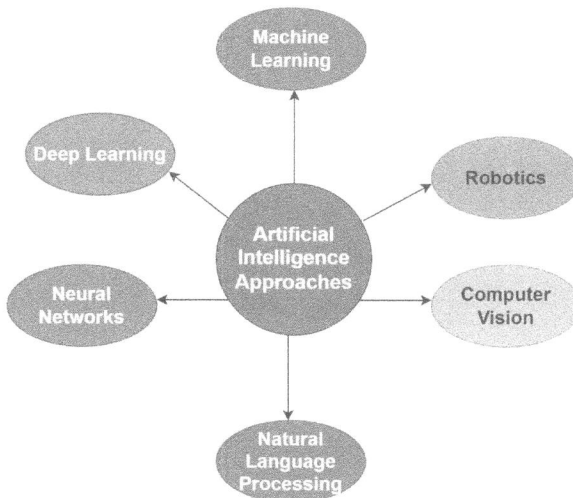

FIGURE 4.5 Techniques of AI.

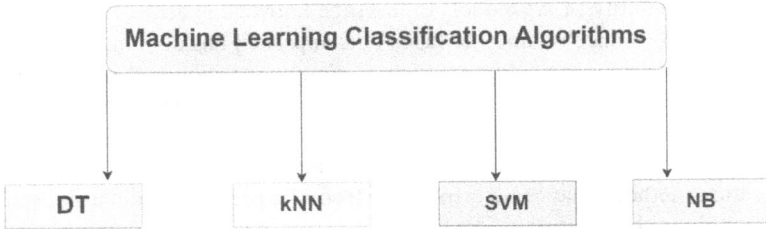

FIGURE 4.6 ML classification algorithms.

technique based on the parameters of labeled data and depends upon the assumption of data. DTs, kNN, and SVM are the non-parametric technique which does not require any labeled data. DTs work on the number of nodes along with the splitting parameter and gives the best results on categorical and numerical values of the dataset. KNN works on the number of nodes and depth. SVM works on high-dimensional data which divides the number of classes on a hyperplane [15]. These four types of ML algorithms are shown in Figure 4.6.

According to Figure 4.7, which compares different machine classifiers, NB is the least accurate because it works on assumptions, which is a vague method, as compared to random forest, DT, and SVM, which are based on different surveys.

4.4.2 DEEP LEARNING

A group of algorithms known as "deep learning" (DL) are built on modeling abstractions at the highest levels made up of several nonlinear modifications. The use of DL allows for the learning of data representations [14]. For instance, there are numerous methods to represent a picture, but not all of them are effective for performing the task of face recognition. Therefore, DL seeks to identify the optimum representation technique for completing this work. An example would be its use in computer vision, to recognize speech and music.

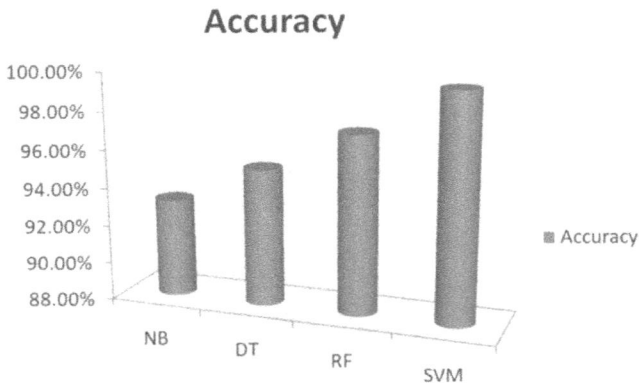

FIGURE 4.7 Comparison of the accuracy of ML classification algorithms.

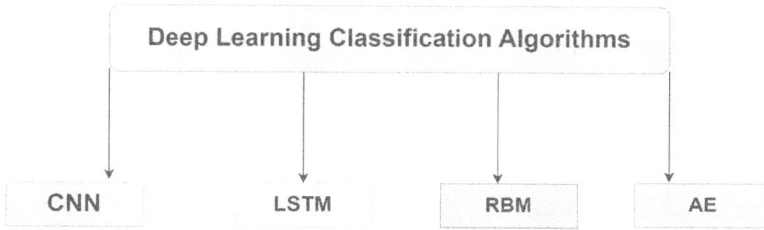

Deep Learning Classification Algorithms

CNN   LSTM   RBM   AE

FIGURE 4.8 DL classification algorithms.

The different types of DL classification algorithms are convolution neural networks (CNN), long short-term memory (LSTM), auto encoders (AE), restricted Boltzmann machine (RBM) [16]. The artificial neural network (ANN), which is a DL classifier, proved the best among all the other classifiers in handling difficult problems. There are several applications of ANN, such as voice recognition, pattern recognition, and handwriting recognition. This model works on data gathering from all the agents and then feeding the data to the ANN. The supervised training of the training phase is presented to an ANN, where both inputs and outputs contain the parameters of events and attacks [11]. The different types of DL algorithms are shown in Figure 4.8.

According to Figure 4.9, the comparison of DL classification algorithms shows that CNN have the highest accuracy and AE have the lowest accuracy, as compared to others.

4.4.3 COMPUTER VISION

Computer vision is one of the subfields of AI. It is a technique that enables computer systems to develop the ability to identify an image and its features. Machine

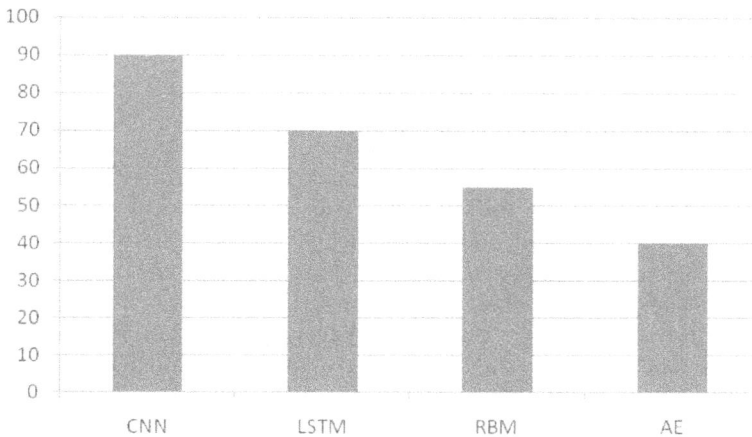

FIGURE 4.9 Comparison of DL classification algorithms.

| Feature Extraction | → | Candidate Generation | → | Classification | → | Performance Evaluation |

Final
Decision

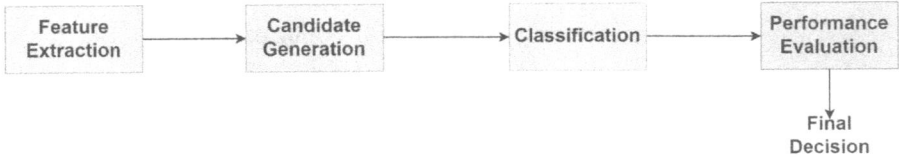

FIGURE 4.10 Computer vision.

understanding of the world is the aim of computer vision [17]. Enabling computers to quickly calculate distances and locations, identify objects based on their appearance, and recognize what they are. But completing this duty is fraught with difficulties. The primary challenge is in modeling the items to ensure the computer can distinguish between multiple kinds, for example between people and cars. However, there are numerous angles from which to view these. Here's the challenge: figuring out if a person is seated, standing, behind a car, or crouching, or even if the car is facing you or not. Along with day and night, there are varying light levels during the day, such as cloudy or sunny. The four steps of computer vision are feature extraction, candidate generation, candidate classification, and performance evaluation. The working of computer vision is shown in Figure 4.10.

4.4.4 NATURAL LANGUAGE PROCESSING

Natural language processing (NLP), also known as mathematical linguistics [8], is a subfield of AI that aims to address the issue of machine comprehension of natural language, or, to put it another way, to enable users to converse with machines more quickly and effectively using natural language as if they were conversing with other people. NLP necessitates an understanding of phrase structure, but it also calls for expertise in the subject content and the context [18]. The study of many points of view is necessary in the vast field of NLP in order to properly process speech. NLP must therefore be explored from a variety of linguistic perspectives [9].

4.4.5 ROBOTICS

Robotics is a multidisciplinary field of engineering and science that encompasses mechanical engineering, electronic engineering, information engineering, computer science, and other related disciplines. It encompasses the disciplines of designing, constructing, operating, and utilizing robots, along with developing computer systems to control them, provide sensory input, and process information. These technologies are utilized to create machines capable of replacing people and imitating human behaviors. Robots are utilized in various scenarios and serve multiple functions. Presently, they are predominantly employed in hazardous settings such as inspecting radioactive materials, detecting and deactivating bombs, and participating in manufacturing processes. Additionally, robots are deployed in environments where human presence is unfeasible, such as outer space, underwater, high-temperature conditions, and the cleanup and containment of dangerous substances and radiation. With ongoing technological progress, it is quite probable that future robots will assume a broader spectrum of responsibilities, potentially being seamlessly integrated into various parts of everyday life and providing aid in personal and household domains. Robotics research is progressing the development of artificial intelligence (AI) systems, allowing robots to independently make decisions and adjust to different surroundings.

4.4.6 NEURAL NETWORKS

Neural networks are a crucial element of artificial intelligence (AI), drawing inspiration from the intricate structure and functioning of the human brain. They serve as the foundation for deep learning, a kind of machine learning in which computers create models of complex concepts in data by utilizing structures consisting of numerous layers of processing. A neural network is composed of interconnected nodes, or "neurons," arranged in layers, with each neuron typically carrying out a basic operation. The input layer receives the initial data, which subsequently traverses one or more hidden layers where the processing occurs via a network of weighted connections. The ultimate layer, also known as the output layer, provides the outcome. Neural networks excel in their capacity to acquire patterns and characteristics from data. The weights of connections are modified by a process called training, which involves adjusting them according to the input received and the output generated. The training process often involves utilizing a predetermined collection of established inputs and outputs, which allows the network to acquire the ability to accurately represent intricate relationships and patterns. Neural networks are very adaptable and potent, employed in several domains such as picture and audio identification, language comprehension, medical diagnosis, financial prediction, and numerous more fields within artificial intelligence. AI research remains highly focused on the ongoing progress and improvement of neural networks, which are crucial for advancing both theoretical knowledge and practical implementations.

4.5   AI-ENABLED IoT

IoT and AI are two fast-developing technologies that have an opportunity to affect numerous sectors and enhance how people live and work. When integrated, both AI and IoT can produce potent synergies which enable cutting-edge applications and service development [19]. AI capabilities are incorporated into IoT systems to enable devices that are connected to gather and analyze data, make intelligent judgments, and automate procedures. This is known as AI-powered IoT. Here are some essential elements of IoT powered by AI. The working of AI in IoT is shown in Figure 4.11.

- *Data gathering and sensor synchronization*: IoT devices come with sensors that gather information about the outside world. The data from these sensors can be processed by AI algorithms to reveal important patterns and insights. To optimize energy use and improve comfort, AI algorithms may analyze data such as relative humidity, temperature, and occupancy collected by sensors in smart homes.
- *Data analysis*: IoT data can be analyzed by AI algorithms in real-time, allowing for quick insights and decisions. AI-powered IoT systems, for

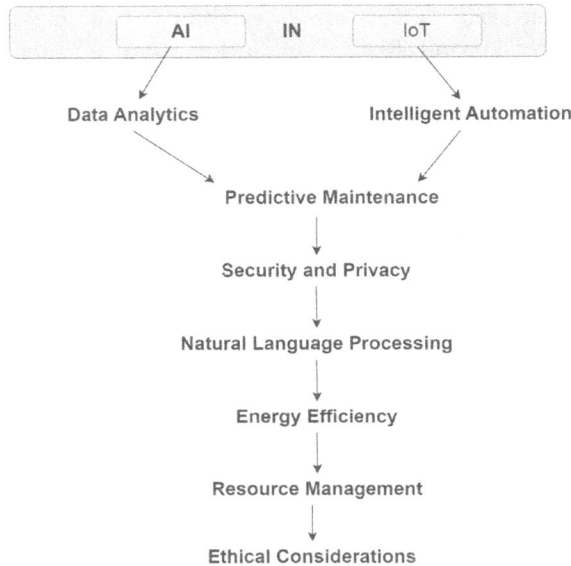

FIGURE 4.11 AI-powered IoT.

instance, may track machinery sensors for inconsistencies in industrial settings and automatically initiate calls for service or closure procedures to
avoid equipment failure.
- *Predictive analytics*: IoT systems may learn from past data to produce
  forecasts and predictions by utilizing ML and AI approaches. This can be
  helpful in several areas, including supply chain optimization, healthcare
  surveillance, and predictive maintenance. AI systems, for instance, can
  analyze data from smart meters to forecast patterns of electricity usage and
  improve distribution of energy [13].
- *Automation*: Based on data analysis and specified rules, AI-powered
  IoT systems might streamline decision-making and control procedures.
  Increased effectiveness and cost savings may result from this. AI algorithms, for instance, can analyze traffic patterns in smart cities, improve
  traffic lights, and dynamically change routes to ease congestion [12].
- *Personalization*: IoT devices with AI capabilities can learn from user
  preferences and behavior to provide individualized experiences. AI algorithms, for instance, can examine data from wearables to comprehend a
  user's fitness objectives and offer personalized exercise recommendations [20].
- *Computing*: Instead of depending only on cloud-based solutions,
  AI-powered IoT frequently uses edge computing, where data processing
  and analysis take place locally on the IoT devices or edge servers. Edge
  computing allows for real-time decision-making in sparsely connected
  locations while lowering latency and enhancing privacy.

FIGURE 4.12 Steps for AI-IoT.

- *Security and privacy*: To safeguard sensitive data and prevent unauthorized access, AI-powered IoT devices need to implement strong security measures. By examining network traffic and device behavior, AI algorithms can help identify anomalies and potential security breaches, improving system security. The pictorial representation of steps in AI-IoT is shown in Figure 4.12.

4.5.1 OPEN CHALLENGES

There are certain open challenges for AI-IoT from different points of view, such as technology, societal impact, and complex systems. These are discussed here.

4.5.1.1 Technology Point of View

In terms of technology, advancements are expected to drive the progress of AI in three key areas: physical devices platform, patterns governing algorithms, and interfaces. The central processing unit (CPU) primarily handles general-purpose computing tasks. While it can also handle AI-related tasks, its performance is relatively subpar. Consequently, there is a growing trend toward developing high-performance AI platforms. Renowned companies like Intel, Google, NVIDIA, Cambrian, and others have devised novel smart technologies, such as the graphics processing unit (GPU). The GPU is poised to supplant the central processing unit (CPU) as the computational and calculation terminal for the next generation. While IoT and AI present many opportunities, there are several open challenges. In the future, as artificial intelligence (AI) faces increasingly complex and challenging tasks, it will require a new computing architecture that can handle malicious requirements. This shift toward designing intelligent platforms based on diverse processors is a prominent trend. Present computer systems are ill-suited for multidimensional and multipath parallel processing. The AI does not work properly in those scenarios where there is a small amount of computation but significant changes, substantial mutual influence, and numerous nonlinear factors. Nonetheless, harnessing the power of quantum computing and real-time multidimensional processing capabilities offers a transformative solution to address the challenges posed by nonlinear multivariate interaction and complex prediction. Consequently, the significant progress in quantum computing technology, specifically the successful development of quantum computers, holds immense potential in establishing a revolutionary computing platform for future AI applications.

4.5.1.2 Society Point of View

Is it truly possible for machines to replicate the minds of human beings and comprehend intensions? The process of judging and evaluating things involves not only common sense but also the influence of personal preferences and emotions in humans. In contrast, AI relies solely on available data to make rational assessments or decisions. This is a complex issue that requires extensive contemplation and research. The advancement and widespread adoption of AI also gives rise to a challenging problem: intelligent machines, apart from being high-tech products, also have an impact on the rules and regulations of human society. Machine behavior remains somewhat opaque, making it susceptible to biases and errors that can result in ethical concerns. When machines propose novel ideas that differ from human beliefs, how should humans respond? As AI increasingly acquires knowledge and skills that were traditionally exclusive to humans, individuals may develop a growing fear of being replaced by machines, leading to reduced motivation for learning and work. These factors can engender psychological anxiety and panic among humans.

The swift advancement of AI has bestowed remarkable convenience and vast prosperity upon society. However, it has also raised numerous concerns regarding various social issues. Who will take responsibility for ensuring the controlled use of AI? What ethical considerations and responsibilities should be attributed to AI? Safety is the fundamental requirement for any technological innovation, including AI. AI depends on various computational algorithms, and extensive datasets. The expansion of the Internet and the proliferation of big data have introduced an element of unpredictability to AI security. On one hand, AI greatly benefits from the abundance of resources offered by big data on the Internet. On the other, the Internet is also a breeding ground for hackers and viruses that pose a substantial threat to AI systems.

While AI greatly enhances social productivity, it also profoundly impacts people's way of life in society, making it a delicate matter for ethicists. The prevention of ethical and liability issues stemming from AI has long been a focus of many professionals. For instance, who should be held accountable for traffic accidents caused by AI? Who should be held accountable in cases of medical mishaps involving smart robotics? Moreover, when it comes to security, AI possesses the capability to self-correct based on knowledge, automation, and independent decision-making. However, humans currently lack complete control over AI, which can potentially result in unforeseen outcomes. In terms of privacy and security, intelligent systems can extract extensive user-related data with the help of mining of seemingly non-related data, allowing for the identification of individual behavioral patterns and even personality traits. The traditional measures for security, related to anonymization of data, prove inefficient against AI's capacity to learn and infer data repeatedly, thereby rendering personal privacy more vulnerable to public exposure.

4.5.1.3   Complex System Point of View

The current implementation of AI in communication networks introduces greater complexity to the system. This complexity arises from the use of case-specific ML techniques, which prioritize achieving a single objective while disregarding other objectives and overall network goals. Consequently, a significant challenge that remains prominent is the focus of research on optimizing one objective, while neglecting important factors such as latency, link quality, storage limitations, and processing overhead in the network as a whole. For instance, a study in [21] proposes increasing spectral efficiency through reinforcement learning but fails to address the potential costs associated with information sharing, storage, and processing when applied to real-world scenarios or large-scale networks. The open challenges are as follows:

* *Interoperability and standardization*: IoT platforms and devices frequently work in contexts with a variety of protocols and standards. Effective integration and cooperation between AI and IoT technologies depend on achieving smooth interoperability and standardization across devices, communication protocols, and data formats [5].
* *Scalability and performance*: Scalability becomes a key difficulty as the number of linked devices increases. Advanced infrastructure, edge computing capabilities, and effective AI algorithms are needed to handle the enormous volume of data created by IoT devices and to process it in real-time while preserving optimal performance and low latency [2].
* *Energy efficiency*: Numerous IoT devices rely on batteries or have constrained power sources. The task of creating AI algorithms that are energy efficient and optimize power usage in IoT devices without sacrificing performance is ongoing. It is necessary to investigate methods like model compression, energy-conscious scheduling, and adaptive power management.
* *Ethical and social implications*: The employment of AI in IoT poses ethical questions about responsibility, fairness, bias, and privacy. Clear standards, laws, and frameworks are necessary to address potential societal and ethical concerns and ensure the ethical development, deployment, and usage of AI-powered IoT devices [22].
* *Regulatory and legal frameworks*: The creation of regulatory and legal frameworks frequently lags behind the quick development of AI and IoT

technology. Industry, legislators, and legal experts must constantly collaborate to address concerns including data ownership, liability, privacy laws,
and ethical standards.

- *Edge to cloud balance*: It can be difficult to choose between running AI
  computations in the cloud or at the edge. While some jobs benefit from
  cloud computing and storage, others necessitate real-time processing and
  decision-making at the edge. For AI-powered IoT systems to be effective
  and efficient, the proper balance must be struck.

4.6 COLLABORATION BETWEEN IoT AND AI

AI and IoT have reached advanced stages of development, and their collaboration
holds tremendous potential for numerous advantages. IoT, often regarded as the cata-
lyst for the Fourth Industrial Revolution (Industry 4.0), has spurred technological
advancements and transformations across diverse domains. It is widely acknowledged that AI is crucial for the progress of IoT, with many experts asserting that AI
is futuristic. In fact, the integration of IoT and AI has been a prevalent practice in
various industries and sectors for quite some time. IoT serves as a data collection
mechanism, generating vast volumes of data, while AI serves as the ideal tool for
effectively interpreting and extracting insights from such extensive data. AI acts as
the decisional engine used to carry out analysis and different data pre-processing
techniques. By leveraging AI, patterns can be understood, leading to more informed
decision-making.

The utilization of ML, combined with the power of big data, has created
new prospects and possibilities within the realm of IoT. Amazon's voice assistant, Alexa [23], exemplifies the process of data collection. However, the subsequent steps of organizing, analyzing, and making informed decisions based on
that data are distinct challenges. It is evident that for AI to enhance its utility in
IoT, there is a need for the development of more precise and efficient algorithms
and tools. By combining IoT with AI, enterprises can leverage the most effective
approach to optimize their store operations and ensure long-term sustainability.
The integration of IoT and AI enables retailers to achieve various benefits, including minimizing theft and maximizing sales through techniques such as cross-
selling. This synergy empowers enterprises to enhance their overall performance
and achieve greater success in the retail industry. The study [24] explores the
interaction between AI and IoT, emphasizing that AI and ML in the realm of
data science go beyond merely applying statistical predictive algorithms to IoT.
The study stated that intelligent system for IoT differs from existing databases,
as it involves specialized techniques tailored to handle time series data, including average methods and similar approaches. To achieve significant outcomes, AI
relies on big data. In fact, AI has the ability to address the challenges associated
with big data analytics.

Due to the vast quantity of data generated by numerous smart devices/objects,
humans face limitations in comprehending and effectively managing such data
using traditional methods. Therefore, it becomes imperative to explore innovative
approaches for analyzing performance data and information. To fully harness the

potential of IoT data, there is a pressing need to significantly enhance the performance. Furthermore, the ongoing progress in AI is leading to a convergence between AI and IoT. The fundamental description of IoT, will ultimately necessitate the intelligence of nearly all devices. In simpler terms, IoT relies on smart devices and machines. The convergence of both is driving the continuous expansion of IoT, influenced by six key factors, with the most influential factor being the emergence of big data and cloud/fog computing.

4.7 AI APPLICATIONS

There exist many applications of AI in different sectors. Some of them are discussed here:

- *Smart industry*: Following automation, electrification, and informationization, AI has emerged as a fundamental pillar of Industry 4.0. Numerous groundbreaking technological advancements in AI have revolutionized various industries, including agriculture, autonomous driving, education, finance, governance, intelligent robotics, manufacturing, the medical industry, the retailing industry, security, and more [24].
- *Smart home*: The concept of automation is explained in [10]. It involves a system that utilizes the controller to enable smart energy conservation. Through a user-friendly Android interface on a smartphone, this system can effectively control and automate various home appliances, such as lights and fans (turning them on or off). The components of the system where a micro-web server is connected via a Local Area Network or Wi-Fi module, allows users to access, monitor, and control devices and appliances using their Android-based smartphones.
- *Self-driving vehicles*: Autonomous driving functions is a robotic system that operates based on three primary technologies—perception, path planning, and control decisions. The implementation of this system lies in intelligent perception, while the remaining components rely heavily on the study, analysis, and implementation of intelligent systems and related fields. To continually enhance driving behavior, autonomous vehicles leverage DL, which proves to be the current most-effective solution. As these vehicles experience diverse traffic conditions and encounter unexpected situations, they generate significant volumes of data and information. This information is then transmitted to a cloud to serve as the samples for training. Through extensive training and learning processes, the vehicles accumulate their own driving experiences, while also benefiting from the training and testing outcomes of different devices [25].
- *Smart education*: In this period, there has been a profound integration of AI and education, leading to significant transformations in the educational landscape. Educational equipment now encompasses both individuals and machines. As a result, the scope of educational AI research has expanded to encompass the study of educational activities and the development of instructional guidelines for both machines and individuals [26].

The fusion of smart systems with that of the education systems has resulted in the emergence of a variety of groundbreaking applications, such as smart robots, smart learning platforms, and sophisticated systems for results evaluation. These advancements alleviate the burdensome tasks faced by teachers and foster a collaborative teaching environment between humans and computers. By merging AI with education, the ultimate goal is to establish an intelligent network learning environment. This involves forging partnerships with government bodies, schools, institutions, enterprises, and other smart network learning platforms, thereby revolutionizing education aims to consistently enhance teaching outcomes and nurture individuals with skills in learning, communication, and innovation.

- *Smart finance system*: The utilization of AI in the financial sector has led to the emergence of digital financial services, thereby infusing the entire industry with increased vitality for further development. The potential applications of AI in finance have garnered widespread recognition, although its implementation is an incremental process. Automation holds significant importance in driving the growth of this sector. It doesn't just enable financial data analysis but also offers valuable services to this field, for example automation provides users with secure services, aids in decision-making for transactions, credit assessment, and financial analytics. Moreover, AI possesses the capability to augment the identification, early detection, prevention, and control abilities across various risks inherent in the following system.

- *Smart governance*: Initially, AI found its first applications in domains characterized by abundant data resources and well-defined scenarios. While it is still relatively new in the realm of intelligent governance, the growing popularity of AI has opened up expansive opportunities for its implementation in various areas, including assistants, smart conferences, robot process automation, processing of any document, and decision-making. The integration of automation into these fields holds great potential, as it has the capacity to enhance efficiency of government, service capabilities, and mitigate labor shortages.

- *Smart manufacturing*: The combination of automation with that of the manufacturing has resulted in enhanced economical and effective productivity, increasing production flexibility, and achieving cost savings. It has facilitated mass customization, improved market forecasting accuracy, and enabled better matching of supply and demand, thereby driving the transformation of manufacturing services and enhancing quality control. However, both present a complex system engineering challenge. People worldwide encounter common issues and hurdles, including AI standardization, internet technology, information security, and the development and implementation of multidisciplinary talents. Examples of areas benefiting from this integration include intelligent products and facilities, smart plants, intelligent management and service systems, intelligent supply chain management, and intelligent monitoring, and decision-making processes [25].

- *Smart healthcare*: The medical industry continues to encounter numerous challenges in its development. However, AI offers a means to effectively combine medicine and technology, contributing to the advancement of medical science and intelligence. Intelligent healthcare leverages advanced networking technologies, notably the widespread adoption of IoT, to establish regional medical information platforms for health records. Moreover, the integration of technology and medicine enables the digitized, electronic, rapid, and accurate processing of medical procedures. For instance, DL techniques are employed for gene prediction, NLP aids in analyzing electronic medical records [27].
- *Smart security system*: The progress and implementation of AI present both pros and cons in the security of information field. First, automation has enhanced protectional capabilities such as security and privacy. Second, it faces privacy concerns, including security related to data, anti-spoofing measures, protection related to privacy, and adapting to dynamic environments. AI enables swift detection of large-scale network data and various security attacks, empowering adaptive security defenses. The important benefit of using automation in network security primarily lies in significantly improving the performance metrics of security monitoring. Given the vast scale of network traffic and the generation of crucial security information provided by security devices and systems, AI can effectively address the challenges of processing and analyzing extensive data. It accurately identifies network attacks, thereby reducing security risks. Examples of AI applications in network security include cyberattacks, prediction of security vulnerabilities, and system failures. The various application areas of AI-IoT are showing in Figure 4.13.

The comparison in Table 4.1 shows the different categories based on Principles, Techniques, and Applications.

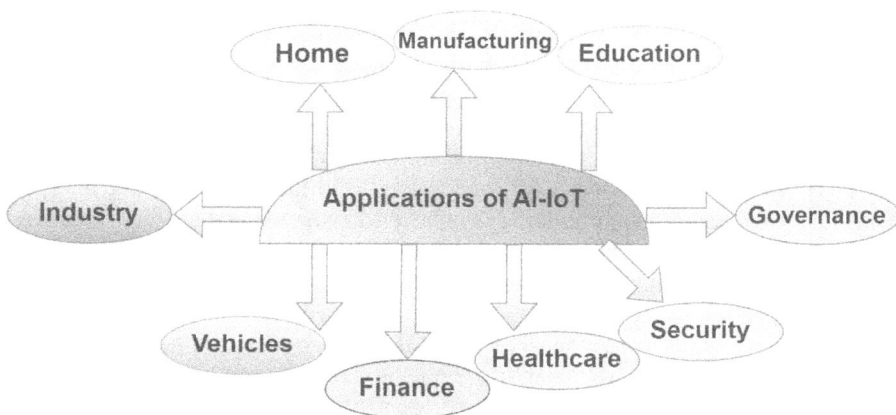

FIGURE 4.13 Applications of AI-IoT.

TABLE 4.1

Comparison Table Based on AI Principles, Techniques and Applications

	Principles	Techniques	Applications
Explanation	The examination of intelligent entities and their actions.	ML, DL, NLP, computer vision, robotics, and expert systems.	Healthcare, finance, education, transportation, agriculture, manufacturing, gaming, and social media.
Emphasis	Gaining insights into the functioning of intelligence and creating agents capable of demonstrating it.	Educating algorithms to identify patterns in data and utilize them to make informed choices.	Enhancing healthcare results, optimizing financial decision-making, tailoring educational journeys, revolutionizing transportation, maximizing agricultural productivity, automating manufacturing operations, enriching gaming encounters, examining social media information.
Problems	Constructing machines with the ability to reason and acquire knowledge akin to humans.	Ensuring the fairness, transparency, and ethicality of AI systems.	Tackling concerns pertaining to privacy, security, prejudice, and the impact on human employment.
Trends	Advancing the capabilities of AI systems to possess human-like abilities in reasoning, planning, and communication.	Incorporating AI into diverse technologies such as blockchain and the IoT.	Venturing into uncharted territories in space, creating intelligent cities, enhancing energy conservation, tackling climate change.

4.8   CONCLUSION AND FUTURE DIRECTIONS

To sum up, the IoT environment is significantly impacted by the fundamentals of AI. Numerous advantages and game-changing prospects result from the integration of AI into IoT systems. IoT installations can achieve increased efficiency, predictive capacities, and automation by utilizing AI techniques like ML, data analytics, and intelligent automation. IoT devices are enabled by AI to gather, analyze, and interpret enormous volumes of data, providing insightful information and enabling reasoned decision-making. It enables IoT systems to adapt, learn, and independently carry out real-time corrections and improvements without the need for human intervention. AI algorithms can also anticipate maintenance requirements, identify security risks, and improve resource management in IoT networks. AI also makes it easier to comprehend natural language, enabling frictionless interactions between people and technology and eventually improving user experiences. Additionally, it is essential for maximizing energy use and improving the performance and scalability of IoT systems as a whole. But as IoT related AI develops, it is essential to address issues like security dangers, ethical quandaries, and privacy concerns. To build trust, protect user data, and reduce potential threats, it is crucial to guarantee that AI is used ethically and responsibly within IoT networks.

In conclusion, the fundamentals of AI provide a solid framework for revolutionizing IoT capabilities. Organizations can unleash new opportunities, encourage innovation, and build intelligent, linked ecosystems that improve productivity, efficiency, and quality of life in general by utilizing AI. The techniques of AI are focused on integrating with other technologies like blockchain and IoT, whereas the principles of AI are focused on creating more sophisticated AI systems that can reason, plan, and communicate like humans. AI is now being used to further scientific research, explore unexplored space, build smart cities, cut greenhouse gas emissions, and boost energy efficiency.

REFERENCES

1. Michalski, R.S., Carbonell, J.G., & Mitchell, T.M. (2013). 'Machine learning: an artificial intelligence approach' (Springer Science & Business Media, Berlin, Germany).
2. Mittal, N., & Singh, S.P. (2019). Healthcare applications of artificial intelligence: A review. In Proceedings of the 3rd International Conference on Computational Intelligence, Communication and Networks Springer (pp. 10–16).
3. Muruganantham, G., & Kamalakkannan, P. (2019). A survey of robotics: Applications, challenges, and recent trends. International Journal of Advanced Science and Technology, 28(16), 202–210.
4. Monostori, L., Kádár, B., Bauernhansl, T., et al. (2016). Cyber-physical systems in manufacturing. CIRP Annals, 65(2), 621–641.
5. Lee, E.A., & Seshia, S.A. (2016). 'Introduction to embedded systems: a cyber-physical systems approach' (MIT Press, Cambridge, Massachusetts).
6. Fortino, G., & Trunfio, P. (2014). 'Internet of things based on smart objects: Technology, middleware and applications' (Springer, New York, USA).
7. Calvillo-Arbizu, J., & Lluch-Lafuente, A. (2019). Artificial intelligence and finance: A review. Frontiers in Robotics and AI, 6, 50.
8. Baheti, R., & Gill, H. (2011). Cyber-physical systems. The Impact of Control Technology, 12, 161–166.
9. Chen, M., Mao, S., & Liu, Y. (2014). Big data: A survey. Mobile Networks and Applications, 19(2), 171–209.
10. Gurav, U., & Patil, G. (2016). IoT based interactive controlling and monitoring system for home automation. International Journal of Advanced Research in Computer Engineering and Technology (IJARCET), 5(9), 2392–2396.
11. Lin, P.H., Wooders, A., Wang, J.T.Y., & Yuan, W.M. (2018). Artificial intelligence, the missing piece of online education? IEEE Engineering Management Review, 46(3), 25–28.
12. Yang, L.T., Di Martino, B., & Zhang, Q. (2017). Internet of everything. Mobile Information Systems, 2017, 1–3.
13. Witten, I.H., & Frank, E. (2016). 'Data mining: Practical machine learning tools and techniques' (Morgan Kaufmann, Burlington, Massachusetts).
14. Rani, S., & Bashir, A.K. (2022, October). Analysis of Machine learning and deep learning intrusion detection system in internet of things network. In 2022 International Conference on Data Analytics for Business and Industry (ICDABI) (pp. 1–9).
15. Schmidhuber, J. (2015). Deep learning in neural networks: An overview. Neural Networks, 61, 85–117.

16. Sharma, A., Rani, S., Shah, S.H., Sharma, R., Yu, F., & Hassan, M.M., (2023). An Efficient Hybrid Deep Learning Model for Denial of Service Detection in Cyber Physical Systems. IEEE Transactions on Network Science and Engineering.
17. Ankita, A., & Rani, S. (2021, July). Machine learning and deep learning for malware and ransomware attacks in 6G network. In 2021 Fourth International Conference on Computational Intelligence and Communication Technologies (CCICT) (pp. 39–44).
18. Chang, C.C., & Lin, C.J. (2011). LIBSVM: A library for support vector machines. ACM Transactions on Intelligent Systems and Technology, 2(3), 1–27.
19. Hassan, Q.F., Khan, A.R., & Madani, S.A. (2017). Internet of things: Challenges, advances, and applications. 'Chapman & Hall/CRC computer and information science series' (CRC Press, Boca Raton, Florida).
20. Linardatos, P., Papastefanopoulos, V., & Kotsiantis, S. (2021). Explainable AI: A review of machine learning interpretability methods. Entropy, 23(1), 18.
21. Sethi, M., Ahuja, S., & Bawa, P. (2021, October). Classification of Alzheimer's disease using neuroimaging data by convolution neural network. In 2021 6th International Conference on Signal Processing, Computing and Control (ISPCC) (pp. 402–406).
22. Bennis, M., Perlaza, S.M., Blasco, P., Han, Z., & Poor, H.V. (2013, July). Self-organization in small cell networks: A reinforcement learning approach. IEEE Transactions on Wireless Communications, 12(7), 3202–3212. doi: 10.1109/TWC.2013.060513.120959
23. Wen, J., Li, Y., Fang, M., Zhu, L., Feng, D., & Li, P. (2023). Fine-Grained and Multiple Classification for Alzheimer's Disease With Wavelet Convolution Unit Network. IEEE Transactions on Biomedical Engineering, 70(9), 2592–2603. doi: 10.1109/TBME.2023.3256042.
24. Bolen, A. (2018). 3 IoT examples from 3 industries: Real world implementation achieving results today. 31. Buntz (2017) Industrial IoT: The top 20 Industrial IoT applications.
25. Grieco, K.A., Rizzo, A., Colucci, S., Sicarri, S., Piro, G., et al. (2014). IoT-aided robotics applications: Technological implications, target domains, and open issue. Computer Communications 54(1): 32–47.
26. Deshpande, A., Pitale, P., & Sanap, S. (2016) Industrial automation using Internet of Things. International Journal of Advanced Research 5(2): 266–269.
27. Friedman, C., Rindflesch, T.C., & Corn, M. (2013). Natural language processing: State of the art and prospects for significant progress, a workshop sponsored by the National Library of Medicine. Journal of Biomedical Informatics, 46(5), 765–773.

5 Blockchain-Based Communication Frameworks for Smart Vehicles

Rakhi Mutha, Neeraj, and Ashu Taneja

5.1 INTRODUCTION

5.1.1 OVERVIEW OF SMART VEHICLES AND THEIR COMMUNICATION REQUIREMENTS

Smart vehicles, including autonomous cars, electric vehicles, and connected vehicles, are revolutionizing the transportation industry. These vehicles rely on advanced communication systems to exchange data with other vehicles, infrastructure components, and backend services. The communication requirements of smart vehicles include real-time data exchange, coordination for safe maneuvering, intelligent routing, efficient charging, and secure transactions, as shown in Figure 5.1. However, existing communication systems face challenges in terms of security, privacy, scalability, and interoperability. In this chapter, we explore how Blockchain technology can address these challenges and provide a robust communication framework for smart vehicles.

Smart vehicles, also known as connected vehicles or intelligent transportation systems (ITS), are automobiles that are equipped with advanced technologies and communication capabilities to enhance safety, efficiency, and overall driving experience [1]. These vehicles leverage various communication technologies to interact with other vehicles, infrastructure, and the surrounding environment.

Here is an overview of smart vehicles and their communication requirements:

Vehicle-to-vehicle (V2V) communication: It enables direct communication between vehicles in close proximity, allowing them to exchange information about their speed, position, acceleration, and other relevant data, as shown in Figure 5.2. This communication helps in improving safety by enabling vehicles to alert each other about potential collisions, hazards, or traffic conditions.

Vehicle-to-infrastructure (V2I) communication: This involves the exchange of information between vehicles and roadside infrastructure such as traffic signals, toll booths, and parking systems, as shown in Figure 5.3. By receiving real-time data from infrastructure, smart vehicles can optimize routes, receive traffic signal prioritization, and access various services such as electronic payment systems.

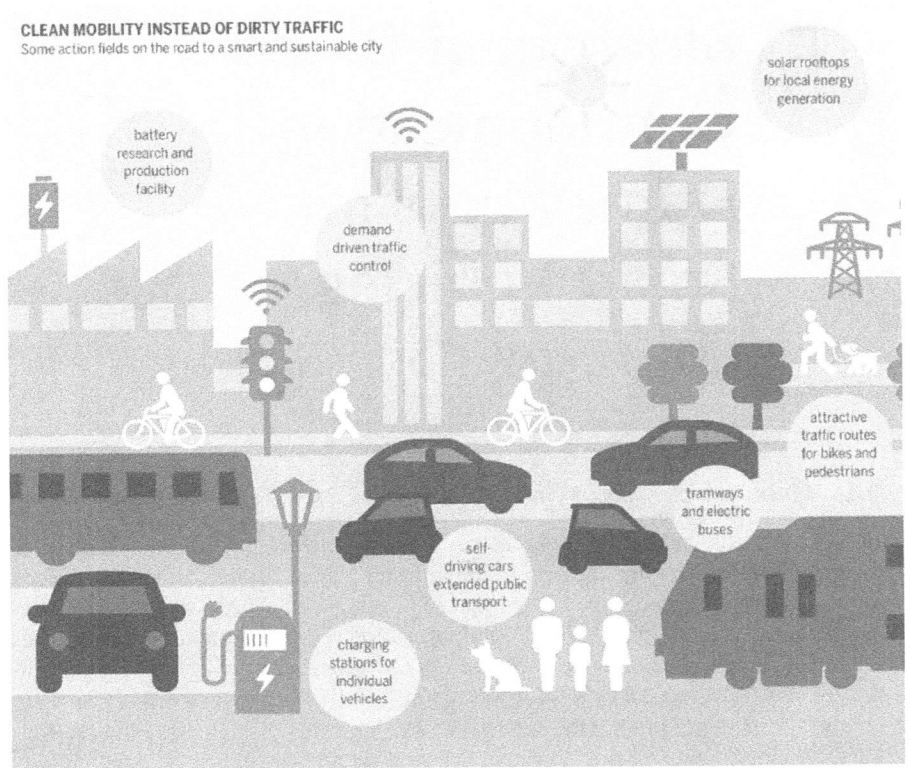

FIGURE 5.1   Smart vehicles for clean mobility.

Vehicle-to-cloud (V2C) communication: This refers to the connection between smart vehicles and cloud-based platforms. This allows vehicles to access a wide range of cloud-based services, such as real-time traffic updates, weather information, software updates, and personalized infotainment content, as shown in Figure 5.4. V2C communication also

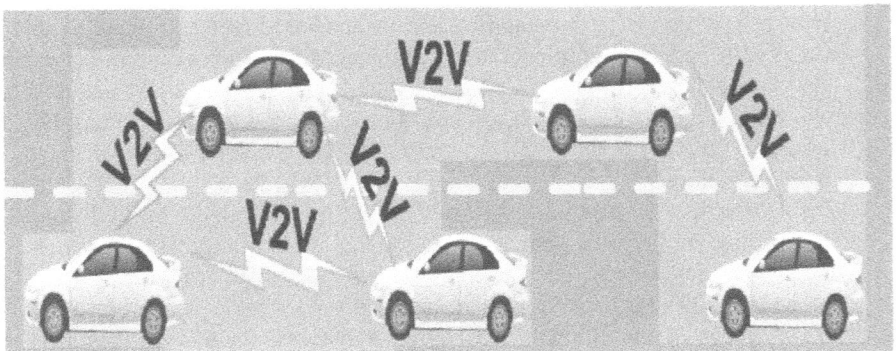

FIGURE 5.2   Vehicle-to-vehicle (V2V) communication.

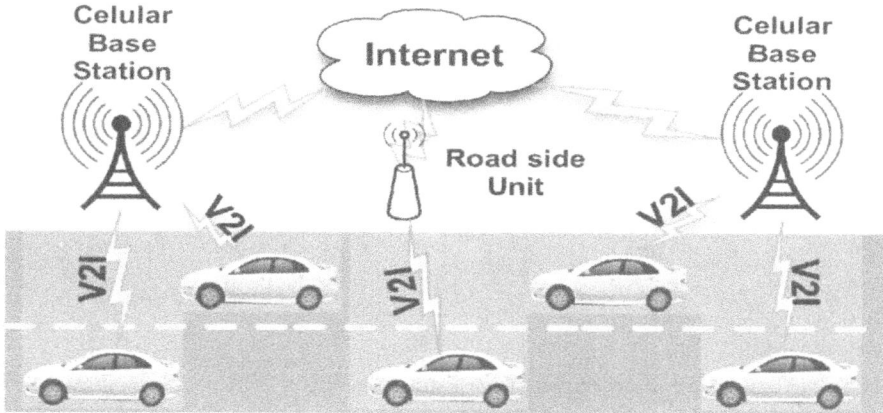

FIGURE 5.3 Vehicle-to-infrastructure (V2I) communication.

enables remote diagnostics and over-the-air software updates for vehicle systems.

Vehicle-to-pedestrian (V2P) communication: It involves the interaction between smart vehicles and pedestrians or vulnerable road users [2]. For example, vehicles can exchange signals with smartphones or wearable devices carried by pedestrians to warn about their presence, reducing the risk of accidents, as shown in Figure 5.5.

Vehicle-to-grid (V2G) communication: It enables smart vehicles to interact with the power grid, as shown in Figure 5.6. Electric vehicles (EVs) can send information about their charging status, battery capacity, and available energy back to the grid [1]. This allows for smart charging, discharging, and

FIGURE 5.4 Vehicle-to-cloud (V2C) communication.

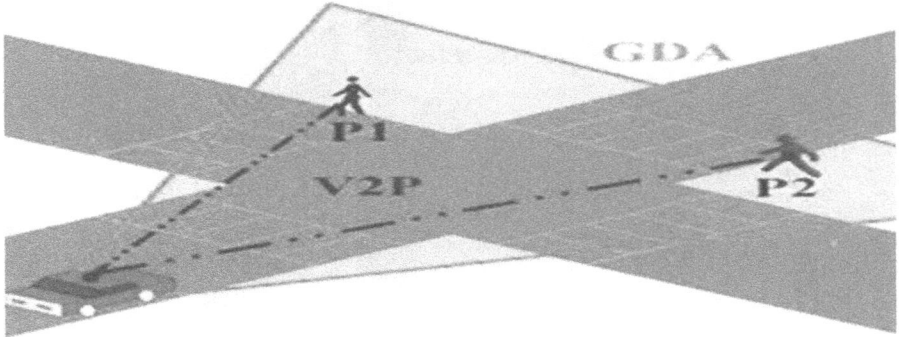

FIGURE 5.5 Vehicle-to-pedestrian (V2P) communication.

load management, facilitating the integration of renewable energy sources
and optimizing energy usage.

Communication technologies: Smart vehicles utilize a combination of communication technologies to enable seamless connectivity. These technologies
include cellular networks (e.g., 4G/5G), dedicated short-range communication (DSRC), Wi-Fi, Bluetooth, and satellite communication. The choice of
communication technology depends on factors such as range, bandwidth,
latency requirements, and the type of information being exchanged.

Security and privacy: With increased connectivity, smart vehicles face security and privacy challenges. Robust security measures, such as encryption, authentication, and intrusion detection systems, are crucial to protect
against cyber threats. Privacy concerns related to the collection and use of
vehicle data also need to be addressed to ensure user trust and compliance
with data protection regulations.

Smart vehicles rely on various communication channels to exchange information
with other vehicles, infrastructure, pedestrians, and the cloud. These communication
capabilities enable enhanced safety, efficiency, and connectivity, paving the way for
a more intelligent and interconnected transportation system.

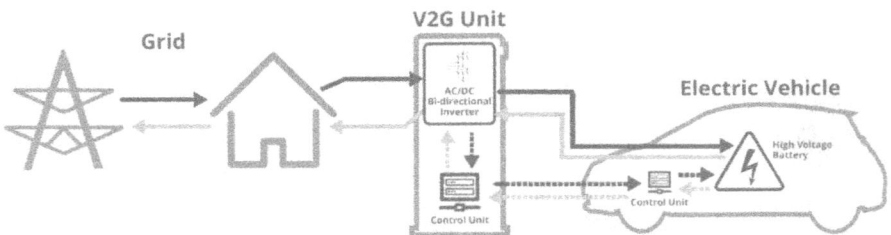

FIGURE 5.6 Vehicle-to-grid (V2G) communication.

5.1.2 Challenges in Current Communication Systems for Smart Vehicles

While communication systems for smart vehicles offer numerous benefits, there are several challenges that need to be addressed to ensure their effective implementation [3]. Here are some of the key challenges in current communication systems for smart vehicles:

Standardization: The lack of standardized protocols and communication frameworks is a significant challenge. Different manufacturers and stakeholders may employ their own proprietary communication technologies, which can create interoperability issues and hinder the seamless communication between vehicles and infrastructure which is crucial for the widespread adoption of smart vehicle communication.

Reliability and latency: Smart vehicles require real-time, reliable communication to ensure timely and accurate exchange of data. However, wireless communication channels can be prone to interference, signal degradation, and latency issues, which can impact the effectiveness of safety-critical applications.

Scalability: As the number of connected vehicles increases, communication networks must be able to handle a growing amount of data traffic. Scalability becomes crucial to support the increasing demand for reliable and high-bandwidth communication among vehicles, infrastructure, and the cloud.

Data privacy and security: Connected vehicles generate and exchange large amounts of data, including sensitive information about vehicle location, driving patterns, and personal preferences. Protecting this data from unauthorized access, cyber threats, and privacy breaches is a critical challenge that requires robust security measures and encryption techniques.

Lack of trust and transparency: Trust is essential for effective communication and coordination among smart vehicles, infrastructure providers, and other stakeholders. However, in current systems, establishing trust is often challenging due to the involvement of multiple parties and the lack of transparency in data exchange and transaction processing.

Infrastructure deployment: To enable efficient communication, a robust and widespread infrastructure deployment is necessary. However, the implementation of infrastructure components, such as roadside units and dedicated communication networks, requires significant investment and coordination among various stakeholders, including governments, service providers, and automotive manufacturers.

Spectrum availability: The limited availability of suitable radio frequency spectrum poses a challenge for communication systems in smart vehicles. The spectrum needs to be efficiently allocated and managed to support the increasing demand for wireless communication without causing interference or congestion.

Longevity and compatibility: Vehicles have a long lifespan, and their communication systems need to be compatible and upgradable over time. Ensuring backward compatibility and providing mechanisms for software and hardware updates are crucial to prevent obsolescence and maintain the effectiveness of communication systems.

Regulatory and legal considerations: The deployment of communication systems for smart vehicles involves compliance with regulations related to spectrum allocation, data privacy, cyber-security, and safety standards. Harmonizing regulations across different regions and ensuring consistent enforcement can be challenging.

Addressing these challenges requires collaboration among automotive manufacturers, communication technology providers, policymakers, and other stakeholders. Standardization efforts, investments in infrastructure, advancements in security measures, and regulatory frameworks all play crucial roles in overcoming these challenges and realizing the full potential of smart vehicle communication systems.

5.1.3 ROLE OF BLOCKCHAIN TECHNOLOGY IN ADDRESSING THESE CHALLENGES

Blockchain technology, originally introduced as the underlying technology for crypto-currencies like Bitcoin, has emerged as a potential solution to address the challenges faced by current communication systems for smart vehicles. Blockchain is a decentralized and tamperproof distributed ledger technology that enables secure, transparent, and immutable record-keeping of transactions and data. Its key features, including decentralization, immutability, transparency, and cryptographic security; make it well-suited for addressing the challenges in smart vehicle communication.

Blockchain technology has the potential to address several challenges in communication systems for smart vehicles. Here we look at the role of blockchain in tackling these challenges:

1. *Data security and privacy*: Blockchain provides a decentralized and immutable ledger that can enhance the security and privacy of data in smart vehicle communication systems. By leveraging cryptography and distributed consensus, blockchain ensures that data exchanged between vehicles and infrastructure remains tamperproof and transparent, as shown in Figure 5.7. It allows for secure and private data sharing without relying on a central authority, reducing the risk of unauthorized access and data breaches.

2. *Interoperability and standardization*: Blockchain can facilitate interoperability among different communication systems by providing a common platform for data exchange. Smart contracts, i.e., self-executing agreements on the blockchain, can help standardize communication protocols and enable seamless interaction between vehicles and infrastructure. By establishing a trustless and decentralized network, blockchain can overcome the challenges of proprietary communication technologies and promote a more standardized approach.

3. *Trust and transparency*: Blockchain's transparent nature enhances trust among participants in smart vehicle communication systems. Every transaction or data exchange recorded on the blockchain is visible to all network

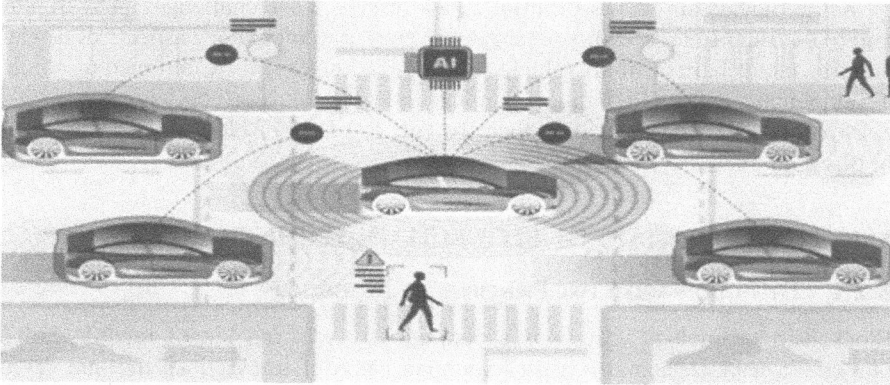

FIGURE 5.7 Role of blockchain technology in addressing these challenges.

participants, fostering transparency and accountability. This can help establish trust between vehicles, infrastructure providers, and other stakeholders, leading to more reliable and secure communication [4].

4. *Scalability and efficiency*: Blockchain technology has seen advancements in scalability solutions, such as off-chain transactions, side chains, and layer-two protocols. These solutions can help address the scalability challenges of communication systems for smart vehicles by reducing the computational and storage requirements on the main blockchain network. By enabling faster and more efficient transactions, blockchain scalability solutions can support the growing data traffic in smart vehicle communication networks.

5. *Smart contracts and automation*: Smart contracts on the blockchain can automate and enforce agreements and protocols in smart vehicle communication systems. These self-executing contracts can facilitate secure and autonomous interactions between vehicles, infrastructure, and other stakeholders. For example, smart contracts can automate toll payments, traffic signal prioritization, or insurance claim settlements, eliminating the need for intermediaries and reducing administrative overheads.

6. *Identity and access management*: Blockchain-based identity solutions can address the challenges of identity verification and access management in smart vehicle communication systems. By storing digital identities and credentials on the blockchain, vehicles and other entities can securely prove their identity and establish trust when communicating with each other. This helps prevent spoofing, unauthorized access, and identity theft.

7. *Micropayments and incentives*: Blockchain technology enables secure and transparent micropayments and incentive mechanisms. In smart vehicle communication systems, this can be leveraged to create decentralized payment systems for services like tolls, parking, or sharing economy platforms. Blockchain-based incentives can encourage data sharing and cooperation among vehicles, infrastructure providers, and users, leading to improved system efficiency and better overall performance [5].

While blockchain shows promise in addressing these challenges, it is important to consider the scalability, energy efficiency, and regulatory aspects of implementing blockchain in smart vehicle communication systems. Continued research, development, and collaboration among industry players, policymakers, and researchers will be crucial to unlock the full potential of blockchain technology in this domain.

5.2 FUNDAMENTALS OF BLOCKCHAIN TECHNOLOGY

5.2.1 DEFINITION AND CORE PRINCIPLES OF BLOCKCHAIN

Blockchain technology maintains a continuously growing list of records called blocks. Each block contains a time-stamped batch of transactions that are cryptographically linked to the previous block, forming a chain of blocks. The core principles of blockchain include decentralization, immutability, transparency, and cryptographic security. These principles ensure that the Blockchain is resistant to tampering and provides a trusted and transparent record of transactions.

Blockchain allows multiple parties to maintain a shared record of transactions or information in a secure and transparent manner. It operates on a peer-to-peer network, where each participant, known as a node, has a copy of the entire blockchain [6].

Core principles of blockchain include the following:

Decentralization: Blockchain operates on a decentralized network where no single entity or central authority has control over the entire system. Instead, the network consists of multiple nodes that participate in the validation and maintenance of the blockchain, as shown in Figure 5.8. This

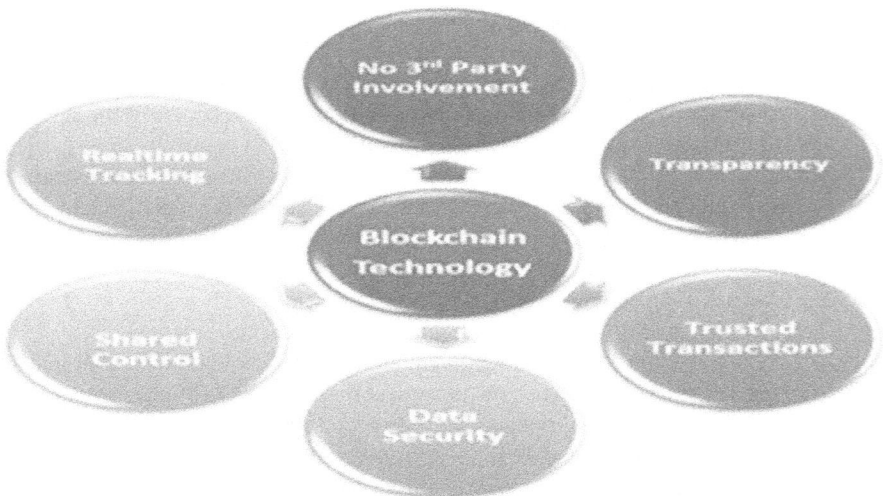

FIGURE 5.8 Core principles of blockchain.

decentralization promotes transparency, resilience, and eliminates the need for intermediaries.

Transparency and immutability: Blockchain maintains a transparent and immutable record of transactions or information. Each transaction, once recorded on the blockchain, becomes a permanent and unchangeable block. All participants in the network can view and verify the transactions, enhancing transparency, accountability, and trust.

Security and cryptography: Blockchain ensures the security of data and transactions through cryptographic techniques. Transactions are encrypted, and each block includes a cryptographic hash, a unique identifier that links it to the previous block, forming a chain. This cryptographic hash ensures the integrity of the data and provides protection against tampering or fraudulent activities.

Consensus mechanisms: Blockchain employs consensus mechanisms to achieve agreement among network participants on the validity of transactions and the order in which they are added to the blockchain. Different consensus algorithms, such as proof of work (PoW), proof of stake (PoS), or practical Byzantine fault tolerance (PBFT), determine how consensus is reached. Consensus mechanisms enable trust and prevent malicious actors from manipulating the blockchain [7].

Smart contracts: Smart contracts are self-executing agreements or protocols stored on the blockchain. They automatically execute predefined conditions and actions when specific conditions are met, as shown in Figure 5.9. Smart contracts enable automation, eliminate the need for intermediaries, and ensure transparency and enforceability of agreements.

Privacy and identity protection: While blockchain provides transparency, it also allows for privacy protection. Cryptographic techniques enable participants to have control over their identity and personal information while still engaging in secure transactions on the blockchain. Private or permission-only blockchains restrict access to authorized participants, ensuring privacy and data protection in specific use cases [8].

These core principles collectively contribute to the trust, security, transparency, and efficiency that blockchain technology offers. They provide the foundation for various applications beyond cryptocurrency, such as supply chain management, healthcare records, voting systems, decentralized finance, and more.

5.2.2 COMPONENTS OF A BLOCKCHAIN: BLOCKS, TRANSACTIONS, AND CONSENSUS MECHANISMS

In a blockchain, each block consists of a header and a body. The header contains metadata such as the block's unique identifier, a timestamp, and a reference to the previous block's identifier. The body of the block contains a set of transactions, which represent the actions or data exchanges recorded on the blockchain. Transactions can include various types of data, such as smart contracts, digital assets, or sensor readings.

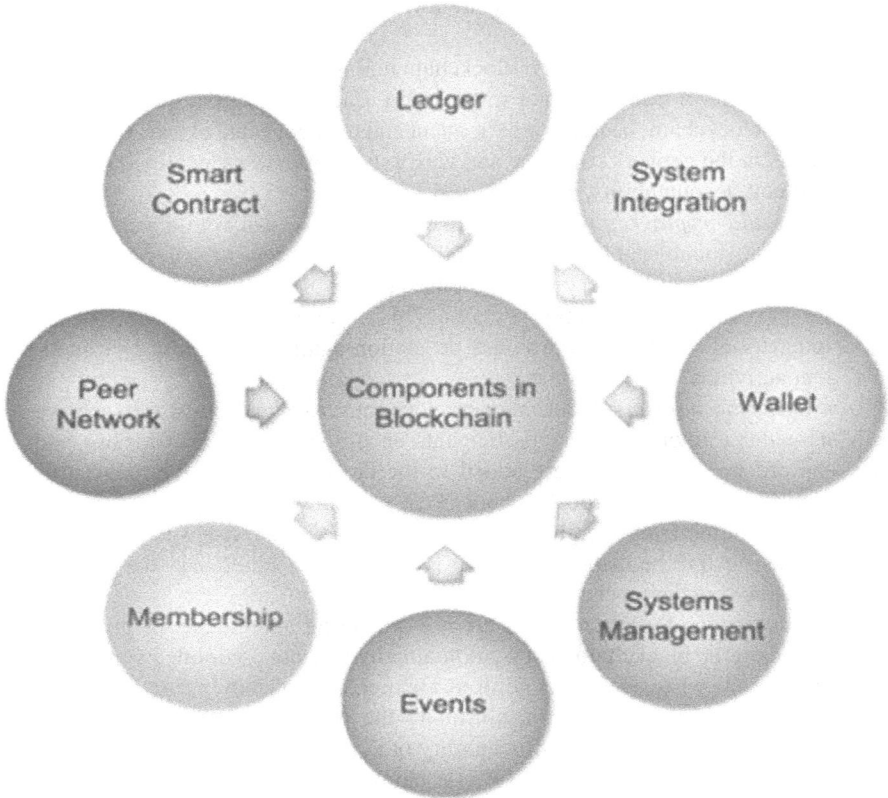

FIGURE 5.9 Components of blockchain.

Components of a blockchain include blocks, transactions, and consensus mechanisms. Let's explore each of these components in more detail:

Blocks: These are the fundamental units of a blockchain. They contain a collection of transactions that are grouped together and added to the blockchain as a single entity. Each block typically includes a unique identifier called a cryptographic hash, a timestamp, and a reference to the previous block in the chain, as shown in Figure 5.10. The cryptographic hash of a block ensures its integrity and links it to the previous block, forming a sequential chain of blocks [9].

Transactions: This represent the actions or information that is recorded on the blockchain. These can include the transfer of digital assets (e.g., cryptocurrencies), the execution of smart contracts, or any other data that needs to be stored on the blockchain. Transactions typically include relevant data, such as sender and recipient addresses, transaction amounts, and additional

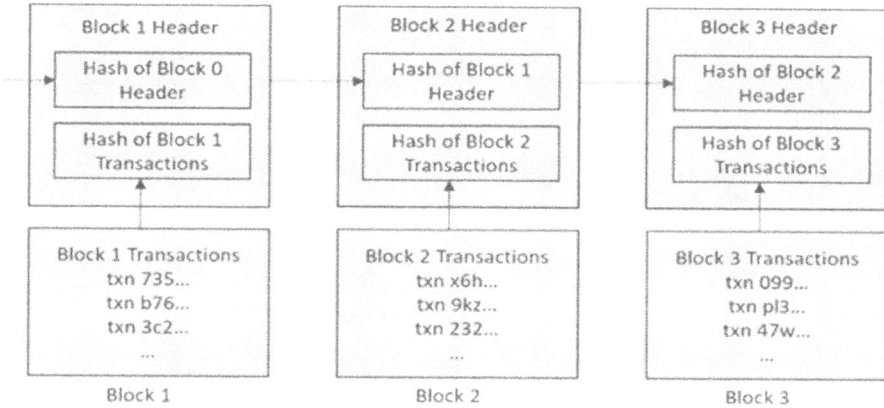

```
┌─────────────────────┐      ┌─────────────────────┐      ┌─────────────────────┐
│    Block 1 Header   │      │    Block 2 Header   │      │    Block 3 Header   │
│  ┌───────────────┐  │      │  ┌───────────────┐  │      │  ┌───────────────┐  │
│  │ Hash of Block 0│  │      │  │ Hash of Block 1│  │      │  │ Hash of Block 2│  │
│  │    Header     │  │      │  │    Header     │  │      │  │    Header     │  │
│  └───────────────┘  │      │  └───────────────┘  │      │  └───────────────┘  │
│  ┌───────────────┐  │      │  ┌───────────────┐  │      │  ┌───────────────┐  │
│  │ Hash of Block 1│  │      │  │ Hash of Block 2│  │      │  │ Hash of Block 3│  │
│  │  Transactions │  │      │  │  Transactions │  │      │  │  Transactions │  │
│  └───────────────┘  │      │  └───────────────┘  │      │  └───────────────┘  │
└─────────────────────┘      └─────────────────────┘      └─────────────────────┘
┌─────────────────────┐      ┌─────────────────────┐      ┌─────────────────────┐
│ Block 1 Transactions│      │ Block 2 Transactions│      │ Block 3 Transactions│
│      txn 735...     │      │      txn x6h...     │      │      txn 099...     │
│      txn b76...     │      │      txn 9kz...     │      │      txn pl3...     │
│      txn 3c2...     │      │      txn 232...     │      │      txn 47w...     │
│         ...         │      │         ...         │      │         ...         │
└─────────────────────┘      └─────────────────────┘      └─────────────────────┘
       Block 1                      Block 2                      Block 3
```

FIGURE 5.10   Blocks, headers, and transactions in a blockchain.

metadata. Once validated and included in a block, transactions become part of the permanent record on the blockchain.

Consensus mechanisms: These are protocols or algorithms that enable agreement among network participants on the validity and order of transactions being added to the blockchain. Consensus is necessary in decentralized systems to ensure that all participants reach a common understanding of the state of the blockchain, as shown in Figure 5.11. Different consensus mechanisms have different approaches to achieving agreement. Some popular consensus algorithms include:

- *Proof of work (PoW)*: In PoW, participants (miners) compete to solve complex mathematical puzzles to validate transactions and add blocks to the blockchain. The solution requires significant computational power, and the miner who finds the solution first is rewarded. This mechanism is used by cryptocurrencies like Bitcoin.
- *Proof of stake (PoS)*: PoS selects block validators based on the number of coins they hold or have staked as collateral. Validators are chosen to create new blocks based on their stake, and the probability of selection is proportional to their stake. PoS consumes less energy compared to PoW and is used by cryptocurrencies like Ethereum.
- *Practical Byzantine fault tolerance (PBFT)*: PBFT is a consensus mechanism used in permissioned blockchains. It requires a predetermined set of trusted validators, and consensus is achieved through multiple rounds of voting. PBFT ensures Byzantine fault tolerance, where a certain number of faulty or malicious nodes cannot disrupt the consensus process.
- *Delegated proof of stake (DPoS)*: DPoS is a variant of PoS where a smaller number of participants, known as "delegates," are elected to validate transactions and produce blocks. Delegates are voted in by coin holders, and they take turns producing blocks in a deterministic order. DPoS is used by cryptocurrencies like EOS [10].

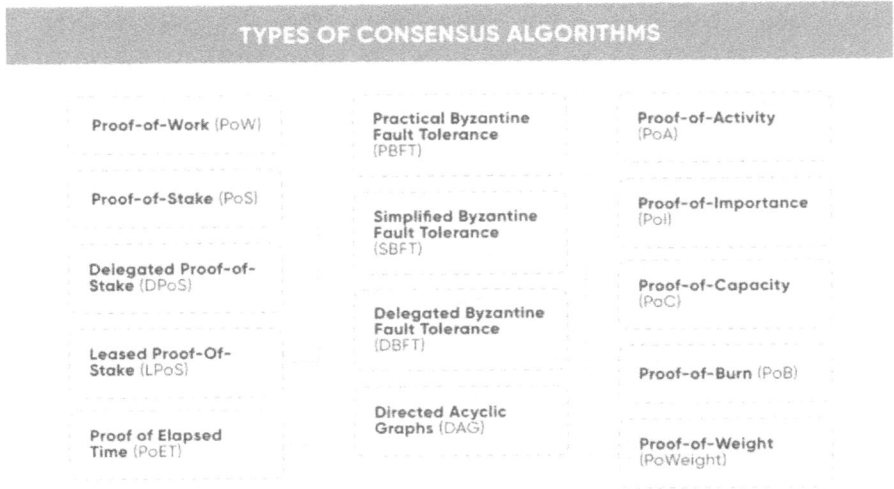

TYPES OF CONSENSUS ALGORITHMS

Proof-of-Work (PoW)

Proof-of-Stake (PoS)

Delegated Proof-of-Stake (DPoS)

Leased Proof-Of-Stake (LPoS)

Proof of Elapsed Time (PoET)

Practical Byzantine Fault Tolerance (PBFT)

Simplified Byzantine Fault Tolerance (SBFT)

Delegated Byzantine Fault Tolerance (DBFT)

Directed Acyclic Graphs (DAG)

Proof-of-Activity (PoA)

Proof-of-Importance (PoI)

Proof-of-Capacity (PoC)

Proof-of-Burn (PoB)

Proof-of-Weight (PoWeight)

FIGURE 5.11   Consensus algorithms.

These components work together to create a secure, transparent, and decentralized system in a blockchain. Blocks store transactions, and consensus mechanisms ensure agreement on the state of the blockchain among network participants, providing the foundation for trust and reliability.

5.2.3 Types of Blockchains: Public, Private, and Consortium

There are different types of blockchains based on their accessibility and permission levels. Public blockchains, like Bitcoin and Ethereum, are open to anyone, and anyone can participate in the consensus process and access the blockchain's data. Private blockchains, on the other hand, are restricted to a specific group of participants who have permission to read, write, and validate transactions. Consortium blockchains are a hybrid approach where a group of organizations collaboratively maintains and governs the blockchain.

There are three main types of blockchains: public, private, hybrid and consortium, as shown in Figure 5.12. Each type has distinct characteristics and use cases. Here's an overview of each type:

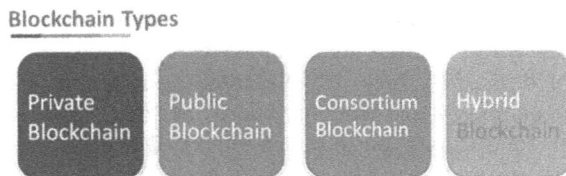

Blockchain Types

| Private Blockchain | Public Blockchain | Consortium Blockchain | Hybrid Blockchain |

FIGURE 5.12   Types of blockchain.

Public blockchain

- Public blockchains are open and permission-less, allowing anyone to participate in the network as a node, validate transactions, and contribute to the consensus process.
- They are decentralized and operate on a peer-to-peer network, where multiple participants maintain a copy of the entire blockchain.
- Public blockchains are typically secured by consensus mechanisms like PoW or PoS.
- Examples of public blockchains include Bitcoin and Ethereum [11].
- Public blockchains are often used for cryptocurrencies, decentralized applications (DApps), and transparent, censorship-resistant systems.

Private blockchain

- Private blockchains, also known as permissioned blockchains, have restricted access and require permission to join the network and participate in the consensus process.
- Participants in a private blockchain are usually known entities, such as specific organizations or individuals, and they have predefined roles and access levels.
- Private blockchains often operate in a centralized or semi-centralized manner, controlled by a single organization or a consortium of organizations.
- Private blockchains prioritize privacy, scalability, and efficiency over full decentralization.
- Use cases for private blockchains include supply chain management, internal record-keeping, and enterprise applications where privacy and control are important.

Consortium blockchain

- Consortium blockchains are a hybrid between public and private blockchains, where a group of organizations or entities form a consortium and jointly participate in maintaining the blockchain network.
- Consortium blockchains are permissioned networks where the consensus process is controlled by a limited number of trusted nodes or validators.
- The participating organizations in a consortium blockchain work together to validate transactions and maintain the blockchain, leveraging the benefits of decentralization and shared governance.
- Consortium blockchains are commonly used in industries where multiple stakeholders collaborate, such as banking, supply chain management, or healthcare, to establish trust, streamline operations, and share data securely.
- It is worth noting that hybrid models and variations of these types also exist, as the blockchain space is continuously evolving and adapting to different use cases. Additionally, interoperability solutions are being developed to enable communication and interaction between different types of blockchains, further expanding their capabilities and potential applications.

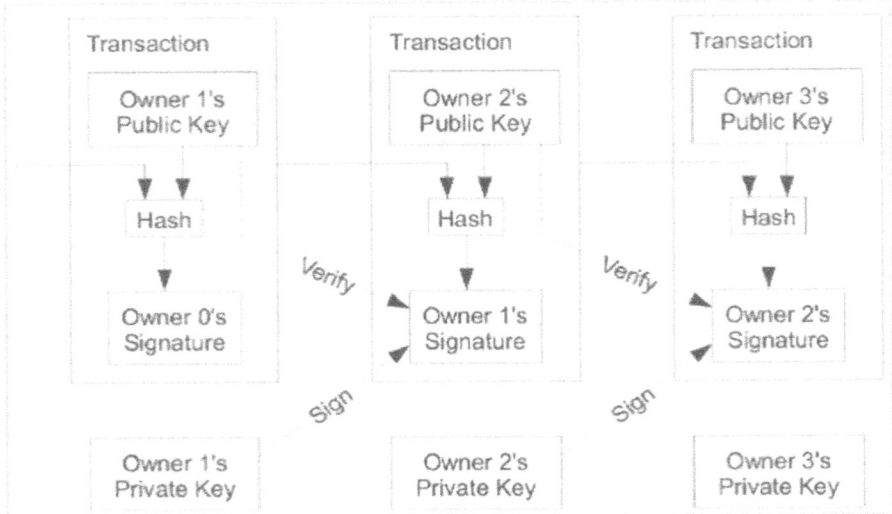

FIGURE 5.13 Cryptographic techniques used in blockchain.

5.2.4 CRYPTOGRAPHIC TECHNIQUES USED IN BLOCKCHAIN

Blockchain employs various cryptographic techniques to ensure the security and integrity of transactions and data. These techniques include cryptographic hash functions, which generate unique identifiers for each block and ensure data integrity, as shown in Figure 5.13. Public-key cryptography is used for secure digital signatures and key management. Consensus mechanisms, such as PoW or PoS, utilize cryptographic algorithms to validate and agree on the order of transactions in the blockchain [12].

Cryptographic techniques play a vital role in ensuring the security and integrity of blockchain systems. Here are some cryptographic techniques commonly used in blockchain:

Hash functions: These are cryptographic algorithms that take an input (data) and produce a fixed-size output called a hash. The output is unique to the input, meaning even a small change in the input will result in a significantly different hash. Hash functions are used extensively in blockchain to generate digital signatures, verify data integrity, and link blocks together in a chain.

Public key cryptography: This is also known as asymmetric cryptography. It involves the use of key pairs: (i) a public key and (ii) a private key.

The public key is freely shared and used to encrypt data or verify digital signatures, while the private key is kept secret and used for decrypting data or creating digital signatures. Public key cryptography is used in blockchain to provide secure communication, establish identity, and enable digital signatures for transaction validation [13].

Digital signatures: These are created using a combination of a private key and a hash function. The private key is used to generate a signature for a specific message or transaction, and the corresponding public key is used to verify the signature's authenticity. Digital signatures ensure that transactions on the blockchain are tamperproof and can be verified by anyone with access to the public key.

Merkle trees: These are also known as hash trees, are hierarchical data structures that allow efficient verification of large sets of data. In a Merkle tree, each leaf node represents a data element, and the intermediate nodes are hash values computed from the concatenation of their child nodes, as shown in Figure 5.14. Merkle trees enable efficient verification of the integrity of large datasets by checking only a small set of hash values instead of the entire dataset.

Zero-knowledge proofs: This allow a party to prove the validity of a statement without revealing any additional information beyond the statement's validity. In the context of blockchain, zero-knowledge proofs can be used to demonstrate knowledge of certain data (such as ownership of a private key) without revealing the actual data. This technique enhances privacy and confidentiality in blockchain transactions.

These cryptographic techniques collectively contribute to the security, privacy, and integrity of blockchain systems. They enable participants to securely interact, validate transactions, and ensure the immutability of the blockchain's data. The proper implementation and use of these cryptographic techniques are essential for maintaining the trust and reliability of blockchain networks.

5.3   BLOCKCHAIN APPLICATIONS IN SMART VEHICLES

5.3.1   SECURE V2V COMMUNICATION

Blockchain technology can enhance the security and privacy of V-2-V communication. By leveraging blockchain, smart vehicles can establish secure and direct communication channels, authenticate each other's identities, and securely exchange critical

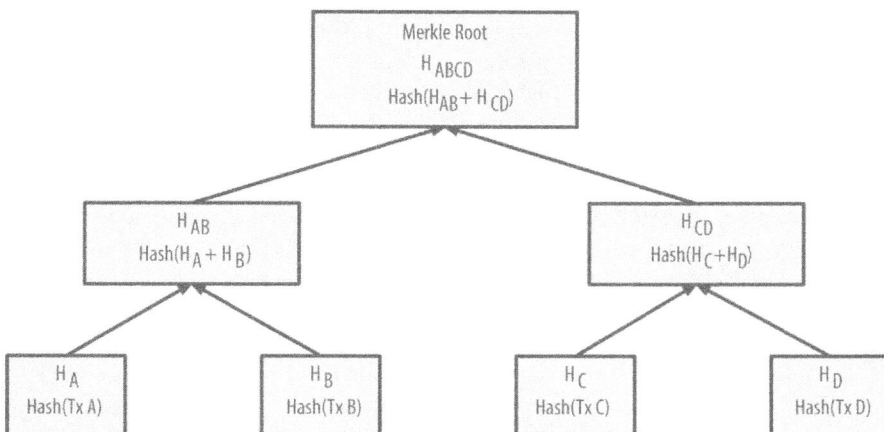

FIGURE 5.14   Merkle tree.

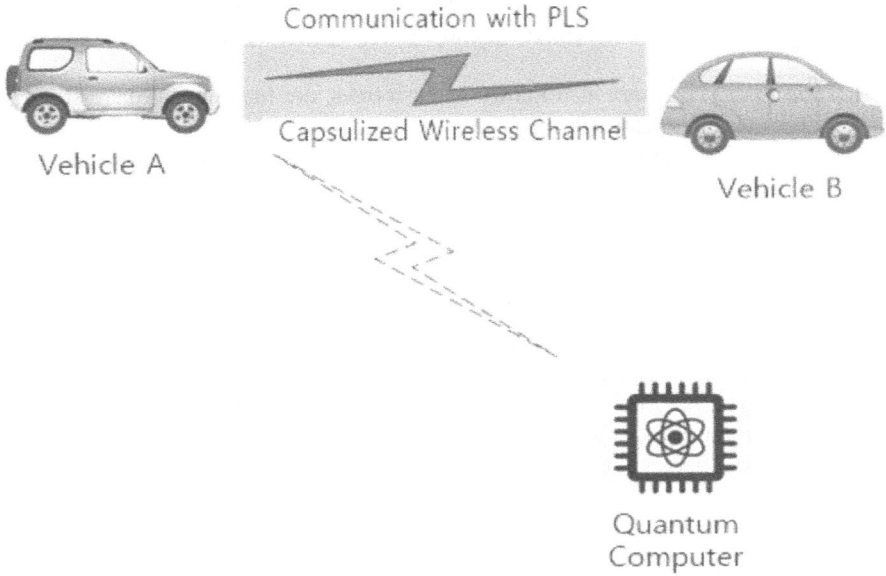

FIGURE 5.15   Secure vehicle-to-vehicle (V2V) communication.

information such as location, speed, and intent, as shown in Figure 5.15. Blockchain ensures that the communication is tamperproof and resistant to malicious attacks, thereby enhancing the safety and reliability of V2V communication [14].

Secure V2V communication refers to the exchange of information between vehicles in a secure and trusted manner. It enables vehicles to share important data, such as location, speed, acceleration, and other relevant information, to enhance safety, improve traffic flow, and enable various cooperative applications.

To achieve secure V2V communication, several measures are typically implemented:

Authentication: Before establishing communication, vehicles need to authenticate each other's identity to ensure that they are communicating with trusted and authorized vehicles. This can be done using digital certificates or cryptographic techniques like public key infrastructure (PKI) to verify the authenticity of the vehicles' digital identities.

Encryption: To protect the confidentiality and integrity of the transmitted data, encryption techniques are employed. Data exchanged between vehicles is encrypted using symmetric or asymmetric encryption algorithms. This ensures that even if the communication is intercepted, the information remains unreadable to unauthorized parties.

Message integrity: To ensure that the received messages have not been tampered with during transmission, message integrity mechanisms are used. This involves the use of cryptographic hash functions or digital signatures to verify the integrity of the received messages and detect any unauthorized modifications.

Privacy preservation: V2V communication should also address privacy concerns. Techniques like pseudonymization or anonymous communication can be employed to prevent the tracking of individual vehicles and protect the privacy of their owners. By using temporary identifiers or random pseudonyms, vehicles can communicate without revealing their true identities [14].

Secure key management: Effective key management is crucial for secure V2V communication. Secure protocols and mechanisms are employed for key generation, distribution, and revocation to ensure that only authorized entities have access to the necessary encryption keys. This helps protect against unauthorized access and maintain the security of the communication channel.

Trust management: Trust management systems can be implemented to assess the trustworthiness of vehicles participating in V2V communication. Trust can be established based on factors such as historical behavior, reputation, and validation by trusted authorities. Trust mechanisms help ensure that vehicles can rely on the information received from other trusted vehicles in making critical decisions.

By implementing these security measures, V2V communication can facilitate cooperative driving, collision avoidance, traffic optimization, and other advanced applications in the realm of connected and autonomous vehicles. These security measures are designed to protect the privacy, integrity, and authenticity of the communication and ensure that vehicles can trust the information received from other vehicles on the road.

5.3.2 Trusted V2I Communication

Blockchain-based communication frameworks enable trusted interactions between smart vehicles and infrastructure components such as traffic lights, road sensors, and charging stations. Blockchain ensures the integrity and authenticity of messages exchanged between vehicles and infrastructure, enabling secure coordination for tasks such as traffic management, intelligent routing, and electric vehicle charging, as shown in Figure 5.16. Additionally, blockchain can facilitate transparent and secure payment mechanisms for services provided by infrastructure providers.

Trusted V2I communication refers to the secure and reliable exchange of information between vehicles and infrastructure components. It enables vehicles to communicate with roadside infrastructure, such as traffic lights, road signs, toll booths, and smart road infrastructure, to facilitate efficient and safe transportation.

Here is an overview of how trusted V2I communication is achieved:

Authentication: V2I communication starts with authenticating the identity of both vehicles and infrastructure components. This authentication process ensures that only authorized and trusted entities can communicate with each other. Digital certificates, cryptographic keys, or secure protocols can be used for authentication, similar to V2V communication.

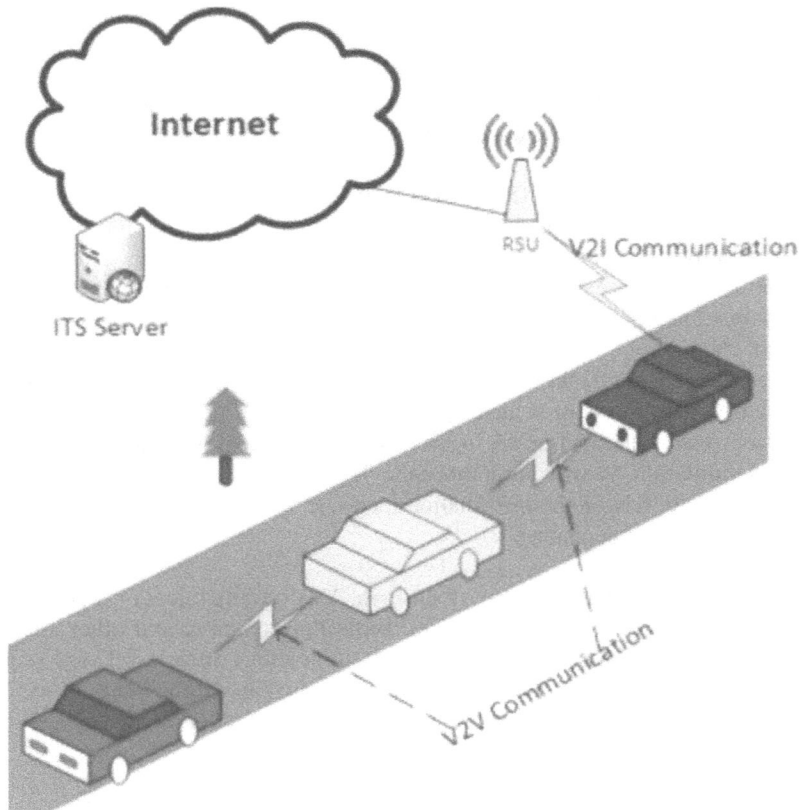

FIGURE 5.16 Trusted vehicle-to-infrastructure (V2I) communication.

Encryption: It is employed to protect the confidentiality and integrity of the data exchanged between vehicles and infrastructure. By encrypting the data using cryptographic algorithms, unauthorized parties are prevented from accessing or modifying the information during transmission. Symmetric or asymmetric encryption techniques can be used, depending on the specific requirements of the communication system.

Message integrity: To ensure the integrity of the messages exchanged between vehicles and infrastructure, mechanisms such as cryptographic hash functions or digital signatures are employed. These techniques verify that the received messages have not been tampered with or altered in transit, providing assurance of data integrity.

Secure communication protocols: Secure communication protocols are utilized to establish and maintain the communication channel between vehicles and infrastructure components. These protocols ensure that the communication is protected against eavesdropping, data manipulation, or unauthorized access. Examples of secure communication protocols commonly used in V2I communication include transport layer security (TLS) and secure socket layer (SSL).

Access control: Access control mechanisms are implemented to regulate which vehicles or entities can access specific infrastructure services or resources. Access control ensures that only authorized vehicles can interact with infrastructure components based on predefined permissions and policies. This helps prevent unauthorized access or misuse of infrastructure resources.

Trust and authentication management: V2I communication may involve the management of trust and authentication. Trust mechanisms can be employed to assess the reliability and trustworthiness of infrastructure components, ensuring that vehicles can rely on the information and services provided by the infrastructure. Similarly, effective management of authentication credentials, such as digital certificates or cryptographic keys, is crucial to maintain the security of the communication.

By implementing these security measures, trusted V2I communication enables vehicles to interact with infrastructure components in a secure and reliable manner. This facilitates the exchange of real-time information, enables intelligent transportation systems, and supports various applications, including traffic management, road safety, and efficient mobility services. The trust and security of V2I communication are crucial for ensuring the effectiveness and safety of connected and autonomous vehicles in a smart transportation ecosystem.

5.3.3 Immutable Data Logging and Tamperproof Records

Blockchain's immutable nature makes it suitable for logging and storing critical vehicle data, such as maintenance records, accident history, or sensor data. By recording these data on the blockchain, a permanent and tamperproof record is created, ensuring data integrity and providing an audit trail for verification purposes. This feature is particularly valuable for establishing trust among vehicle owners, service providers, and regulatory authorities [15].

Immutable data logging and tamperproof records refer to the concept of creating and maintaining data that cannot be modified or tampered with after it has been recorded. This concept is crucial in various fields where data integrity, audit ability, and transparency are essential, such as blockchain, financial transactions, legal records, supply chain management, and more.

Here is an overview of how immutable data logging and tamperproof records are achieved:

Cryptographic hash functions, immutable data logging often relies on cryptographic hash functions. A cryptographic hash function takes input data and produces a fixed-size output called a hash or digest, as shown in Figure 5.17. These functions have several important properties: they are deterministic (the same input always produces the same output), quick to compute, and computationally infeasible to reverse-engineer the original input from the hash. This property makes them suitable for verifying data integrity.

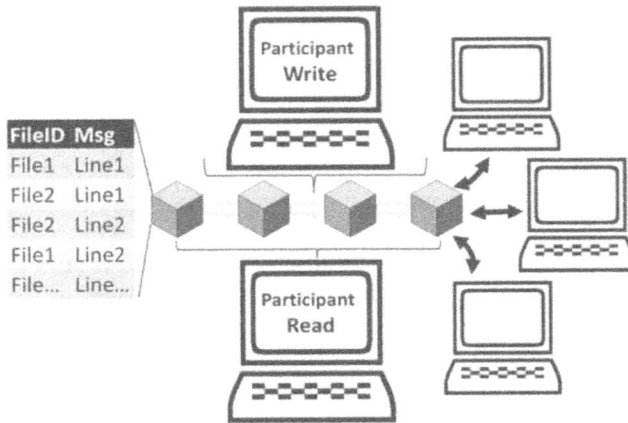

FIGURE 5.17   Immutable data logging and tamperproof records.

Hash pointers are used to create a tamperproof record, a hash pointer can be used. A hash pointer is a reference to a specific data block or record, along with the hash of that data block. By including the hash of the previous block in the hash pointer of a new block, a chain of linked blocks is created. This ensures that any modification in a block's content will result in a different hash, breaking the chain and indicating tampering.

Distributed ledger technologies (DLT), such as blockchain, are designed to create tamperproof and immutable records. In a blockchain, data is stored across a network of decentralized nodes, and each block contains a hash pointer linking it to the previous block, forming a chain. The decentralized and consensus-driven nature of blockchain makes it extremely difficult to alter previously recorded data without the consensus of the network.

Data auditing and verification, immutable data logging allows for easy auditing and verification of records. By comparing the hash of a stored record with its original hash, one can verify if the record has been tampered with. This verification process can be automated and performed by any party with access to the original data and the corresponding hash.

Timestamping plays a crucial role in ensuring the integrity of data records. Each record or block in the chain can be associated with a timestamp indicating when it was created or added to the system. Timestamping helps establish the order of events and prevents retroactive modification of records.

Encryption and digital signatures, in addition to the above techniques, encryption and digital signatures can be used to provide an extra layer of security to the data. Encryption protects the confidentiality of the data, ensuring that only authorized parties can access it. Digital signatures provide a means to authenticate the origin and integrity of the data, allowing recipients to verify that the data has not been tampered with.

By employing these techniques, immutable data logging and tamperproof records offer enhanced data integrity, auditability, and transparency. They provide a robust framework for recording and storing sensitive information, ensuring that it remains secure and unaltered over time.

5.3.4 DECENTRALIZED IDENTITY MANAGEMENT FOR VEHICLES

Blockchain enables decentralized identity management for vehicles, allowing them to have unique and verifiable digital identities. Each vehicle can have a digital certificate stored on the blockchain, which securely associates its identity with relevant attributes and permissions. This decentralized identity management eliminates the need for a centralized authority and provides a secure and reliable mechanism for verifying the identity of vehicles participating in the communication network.

Decentralized identity management for vehicles involves the management and control of digital identities of vehicles in a decentralized and distributed manner. It allows vehicles to have self-sovereign identities, where they have control over their own identity information and can securely interact with other entities without relying on a centralized authority.

Here's an overview of decentralized identity management for vehicles:

Self-sovereign identity: This refers to the concept where individuals or entities have complete control over their identity information. In the context of vehicles, self-sovereign identity enables vehicles to manage their own identity attributes, such as vehicle identification number (VIN), registration details, ownership information, and other relevant data.

Decentralized identity infrastructure: The management relies on distributed ledger technologies (DLT) or blockchain to create a decentralized identity infrastructure. The blockchain serves as a secure and immutable ledger for recording and validating vehicle identities and associated attributes. It eliminates the need for a centralized identity provider or authority, enhancing privacy, security, and control.

Digital identity wallets: Vehicles can have digital identity wallets, similar to digital wallets used for cryptocurrencies. These wallets store and manage the vehicle's identity credentials, such as digital certificates, cryptographic keys, and other relevant information [16]. The wallets provide secure storage and enable the vehicle to present and authenticate its identity when interacting with other vehicles, infrastructure, or service providers [15].

Privacy and selective disclosure: Decentralized identity management allows vehicles to selectively disclose identity attributes or information as required. Instead of sharing all identity details with every entity, vehicles can share only the necessary information on a need-to-know basis. This ensures privacy and minimizes the exposure of sensitive information.

Interoperability and standards: To enable seamless interaction and interoperability, standards and protocols are necessary for decentralized identity management. Organizations such as the Decentralized Identity Foundation (DIF) and the World Wide Web Consortium (W3C) are working on

developing standards and specifications for decentralized identity, ensuring compatibility and interoperability across different systems and platforms.
Trust and verification: Trust mechanisms are employed in decentralized identity management to establish the trustworthiness of vehicle identities. Trust can be built through reputation systems, attestations from trusted authorities, or through consensus mechanisms within the decentralized identity infrastructure. Verification of vehicle identities and attributes can be performed by other vehicles, infrastructure components, or third-party verifiers using cryptographic techniques.

Decentralized identity management for vehicles enhances security, privacy, and autonomy in managing identity information. It enables vehicles to securely interact with other vehicles, infrastructure, and service providers, while maintaining control over their identity data. By leveraging blockchain technology and decentralized principles, vehicles can establish trusted and self-sovereign identities in a decentralized ecosystem.

5.3.5 SMART CONTRACTS FOR AUTOMATED TRANSACTIONS

Smart contracts, self-executing agreements recorded on the blockchain, can automate transactions and enforce predefined rules without relying on intermediaries. In the context of smart vehicles, smart contracts can facilitate automated transactions such as toll payments, insurance claims, or electric vehicle charging. By eliminating the need for intermediaries, smart contracts enhance efficiency, transparency, and trust in these transactions.

Smart contracts are self-executing contracts with the terms of the agreement directly written into code. They are designed to automatically facilitate, verify, and enforce the performance of a contract without the need for intermediaries, as shown in Figure 5.18. Smart contracts are commonly associated with blockchain

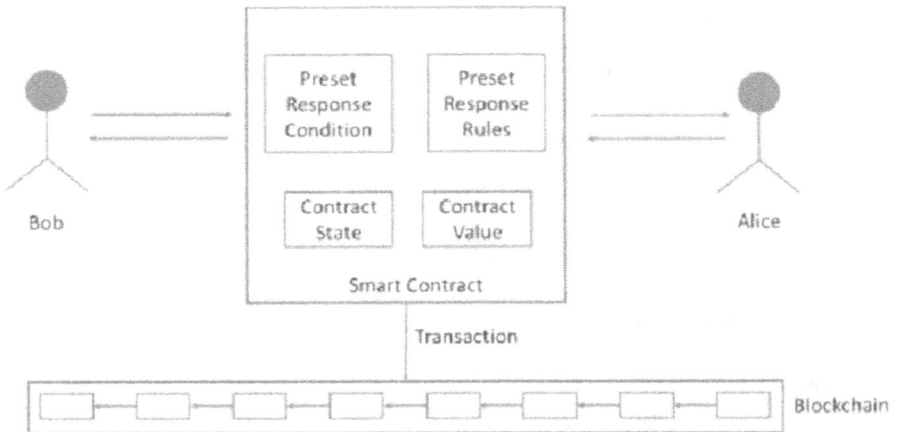

FIGURE 5.18   Smart contract.

technology, although they can also be implemented in other distributed ledger technologies.

Here is an overview of smart contracts for automated transactions:

Contract logic in code: Smart contracts encode the terms and conditions of a contract into computer code. This code contains predefined rules, conditions, and actions that are triggered based on specified events or inputs. The code serves as an agreement between the involved parties and is executed automatically when certain conditions are met.

Decentralized execution: Smart contracts are typically deployed on a blockchain or distributed ledger, enabling decentralized execution. The decentralized nature ensures that the contract is executed across multiple nodes in the network, providing transparency, immutability, and resilience to single points of failure.

Automation and self-execution: Once deployed on the blockchain, smart contracts operate autonomously, automatically executing the predefined actions based on predefined triggers or inputs. This eliminates the need for manual intervention or reliance on intermediaries, streamlining and automating the transaction process.

Trust and security: Smart contracts rely on the underlying blockchain's security mechanisms, such as cryptographic algorithms and consensus protocols, to ensure the integrity and trustworthiness of the contract execution. The transparency and immutability of the blockchain provide auditability and prevent tampering or unauthorized modification of the contract's logic.

Conditional triggers and actions: Smart contracts can be programmed to execute specific actions based on predefined conditions or triggers. For example, in a supply chain smart contract, a payment can be automatically released to the supplier when certain delivery milestones are met. These conditions are encoded in the contract code and executed without the need for manual intervention.

Real-time validation and verification: Smart contracts enable real-time validation and verification of transactions. The predefined rules and conditions within the contract's code ensure that transactions are validated and verified automatically, providing instant confirmation and reducing the need for time-consuming manual processes.

Multiparty agreements: Smart contracts can facilitate complex multiparty agreements by automating and enforcing the obligations of each party involved. The code governs the interactions and ensures that all parties fulfill their responsibilities as agreed upon, minimizing disputes and enhancing efficiency.

Immutable record keeping: The execution and outcome of smart contracts are recorded on the blockchain, creating an immutable and transparent record of the transactions. This provides an auditable trail and a historical reference of all executed smart contract transactions.

Smart contracts have the potential to revolutionize various industries by automating and streamlining transactions, reducing costs, enhancing trust, and enabling new business models. They are particularly well-suited for scenarios involving repetitive or complex transactions, where efficiency, accuracy, and trust are crucial.

5.4 DESIGN CONSIDERATIONS FOR BLOCKCHAIN-BASED COMMUNICATION FRAMEWORKS

5.4.1 SCALABILITY AND THROUGHPUT CHALLENGES

Blockchain technology faces inherent scalability and throughput limitations due to the consensus mechanisms and the need to maintain a full copy of the blockchain across all network participants. Addressing these challenges requires exploring innovative solutions such as sharding, sidechains, or off-chain transactions to enhance the performance of blockchain-based communication frameworks for smart vehicles [17].

5.4.2 CONSENSUS MECHANISMS SUITABLE FOR SMART VEHICLE NETWORKS

Consensus mechanisms play a crucial role in maintaining the integrity and security of the blockchain network. However, the resource-intensive nature of traditional consensus algorithms like PoW may not be suitable for resource-constrained smart vehicles. Alternative consensus mechanisms such as PoS, DPoS, or PBFT should be considered to achieve a balance between security and resource efficiency.

5.4.3 PRIVACY AND CONFIDENTIALITY OF VEHICLE-RELATED DATA

Preserving privacy and confidentiality is paramount in smart vehicle communication. Blockchain inherently provides transparency, but privacy-enhancing techniques like zero-knowledge proofs, ring signatures, or homomorphic encryption can be applied to protect sensitive vehicle data while still maintaining the integrity and auditability of the blockchain.

5.4.4 INTEROPERABILITY WITH EXISTING COMMUNICATION PROTOCOLS

To ensure seamless integration with existing communication protocols and infrastructure, Blockchain-based communication frameworks need to be designed with interoperability in mind. Standards and protocols that facilitate interoperability, such as the interledger protocol (ILP) or the vehicle-to-everything (V2X) communication standards, should be considered to enable compatibility between blockchain networks and traditional communication systems.

5.4.5 ENERGY EFFICIENCY AND RESOURCE CONSTRAINTS

Smart vehicles have limited computing resources and energy capacities. Blockchain-based communication frameworks must be optimized to minimize the energy consumption and computational overhead of participating vehicles. Techniques like lightweight consensus algorithms, data compression, and efficient data synchronization mechanisms can help mitigate resource constraints and enable practical implementation in smart vehicle networks [18].

5.5 CASE STUDIES AND IMPLEMENTATIONS

5.5.1 EXAMPLE BLOCKCHAIN-BASED COMMUNICATION FRAMEWORKS FOR SMART VEHICLES

This section presents real-world examples of blockchain-based communication frameworks developed for smart vehicles. It explores projects that have leveraged Block chain technology to enhance security, data integrity, and interoperability in smart transportation systems [19]. Case studies can include initiatives like MOBI (Mobility Open Blockchain Initiative) and various pilot projects conducted by automotive manufacturers, transportation authorities, and technology companies.

5.5.2 REAL-WORLD PILOT PROJECTS AND THEIR OUTCOMES

Highlighting the outcomes of real-world pilot projects provides insights into the practical implementation of blockchain-based communication frameworks for smart vehicles. It examines the challenges faced, lessons learned, and the impact on enhancing the efficiency, safety, and sustainability of smart transportation systems. Examples can include trials of blockchain-based V2V communication, decentralized charging infrastructure, or supply chain management for autonomous vehicles [20].

5.5.3 LESSONS LEARNED AND BEST PRACTICES FOR IMPLEMENTING BLOCKCHAIN IN SMART TRANSPORTATION SYSTEMS

Based on the experiences gained from case studies and pilot projects, this section discusses the key lessons learned and best practices for implementing Blockchain in smart transportation systems. It covers aspects such as system architecture, network governance, security measures, data management, and collaboration among stakeholders. These insights will aid future implementations and help optimize the benefits of blockchain technology in smart vehicle communication.

5.6 BENEFITS AND CHALLENGES OF BLOCKCHAIN-BASED COMMUNICATION FRAMEWORKS

5.6.1 ENHANCED SECURITY AND RESISTANCE TO CYBER-ATTACKS

Blockchain's decentralized and immutable nature enhances the security of smart vehicle communication. By removing single points of failure and providing tamper-proof records, blockchain-based communication frameworks can mitigate the risks of cyber-attacks, data manipulation, and unauthorized access. However, it is essential to continually evaluate and address emerging security threats and vulnerabilities specific to blockchain implementations.

5.6.2 Improved Data Integrity and Trustworthiness

Blockchain technology ensures data integrity by creating a transparent and auditable record of all transactions and communication events. Smart vehicles can rely on the blockchain as a single source of truth, fostering trust among network participants and enabling secure and reliable data exchange. Nevertheless, the accuracy and reliability of data feeding into the blockchain need to be ensured, considering the potential for malicious or faulty data inputs.

5.6.3 Increased Efficiency and Reduced Costs

Blockchain-based communication frameworks have the potential to streamline and automate various processes in smart transportation systems, resulting in increased operational efficiency and reduced costs. For instance, smart contracts executed on the blockchain can automate payment settlements, toll collection, or insurance claims, eliminating the need for intermediaries and reducing administrative overhead. However, the performance and efficiency of blockchain networks should be continuously optimized to deliver real-time responsiveness required for smart vehicle communication.

5.6.4 Potential Challenges and Limitations of Blockchain Adoption in Smart Vehicles

Although blockchain offers significant advantages, it is essential to acknowledge its limitations and potential challenges when implementing it in smart vehicles. These challenges include the computational and storage requirements of blockchain networks, regulatory considerations, standardization efforts, and the need for widespread adoption to achieve the desired network effects. Additionally, scalability issues and the trade-off between decentralization and performance should be carefully managed.

5.7 FUTURE DIRECTIONS AND RESEARCH OPPORTUNITIES

5.7.1 Integration of Blockchain with Emerging Technologies (e.g., IoT, AI)

The integration of blockchain with other emerging technologies such as the Internet of Things (IoT) and artificial intelligence (AI) opens up new avenues for innovation in smart vehicle communication. Research should focus on exploring synergies between these technologies to enhance the security, privacy, and intelligence of blockchain-based communication frameworks. Examples include combining blockchain with edge computing, machine learning algorithms, or secure device-to-device communication protocols [21].

Integration of blockchain with emerging technologies, such as IoT and AI, can offer several benefits and open up new possibilities. Here are some ways blockchain can be integrated with these technologies:

Block chain and IoT integration

a. *Data integrity and security*: Blockchain can provide a tamper-proof and transparent ledger for recording IoT device data. Each data transaction can be securely recorded on the blockchain, ensuring data integrity and preventing unauthorized modification.

b. *Trusted device identity*: Blockchain-based identity management can enable secure and decentralized device identity verification, authentication, and access control for IoT devices.

c. *Device coordination and automation*: Smart contracts on the blockchain can facilitate automatic coordination and interaction between IoT devices without the need for centralized control, as shown in Figure 5.19. This can enable decentralized decision-making and autonomous IoT networks.

d. *Supply chain transparency*: Blockchain can enhance supply chain management in IoT applications by providing an immutable and transparent record of every step in the supply chain, ensuring traceability, authenticity, and accountability of IoT devices and components.

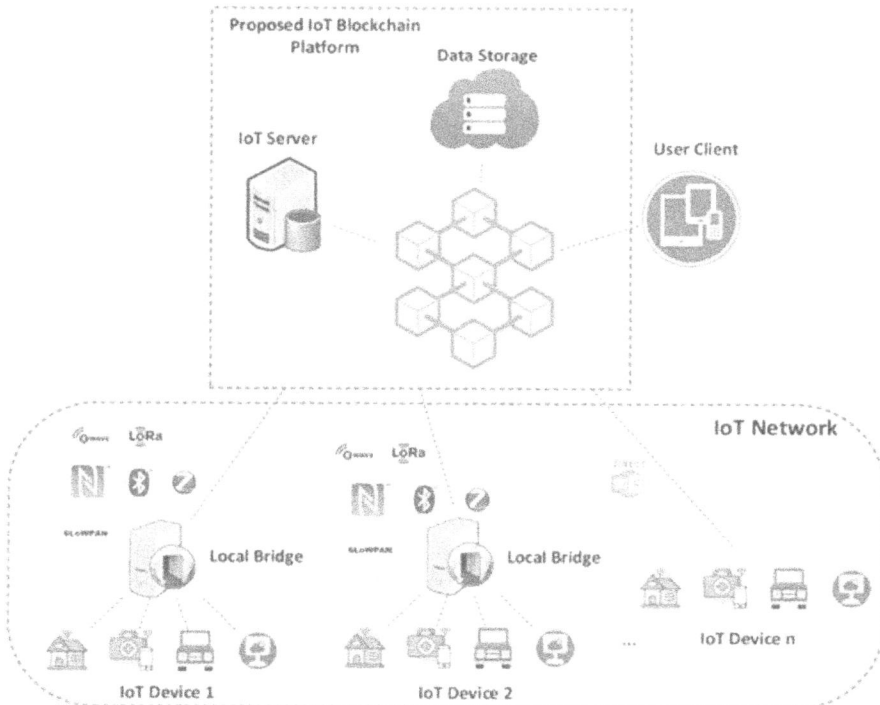

FIGURE 5.19 Blockchain and IoT integration.

Blockchain and AI Integration

a. *Data privacy and ownership*: Blockchain can enable secure data sharing and collaboration in AI applications while ensuring privacy and ownership rights. It allows individuals to have control over their data and grant access to AI models or algorithms without compromising data security.

b. *AI model verification and transparency*: Blockchain can be used to record the training data, algorithms, and models used in AI systems, ensuring transparency and traceability. This helps in verifying the integrity and authenticity of AI models, promoting trust and accountability.

c. *Decentralized AI marketplaces*: Blockchain-based platforms can facilitate decentralized AI marketplaces, where individuals and organizations can securely buy, sell, and trade AI models, data, or computational resources. Smart contracts can automate the transactions and ensure fair and transparent exchanges.

d. *Federated learning and data collaboration:* Blockchain can enable secure and privacy-preserving federated learning, where AI models are trained on distributed data sources without sharing the raw data, as per Figure 5.20. Blockchain provides a trust layer for coordinating the learning process and aggregating the model updates.

Integrating blockchain with IoT and AI can enhance data security, privacy, transparency, and trust in these emerging technology domains. It enables new business models, decentralized decision-making, and collaborative ecosystems while addressing the challenges of data integrity, ownership, and interoperability.

5.7.2 STANDARDIZATION EFFORTS AND REGULATORY CONSIDERATIONS

Standardization bodies, industry consortia, and regulatory authorities play a crucial role in establishing interoperability, security standards, and legal frameworks

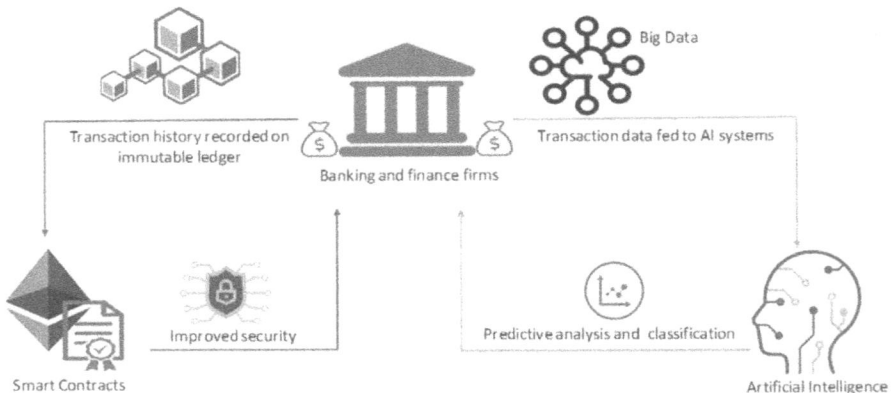

Transaction history recorded on immutable ledger

Banking and finance firms

Big Data

Transaction data fed to AI systems

Improved security

Predictive analysis and classification

Smart Contracts

Artificial Intelligence

FIGURE 5.20 Blockchain, smart contracts and AI integration.

for blockchain-based communication frameworks in smart transportation systems. Collaboration among stakeholders is necessary to develop consistent and widely accepted protocols, ensuring a smooth integration of blockchain into existing infrastructures and regulatory frameworks.

Standardization efforts and regulatory considerations play a crucial role in the successful adoption and implementation of blockchain technology. Here are some key points to consider:

Standardization efforts

a. *Interoperability*: Standardization helps ensure interoperability between different blockchain platforms, protocols, and applications. Efforts are underway to develop common standards for data formats, communication protocols, smart contracts, and consensus mechanisms.

b. *Data exchange and integration*: Standardized formats and protocols enable seamless data exchange and integration between blockchain systems and existing IT infrastructure, promoting compatibility and reducing integration complexities.

c. *Security and privacy*: Standards related to cryptographic algorithms, key management, identity management, and secure coding practices help establish best practices for ensuring the security and privacy of blockchain implementations.

d. *Governance and compliance*: Standardization efforts address governance models, regulatory compliance, and legal frameworks associated with blockchain technology. This includes guidelines for anti-money laundering (AML), know your customer (KYC) requirements, and data protection regulations.

Regulatory considerations

a. *Jurisdiction and compliance*: Blockchain-based applications may operate across different jurisdictions, requiring compliance with local laws and regulations. Regulatory considerations include data privacy, consumer protection, financial regulations, securities laws, and intellectual property rights.

b. *Smart contracts and legal frameworks*: As smart contracts gain prominence, legal frameworks need to adapt to accommodate their use. Definitions of enforceability, contractual obligations, and dispute resolution mechanisms may need to be revisited to align with the unique characteristics of blockchain-based contracts.

c. *Anti-money laundering (AML) and counter-terrorism financing (CTF)*: Blockchain-based systems may need to comply with AML and CTF regulations. Regulatory frameworks may require identity verification, transaction monitoring, and reporting mechanisms to address these concerns.

d. *Tokenization and securities regulations*: If blockchain-based assets are considered securities or involve tokenization, they may fall under existing securities regulations. Compliance with securities laws, such as registration, investor protection, and trading regulations, should be considered.

Governments, industry consortia, and standardization bodies are actively working on developing guidelines, best practices, and regulatory frameworks for blockchain technology. Collaboration among stakeholders, including technology providers, regulators, legal experts, and industry participants, is vital to establishing a balanced regulatory environment that fosters innovation, protects consumers, and ensures compliance with legal and regulatory requirements.

5.7.3 EXPLORING HYBRID COMMUNICATION MODELS COMBINING BLOCKCHAIN AND OTHER APPROACHES

Hybrid communication models that combine blockchain with other communication approaches, such as peer-to-peer networks, centralized cloud services, or edge computing, can address the limitations and scalability challenges of blockchain technology. Research should investigate the potential of hybrid models to strike a balance between the benefits of blockchain and the efficiency of alternative communication architectures.

Hybrid communication models that combine blockchain with other approaches can leverage the strengths of both technologies and address specific use cases more effectively. Here are a few examples:

Blockchain and centralized communication: In some scenarios, a hybrid model can be designed where blockchain is used for certain aspects of communication, such as decentralized identity management, data integrity, and audit ability, while centralized communication channels are utilized for faster and more efficient data transfer. This approach combines the immutability and transparency of blockchain with the scalability and low-latency capabilities of centralized systems.

Blockchain and peer-to-peer (P2P) communication: Block chain can be integrated with peer-to-peer communication protocols to create a hybrid model. P2P communication enables direct and secure data exchange between nodes, while blockchain ensures trust, consensus, and data integrity. This combination can be beneficial in decentralized applications, IoT networks, and supply chain management, where direct peer-to-peer communication is necessary, but the blockchain provides an immutable and transparent record of interactions.

Blockchain and cloud communication: Hybrid models can also integrate blockchain with cloud communication services. Blockchain can be used to store critical data and transactions, while cloud-based communication services provide efficient and scalable messaging, real-time data synchronization, and computing resources. This combination is suitable for applications requiring a balance between data security, decentralization, and scalable communication.

Blockchain and Internet of Things (IoT) gateways: IoT gateways can serve as intermediaries between IoT devices and the blockchain network. The gateway collects and validates data from IoT devices, performs necessary processing or filtering, and securely submits the data to the blockchain for storage or execution of smart contracts. This hybrid approach enables efficient data aggregation, while blockchain ensures data integrity and transparency.

Blockchain and off-chain communication: Off-chain communication refers to the transfer of data or transactions outside the blockchain network. In a hybrid model, certain noncritical or high-frequency data can be communicated off-chain, while important data or transactions are recorded on the blockchain. This approach reduces blockchain congestion and improves scalability while leveraging the security and immutability of the blockchain for critical operations.

The choice of hybrid communication models depends on the specific requirements, scalability needs, latency constraints, and security considerations of the use case. By combining blockchain with other communication approaches, organizations can optimize their solutions and strike a balance between decentralization, scalability, efficiency, and data integrity.

5.7.4 ETHICAL AND LEGAL IMPLICATIONS OF BLOCKCHAIN-BASED SMART VEHICLE COMMUNICATION

The adoption of blockchain technology in smart vehicles raises ethical and legal considerations regarding data privacy, ownership, consent, and liability. Further research is needed to explore frameworks and policies that address these concerns and ensure compliance with legal requirements, consumer rights, and societal expectations.

The adoption of blockchain-based smart vehicle communication raises several ethical and legal implications that need to be carefully addressed. Here are some key considerations:

Privacy and data protection: Blockchain-based smart vehicle communication involves the collection, storage, and sharing of sensitive data, including vehicle and driver information. Ethical considerations include obtaining informed consent for data collection, ensuring data protection, implementing strong encryption and access controls, and providing transparency to users regarding data usage and sharing practices. Compliance with relevant data protection regulations, such as the General Data Protection Regulation (GDPR), is essential.

Security and integrity: Blockchain-based systems are designed to be secure and tamperproof. However, vulnerabilities in smart contracts, consensus mechanisms, or implementation can lead to security breaches [22]. Ethical considerations involve implementing robust security measures, conducting regular audits and vulnerability assessments, and promptly addressing any identified weaknesses. Protecting the integrity of data and ensuring the accuracy of smart contract execution is crucial for maintaining trust in blockchain-based smart vehicle communication.

Liability and accountability: The use of blockchain-based smart vehicle communication can raise questions of liability and accountability. If a smart contract or automated transaction leads to a malfunction or accident, it becomes essential to establish responsibility. Ethical considerations involve determining legal frameworks for liability, ensuring transparency in the execution of smart contracts, and providing mechanisms for dispute resolution and accountability in case of errors or failures.

Access and digital divide: Blockchain-based smart vehicle communication relies on internet connectivity and access to technology infrastructure. Ethical considerations include ensuring equitable access to technology and infrastructure, particularly in underserved or marginalized communities. Efforts should be made to bridge the digital divide and prevent exclusion or discrimination based on access to blockchain-based communication systems.

Transparency and auditing: Blockchain is often associated with transparency and audit ability. Ethical considerations involve providing users with visibility into the functioning of the blockchain-based smart vehicle communication system, ensuring transparency in decision-making processes, and allowing users to verify the accuracy of data and transactions. Implementing mechanisms for auditing and third-party verification can enhance trust and accountability.

Social and environmental impact: The adoption of blockchain-based smart vehicle communication can have broader social and environmental implications. Ethical considerations involve assessing the environmental impact of blockchain networks, such as energy consumption, and considering the potential social implications, such as job displacement due to automation. Efforts should be made to mitigate negative impacts and ensure the technology is aligned with sustainable and socially responsible practices.

Regulatory compliance: Compliance with existing legal and regulatory frameworks is crucial in blockchain-based smart vehicle communication. Ethical considerations include ensuring compliance with relevant transportation, data protection, cybersecurity, and consumer protection regulations. Collaborating with regulatory authorities and industry stakeholders to establish appropriate regulations and standards can help address ethical and legal concerns effectively.

Addressing these ethical and legal implications requires a multidisciplinary approach involving technology developers, policymakers, regulators, and other stakeholders. By proactively addressing these considerations, blockchain-based smart vehicle communication can be developed and deployed in a manner that prioritizes ethical values, protects user rights, and ensures compliance with legal obligations.

5.8 CONCLUSION

Blockchain technology holds significant promise in establishing a robust communication framework for smart vehicles. By leveraging the decentralized and tamperproof nature of blockchain, smart transportation systems can enhance security, data integrity, and efficiency while fostering trust among network participants. However, there are design considerations, challenges, and research opportunities that need to be addressed to realize the full potential of blockchain-based communication frameworks. Continued collaboration among researchers, industry stakeholders, and policymakers is crucial to drive the adoption and evolution of blockchain in smart vehicle communication and shape the future of intelligent transportation systems.

Blockchain-based smart vehicle communication holds immense potential to revolutionize the automotive industry by enabling secure, transparent, and efficient communication among vehicles, infrastructure, and other stakeholders. By leveraging the inherent benefits of blockchain, such as decentralization, immutability, and transparency, smart vehicles can enhance safety, efficiency, and data integrity in transportation systems.

However, there are several challenges and considerations that need to be addressed to fully realize the potential of blockchain-based smart vehicle communication. These include scalability, privacy, security, regulatory compliance, and interoperability. Ongoing research, industry collaborations, and standardization efforts are crucial to overcome these challenges and establish robust frameworks for blockchain integration in the automotive sector.

The future scope of blockchain-based smart vehicle communication is promising. As technology advancements continue, we can expect to see increased integration of blockchain with emerging technologies like IoT, AI, and edge computing. This integration will further enhance the capabilities of smart vehicles, enabling real-time data exchange, autonomous decision-making, and secure, trusted interactions among vehicles and infrastructure.

Moreover, the application of blockchain in smart vehicle communication is not limited to individual vehicles. It extends to broader use cases such as mobility-as-a-service (MaaS), electric vehicle charging networks, supply chain management, and insurance systems. Blockchain-based solutions have the potential to streamline these areas, enhance efficiency, and promote trust among participants.

Blockchain-based smart vehicle communication represents a transformative shift in the automotive industry. By addressing the ethical, legal, and technical considerations and embracing collaborative efforts, we can shape a future where smart vehicles communicate seamlessly, improving safety, efficiency, and sustainability in transportation systems. Continued research, innovation, and industry-wide adoption will drive the evolution of blockchain-based smart vehicle communication and unlock its full potential.

REFERENCES

1. Li, C., Qiao, J., Deng, L., Zhang, R., & Vasilakos, A. V. (2020). A Survey on Block Chain-Based Communication Systems for IoT. IEEE Communications Surveys & Tutorials, 22(1), 674–705.
2. Yao, Y., Li, J., Zhang, K., & Zhang, Y. (2020). A Block chain-Based Communication Framework for Connected Vehicles in Smart City. In 2020 IEEE International Conference on Artificial Intelligence and Computer Applications (ICAICA) (pp. 408–413). IEEE.
3. Rahman, M. H., Basu, A., & Dey, K. C. (2019). Block chain-based Secure and Trustworthy Communication Framework for Connected Vehicles. In 2019 12th International Conference on Developments in eSystems Engineering (DeSE) (pp. 271–276). IEEE.

4. Gupta, D., Rani, S., Ahmed, S. H., Garg, S., Piran, M. J., & Alrashoud, M. (2021). ICN-Based Enhanced Cooperative Caching for Multimedia Streaming in Resource Constrained Vehicular Environment. IEEE Transactions on Intelligent Transportation Systems, 22(7), 4588–4600.
5. Rani, S., Gupta, D., Herencsar, N., & Srivastava, G. (2023). Blockchain-Enabled Cooperative Computing Strategy for Resource Sharing in Fog Networks. Internet of Things, 21, 100672.
6. Alotaibi, M., Alshehri, A., Alharthi, M., Alabdulkarim, R., & Almuhaideb, N. (2021). A Block Chain-Based Communication Framework for Intelligent Transportation Systems. Sensors, 21(1), 107.
7. Arora, G. K., Mutha, R., Sangari, M. S., Aswal, U., Bhattacherjee, A., & Agarwal, A. (2023). An Analysis of the Effects and Interaction of Hyper Parameters in Convolutional Neural Networks.2023 Second International Conference on Electronics and Renewable Systems (ICEARS)(pp. 1057–1063). IEEE Explore. https://doi.org/10.1109/ICEARS56392.2023.10085483
8. Mutha, R., Pathak, D. N., Ahuja, V., Bhandari, P. R., Rahi, P., & Mamodiya, U. (2023). Medical Image Fusion Solid Works Supporting Multiple Techniques with Feature-Level Transforms.2023 International Conference on Sustainable Computing and Data Communication Systems (ICSCDS)(pp. 937–942). IEEE Explore. https://doi.org/10.1109/ICSCDS56580.2023.10105129
9. Hou, Z., Li, J., & Duan, Q. (2020). Block Chain-Based Secure Communication Architecture for Cooperative Connected Vehicles. IEEE Access, 8, 23242–23251.
10. Yu, S., Li, J., Yang, X., & Zhang, Y. (2020). Blockchain-Based Communication Framework for Secure Data Sharing in Smart Transportation Systems. IEEE Access, 8, 141364–141374.
11. Thakur, A., & Ranjan, R. (2023). Evaluate the Performance of Deep CNN Algorithm Based on Parameters and Various Geometrical Attacks. Wireless Personal Communications, 132(4), 1–16.
12. Mutha, R. Dr.(2021). Block Chain Based Financial Transactions Positive and Negative Aspects, International Journal Of Engineering Research & Technology, 10(12), 481–484.
13. Mutha, R., & Keswani, B. (15 June 2023). Digital Asset Management Using Block Chain. AIP Conference Proceedings, 2782(1), 020028. https://doi.org/10.1063/5.0154853
14. Keswani, B., Keswani, P., & Purohit, R. (2020). History and Generations of Security Protocols. In Design and Analysis of Security Protocol for Communication (eds D. Goyal, S. Balamurugan, S.-L. Peng and O.P. Verma). Wiley Publication. https://doi.org/10.1002/9781119555759.ch1
15. Mutha, R., Lavate, S., Limkar, S. et al. (2023). HDFRMAH: Design of a High-Density Feature Representation Model for Multidomain Analysis of Human Health Issues. Soft Computing, 27, 8493–8503. https://doi.org/10.1007/s00500-023-08311-9
16. Li, J., Hou, Z., Zhang, Y., & Zhang, K. (2020). A Block chain-Based Secure Communication Framework for Connected Vehicles in Smart City. IEEE Transactions on Intelligent Transportation Systems, 10, 20995–21031. https://doi.org/10.1109/ACCESS.2022.3149958
17. Mutha, R., Pawar, M. E., Limkar, S. et al. (2023). MPCITL: Design of an Efficient Multimodal Engine for Pre-Emptive Identification of CKD via Incremental Transfer Learning on Clinical Data Samples. Soft Computing. https://doi.org/10.1007/s00500-023-08774-w
18. Taneja, A., Saluja, N., & Rani, S. (2022). An Energy Efficient Dynamic Framework for Resource Control in Massive IoT Network for Smart Cities. Wireless Networks. https://doi.org/10.1007/s11276-022-03047-0

19. Xie, S., Zhou, Y., Wang, X., & Ma, J. (2020). A Block chain-Based Secure Communication Framework for Intelligent Transportation Systems. In 2020 IEEE 5th Information Technology and Mechatronics Engineering Conference (ITOEC) (pp. 124–128). IEEE.
20. Wang, J., Chen, Y., Li, M., & Zhao, Q. (2021). Block Chain-Based Secure Communication Framework for Connected Vehicles. In 2021 IEEE International Conference on Information and Automation (ICIA) (pp. 504–508). IEEE.
21. Duan, Q., Li, J., Zhang, K., & Wang, X. (2021). Block Chain-Based Secure Communication Framework for Cooperative Vehicular Systems. IEEE Access, 9, 13332–13343.
22. Jain, S., Sharma, C., Das, P., Shambhu, S., & Chen, H. Y. (2023). Blockchain and Cryptocurrency: A Bibliometric Analysis. Journal of Advanced Computational Intelligence and Intelligent Informatics, 27(5), 822–836.

6 A Sustainable IoT-Based Smart Transportation System for Urban Mobility

Mahendra Balkrishna Salunke, Shailesh V. Kulkarni, Shraddha Ovale, Bhushan M. Manjre, Suresh Limkar, Farhadeeba Shaikh, and Vyasa Sai

6.1 INTRODUCTION

In today's metropolitan centers across the world, commuting in cities is a significant problem. The exponential rise in the number of cars on the road and the increasing urbanization have all worsened traffic congestion, air pollution, and ineffective transportation systems [1]. These problems have an adverse effect on the quality of life for city people, as well as having a substantial influence on the economy and the environment. Modern transportation systems need to be improved since they are too dependent on static infrastructure and insufficient data to handle the complexity of urban mobility [2]. As the Internet of Things (IoT) expands, our perspective and approach toward transportation systems will shift fundamentally. Utilizing IoT technology [3], such as sensors, wireless connectivity, and cloud computing, it is possible to create intelligent transportation systems that are better for the environment, use less energy, and can adapt to the demands of city dwellers. IoT-based smart transportation systems (Figure 6.1) must be able to receive and understand real-time data from a variety of sources, including cars, infrastructure, and users, in order to fully achieve the potential of these systems and reap the benefits [4]. With the use of this data-driven method, more understanding of traffic patterns, road conditions, and user behavior can be attained. As a consequence, those who manage transport policy will be better able to decide how to increase system-wide efficiency and lessen environmental impact.

Because of the technology supporting the IoT, these systems may offer customers timely and pertinent information. Possibly the largest benefit of these systems is that users will have access to real-time information on traffic congestion, parking availability, and public transit timetables thanks to the installation of IoT sensors in cars, traffic lights, and infrastructure. By providing users with this knowledge, we can enable them to make more intelligent decisions regarding their schedules, modes of transportation, and routes, ultimately cutting down on travel time and increasing productivity [5]. Intelligent transport systems based on the IoT may make proactive traffic flow control practicable. Algorithms using machine learning (ML) capabilities

DOI: 10.1201/9781003437079-6

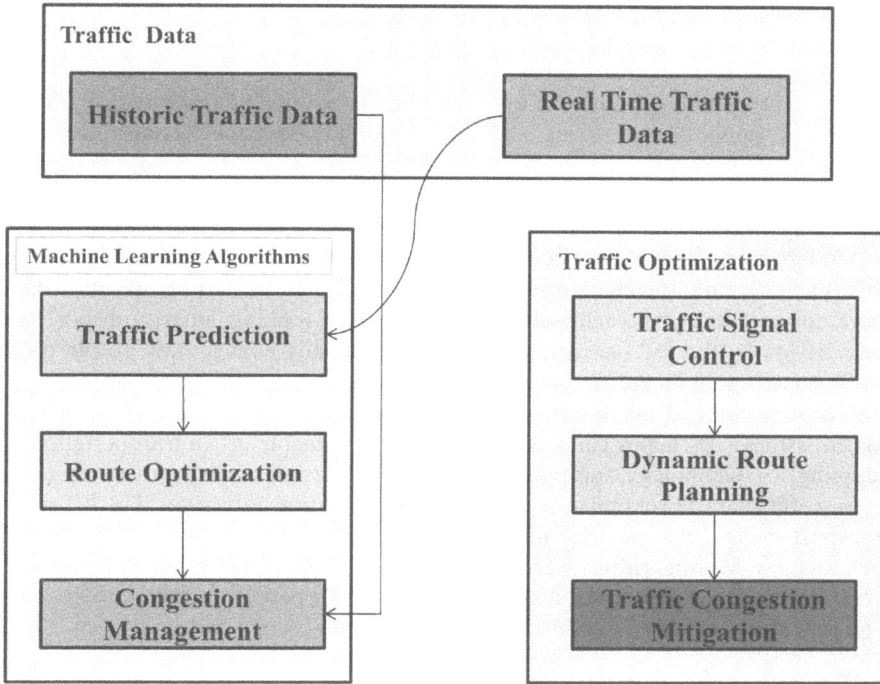

FIGURE 6.1   IoT-based smart transportation system.

from IoT-connected devices can control traffic more effectively by spotting patterns and making changes in real time. This dynamic traffic management reduces congestion, cuts travel times to a minimum, and improves system efficiency. IoT-based smart transport solutions not only assist in keeping traffic flowing smoothly, but they also make it simpler for individuals to move around cities sustainably [6]. These programmers encourage people to utilize shared mobility services and public transportation, instead of driving their own cars, in an effort to reduce the number of private automobiles on the road. The result is a noticeable reduction in traffic and air pollution on the roads. Additionally, IoT-connected sensors may be able to monitor the air quality and offer information about how transportation-related activities affect the environment. It is therefore possible that more individuals will select green modes of transportation [7].

Design concepts must be carefully considered while developing an IoT-based smart transportation system. It is essential that these resources be used as effectively as possible in order to guarantee that the system will operate with the least amount of available energy resources. This objective may be attained by using energy-efficient IoT devices, energy-harvesting techniques, and optimization algorithms that reduce unnecessary data transfer [8]. The system's architecture must prioritize the use of cleaner forms of transportation, such as public transportation and electric automobiles, in order to minimize emissions. If smart transportation systems incorporated electric vehicle charging stations and financial incentives to purchase them, greenhouse gas emissions may be significantly decreased. The cornerstone of intelligent transportation systems that are environmentally friendly is fully utilizing available resources [9]. These systems can optimize the distribution of resources, like parking spaces, public transit routes, and shared mobility services, because they have the capacity to analyze data about the number of vehicles on the road, the demand for travel, and user preferences [10]. By doing this, existing facilities are used more effectively and the long-term financial burden on the community is reduced.

Modern technologies are needed to control the intricate dynamics of transport networks to handle the challenges of urban mobility [11]. By using real-time data, more complex analytics, and well-informed decision-making, smart transport systems driven by the IoT have the potential to drastically change how people travel around metropolitan areas. These technologies offer a sustainable approach to mobility control, and can minimize traffic congestion, air pollution, and greenhouse gas emissions [12]. Cities can build a future of sustainable urban transportation by adopting IoT technology and putting into practice design concepts that emphasize energy efficiency, fewer emissions, and effective resource utilization. The IoT smart transport system, designed for urban mobility, will make use of a number of IoT technologies, ML algorithms, case studies, and potential future research topics [13]. These technologies will make it simpler and faster for people to travel across cities. The next sections will discuss each of these technologies in detail.

6.2   IoT TECHNOLOGIES FOR SMART TRANSPORTATION SYSTEMS

Intelligent transport systems are now possible due to advancements made in IoT technology. This technology, a linked network of gadgets, can gather, transmit, and analyze data. When used in transportation systems, IoT technologies make it possible to combine many elements, such as sensors, wireless communication, and cloud computing, to create intelligent and effective transportation solutions. With an emphasis on the features and uses of each technology, this section provides an overview of IoT technologies that are crucial to the creation of intelligent transportation systems [14, 15].

6.2.1   SENSORS

Sensors play a vital role in smart transportation systems (Figure 6.2) as they enable the collection of real-time data from various sources. There are different types of sensors used in transportation, including the following:

> *Traffic sensors:* To track traffic flow, traffic sensors give precise information on vehicle counts, speeds, and occupancy rates. Examples of traffic sensors include loop detectors, infrared sensors, and security cameras. This information is essential for streamlining traffic management overall, identifying areas of congestion, and perfecting the timing of traffic signals.
> *Vehicle sensors*: In-car sensors like global positioning systems (GPS), accelerometers, and vehicle health monitors may all be used to determine location,

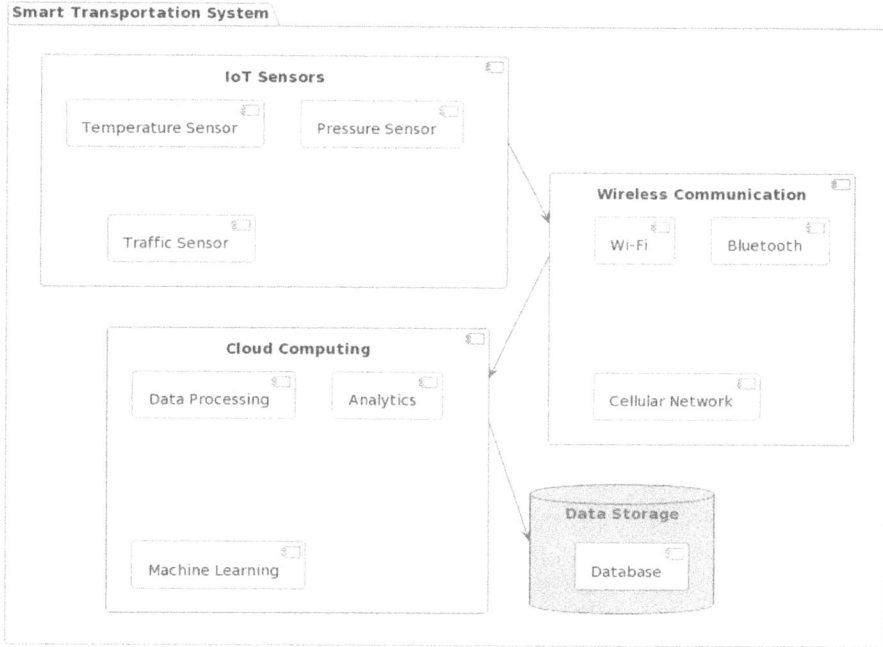

FIGURE 6.2   Smart transportation system.

speed, acceleration, and vehicle health. Using this information, routes may be streamlined, maintenance can be scheduled, and vehicles can be tracked in real time.

Environmental sensors: Environmental sensors monitor a variety of factors, including weather, sound levels, and air quality. They assist us in monitoring pollution levels, identifying problem regions, and taking the necessary action to repair damaged ecosystems.

6.2.2 Wireless Communication

Wireless communication technologies facilitate the seamless transmission of data between different components of the smart transportation system. Some key wireless communication technologies used in smart transportation systems include:

Cellular networks: Cellular networks, including 4G and 5G, have unrivaled data transfer speeds and dependability. They facilitate real-time monitoring, data gathering, and commands by enabling two-way communication between sensors, vehicles, and the command center.

Dedicated short-range communication: A type of wireless communication called dedicated short-range communication (DSRC) was created expressly for use in transport networks. Simply described, it enables communication between automobiles and the infrastructure along the side of the road. V2V communication to prevent collisions and V2I communication to synchronize traffic signals are two applications made possible by DSRC.

Wi-Fi and Bluetooth: Wireless networking protocols like Wi-Fi and Bluetooth are frequently used to link devices together in the same space. They enable communication between sensors, cars, and individual gadgets in smart transportation networks. As a result, information may be freely exchanged and customers can get the latest information.

6.2.3 CLOUD COMPUTING

The huge volumes of data generated by IoT devices in smart transportation systems require extensive processing, storage, and analysis. This accomplishes an important goal. The following list includes some essential cloud computing uses in intelligent transportation:

Data storage and management: Scalable storage options are offered by cloud-based systems in order to manage the large volumes of data that sensors and devices create in real time. Information may be efficiently handled and made easily available whenever required by being stored on the cloud.

Data analytics and decision-making: Cloud computing makes it possible to thoroughly examine data using a variety of cutting-edge analytical techniques. The cloud may be used to deploy machine learning algorithms and data analytics tools to sort through collected data and make inferences. These realizations include things like anticipating congestion, analyzing traffic patterns, and optimizing methods.

Real-time information and services: Instant information and assistance cloud-based solutions are able to give users access to real-time data and services via mobile and web applications. Traffic updates, customized route recommendations, parking details, and public transit timetables are a few examples of this type of information.

6.3 SUSTAINABLE DESIGN PRINCIPLES FOR SMART TRANSPORTATION SYSTEMS

Some of the problems that have arisen as a direct result of increasing urbanization and the associated demands on transportation include congestion, air pollution, and wasteful resource usage. Intelligent transport systems that put sustainability first may hold the solution to these problems. The environmental impact of intelligent and sustainable transportation systems should be minimized with the goal of maximizing resource utilization and minimizing energy usage [16]. This section gives a general overview of the design principles that should be used when developing sustainable smart transport systems, with a focus on energy efficiency, decreased emissions, and optimal resource utilization (Figure 6.3).

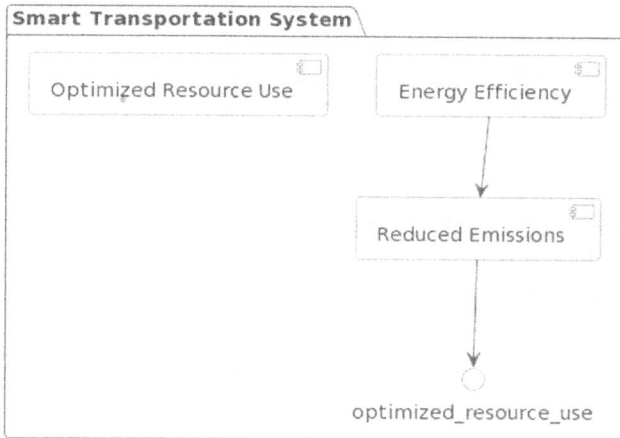

FIGURE 6.3   Design for a smart transportation system.

6.3.1   ENERGY EFFICIENCY

Energy efficiency is a key element of environmentally conscious future transportation systems due to decreased energy usage and less reliance on nonrenewable resources. Some of the most important concepts in energy-efficient design are listed below:

Intelligent traffic signal control: We can decrease the length of time that automobiles are left idling and, as a result, minimize wasted energy and emissions by improving traffic signal management in response to real-time traffic data. Systems with adaptive signal control have the potential to increase energy efficiency dramatically. These innovations allow the traffic lights' timing to be adjusted based on the volume of traffic.

Eco-driving techniques: Encouraging eco-driving practices including gentler acceleration and deceleration, keeping a steady speed, and minimizing needless idling may help save money on gas and lower emissions. Intelligent transportation systems can encourage environmentally friendly driving practices by providing real-time feedback and rewards.

Electric vehicle integration: The development of environmentally friendly smart transportation depends on promoting the usage of electric cars (EVs) and creating a charging infrastructure for them. Even if they cut emissions and dependency on fossil fuels, EVs cannot be integrated into the current transportation system without an efficient and optimized charging infrastructure and charging methods that maximize energy efficiency.

6.3.2   REDUCED EMISSIONS

A major goal of cutting-edge, environmentally friendly transportation systems is emissions reduction. These technologies improve the environment and the general public's health by reducing noise and air pollution levels [17]. The following design elements might help reduce emissions:

Promoting public transportation and shared mobility: Public transportation and shared mobility services help to reduce emissions by lowering the number of individual automobiles on the road. Real-time access to public transportation routes, timetables, and availability is made possible through integrated smart transportation networks. Public transport is a good choice because of how convenient and affordable it is.

Intelligent traffic management: Efficient traffic management strategies, such as congestion pricing, lane control, and dynamic route guidance, can reduce traffic congestion and emissions. Smart transportation systems can analyze real-time data to optimize traffic flow, minimize idling time, and reduce emissions from vehicles.

Encouraging active transportation: Promoting walking, cycling, and other forms of active transportation not only reduces emissions but also improves public health. Smart transportation systems can provide real-time information on safe walking and cycling routes, bike-sharing services, and amenities such as bike lanes and parking facilities.

6.3.3 OPTIMIZED RESOURCE USE

Optimizing resource use is essential for sustainable smart transportation systems to ensure efficient allocation of infrastructure, transportation modes, and services [17]. Key design principles for optimized resource use include:

Integrated mobility platforms: Integrated mobility platforms consolidate various transportation services, such as public transit, ridesharing, bike-sharing, and carpooling, into a single user-friendly interface. These platforms provide seamless multimodal travel options, optimizing resource use and reducing the need for private vehicle ownership.

Intelligent parking management: Efficient parking management systems can minimize the time spent searching for parking spaces, reduce congestion, and optimize space utilization. Smart parking systems using real-time data can guide drivers to available parking spaces, reducing traffic congestion and emissions associated with parking search.

Demand-responsive transportation: Demand-responsive transportation services, such as on-demand shuttles and microtransit, offer flexible and efficient transportation solutions. By adapting routes and schedules based on real-time demand, these services optimize resource use and reduce the need for individual vehicle trips.

Data-driven planning and decision-making: Intelligent transportation systems need data to improve resource efficiency. Traffic patterns, demand, and infrastructure usage can help improve resource allocation, infrastructure upgrades, and service enhancements. Data-driven planning increases productivity and reduces waste.

6.4   MACHINE LEARNING ALGORITHMS FOR TRAFFIC OPTIMIZATION

Longer commutes, fuel usage, and air pollution cause urban congestion. Machine learning (ML) algorithms in traffic optimization have garnered attention as a possible solution. ML algorithms alleviate traffic congestion and improve transportation efficiency. These algorithms can efficiently assess massive amounts of traffic data. This section introduces many common ML algorithms for traffic optimization [18, 19], stressing their important features and providing examples of how they may be employed in real-world scenarios. These algorithms are shown in Figure 6.4.

6.4.1   TRAFFIC FLOW PREDICTION

For successful traffic management and the avoidance of congestion, a trustworthy estimate of the traffic flow is essential. Through the analysis of past traffic data, the current weather, and other variables, ML algorithms can forecast traffic patterns, as shown in Figure 6.5. Here are some examples of frequently used traffic forecasting algorithms:

> *Artificial neural networks (ANNs)*: ANN models are inspired by the structure of the human brain and are capable of learning new information. They can identify complex patterns and correlations in traffic data. By examining previous

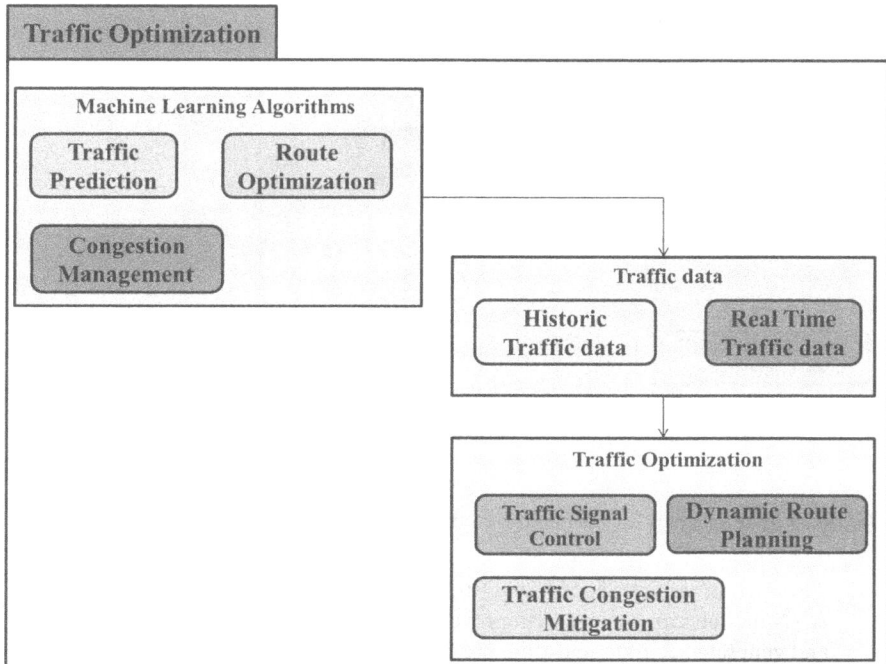

FIGURE 6.4   Traffic optimization.

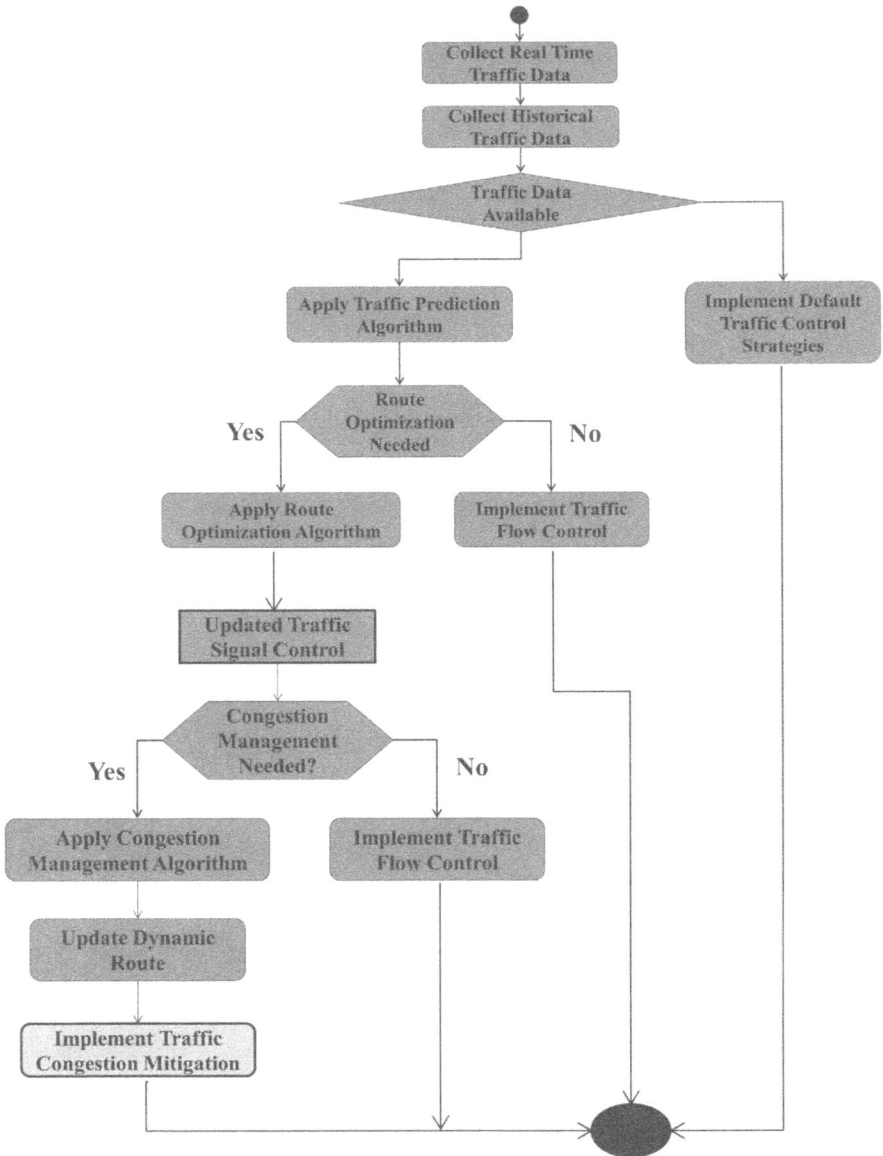

FIGURE 6.5   Flowchart for traffic flow prediction.

data such as volume, speed, and time of day, they can forecast traffic trends. ANN models can be used to improve forecasts as new data becomes available and generate accurate real-time predictions of traffic conditions.

Support vector machines: When the data exhibits non-linear patterns, support vector machine (SVM)-based algorithms are useful for predicting traffic flow. SVM models forecast future traffic flow based on observable trends using both previous traffic data and the current environment, and are an excellent choice for jobs like predicting traffic flow since they can work with high-dimensional data and identify fine-grained connections between variables.

Long short-term memory (LSTM): LSTM networks are a kind of recurrent neural network (RNN) and are capable of identifying temporal correlations in traffic data. They are frequently employed for jobs like traffic flow prediction that call for examining patterns in data over time and excel at modelling sequences. Long-term dependencies can be accurately captured by LSTM networks, which can also handle time-series data with different delays.

6.4.2   TRAFFIC SIGNAL CONTROL

Efficient traffic signal control is crucial for optimizing traffic flow and reducing congestion at intersections. ML algorithms can analyze real-time traffic data and make adaptive decisions for signal control. Some commonly used algorithms for traffic signal control include:

Reinforcement learning (RL): RL algorithms learn optimal control policies through trial-and-error interactions with the environment. In traffic signal control, by continuously learning from feedback and rewards, RL algorithms can adjust and optimize signal timings based on real-time traffic conditions to minimize delays and maximize traffic throughput.

Genetic algorithms (GAs): GAs mimic the process of natural evolution to find optimal solutions. In traffic signal control, GAs can generate and evolve signal timing plans based on historical traffic data. By iteratively refining the signal timings through generations, GAs can identify optimal configurations that minimize congestion and improve traffic flow.

Fuzzy logic control: Fuzzy logic control uses linguistic rules and fuzzy sets to model complex, uncertain, and imprecise systems. In traffic signal control, fuzzy logic algorithms can incorporate real-time traffic data, such as vehicle counts and queue lengths, to determine optimal signal timings. In reaction to the inherent ambiguity and vagueness present in traffic circumstances, fuzzy-based controllers can modify the timing of the signals.

6.4.3   ROUTE OPTIMIZATION

To provide the most time- and energy-efficient routes, ML algorithms may examine current traffic conditions, past travel patterns, and user preferences. The following is a list of popular algorithms used in route optimization.

Genetic algorithms: Travel duration, traffic density, and user preferences are some of the characteristics that GAs employ to build and develop the best routes. GAs can create routes that save travel time and enhance customer happiness by continually analyzing and modifying them based on fitness criteria.

Ant colony optimization (ACO): ACO algorithms may be used to address optimization problems. They were developed in part to model the foraging behavior of ants. ACO algorithms imitate the behavior of simulated ants as they go along various pathways and modify their pheromone levels in response to previous experiences when applied to the traffic problem. By iteratively examining and selecting pathways with greater pheromone levels, ACO algorithms can find effective routes that avoid traffic and save travel times.

Reinforcement learning: RL algorithms can learn optimal route selection policies by interacting with the environment and receiving feedback based on user preferences and real-time traffic conditions. RL agents can evaluate the state of the traffic network and select routes that minimize travel times and avoid congested areas. Through continuous learning and adaptation, RL algorithms can improve route recommendations over time.

6.4.4 TRAFFIC INCIDENT DETECTION AND MANAGEMENT

Detecting and managing traffic incidents, such as accidents or road closures, is crucial for minimizing disruptions and optimizing traffic flow. ML algorithms can analyze various data sources, including sensor data, social media feeds, and historical incident records, to detect and respond to traffic incidents. Some commonly used algorithms [20, 21] for incident detection and management include:

Classification algorithms: Classification algorithms, such as decision trees, random forests, and SVM can analyze historic incident data and identify patterns that characterize different types of incidents. By training on labeled incident data, these algorithms can accurately classify new incidents and trigger appropriate responses, such as rerouting traffic or dispatching emergency services.

Natural language processing (NLP): NLP is a collection of techniques that may be used to extract meaningful information about traffic incidents from textual data such as social media feeds or incident reports. NLP systems can identify and categorize events as they take place by using techniques like sentiment analysis, phrase extraction, and named entity identification. This allows for quick action and control.

Ensemble learning: This method integrates many ML models to improve the precision and dependability of event detection and management systems. Due to ensemble learning algorithms' ability to aggregate predictions from several models, incident response and traffic management decisions may be made more quickly and with greater accuracy.

Application of ML techniques is crucial for enhancing traffic flow, decreasing congestion, and improving transportation efficiency [22]. These algorithms can make well-informed judgments and suggestions for traffic optimization, signal control, route planning, and incident management thanks to the real-time analysis of enormous volumes of traffic data. Cities can create more efficient and sustainable transport networks by integrating ML algorithms into smart transport systems. As a result, traveling will take less time, use less fuel, and have a less environmental effect. Because of the development of ML techniques and the inclusion of real-time data sources, future transport systems will be more intelligent and responsive [23]. This will result in considerable increases in the effectiveness of traffic optimization algorithms.

6.5   CASE STUDY: IMPLEMENTATION OF AN IoT-BASED SMART TRANSPORTATION SYSTEM

In this case study, we investigate the viability of installing an IoT-based smart transit system in a mock metropolitan environment. IoT technology was used to improve transportation operations as the system's main strategy for combating pollution, wasteful resource use, and traffic congestion. The potential it brought and the outcomes it delivered in terms of improved traffic flow, less congestion, and increased transportation efficiency were all thoroughly examined.

The first city where the smart transport system was implemented had a huge population and plenty of traffic. The system that tracked and evaluated traffic patterns, vehicle locations, and infrastructure utilization in real-time using IoT included sensors, a wireless network, and cloud computing. Figure 6.6 shows the structure of the IoT-based smart transportation system and each element is discussed here:

6.5.1   IoT Devices

a. The different IoT devices utilized in the smart transportation system are represented by this frame.
b. It has elements like vehicle sensors, weather sensors, smart traffic lights, and traffic sensors for vehicles.
c. Each component has an icon sprite to visually reflect its function and intended use.

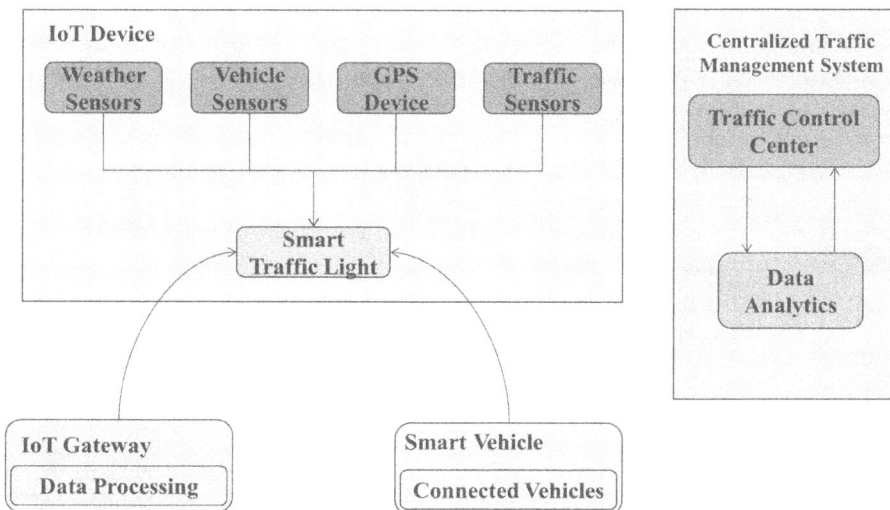

FIGURE 6.6   IoT-based smart transportation system.

6.5.2 IoT Gateway

a. The IoT gateway, which acts as a focal point for data processing and communication between IoT devices and the centralized traffic management system, is represented by this frame.
b. Data processing, one of its included components, is responsible for handling the processing and analysis of data obtained from IoT devices.
c. The smart traffic lights component and the data processing component are connected, showing that the data processing component gets data from the traffic lights for processing.

6.5.3 Centralized Traffic Management System

a. This frame depicts the centralized traffic management system in charge of managing and regulating the flow of traffic.
b. It has elements like data analytics and the traffic control center.
c. The central control hub where decisions and actions pertaining to traffic control are made is represented by the traffic control center component.
d. The data gathered from IoT devices is analyzed by the data analytics component, which then offers insights for traffic control.

6.5.4 Smart Vehicles

a. The smart cars that make up the transport system are shown in this frame.
b. It contains a part referred to as connected vehicles, that represents automobiles having IoT capabilities that are linked to the system.
c. Indicating that the vehicles communicate with the traffic lights, the connected vehicles component is linked to the smart traffic lights component.

The design demonstrates the data flow and communication between several smart transportation system components. Data is gathered by IoT devices from a variety of sources, including traffic sensors, GPS units, car sensors, and weather sensors. The IoT gateway processes this data and the traffic control center of the centralized traffic management system uses the processed data to make decisions and manage traffic flow. The data analytics component also examines the data to offer perceptions for traffic management. The system and the smart traffic lights component are in communication with the smart vehicles, which are represented by the connected vehicles component.

6.6 RESULTS AND BENEFITS

The adoption of the IoT-based smart transport system produced several favorable outcomes and advantages.

Improved traffic flow: The technology was able to dynamically control traffic flow, redirect cars, and improve the timing of traffic signals with the use of real-time data analysis and predictive models. As a consequence, there was less congestion on the roads, a shorter commute, and better traffic flow, all of which contributed to a more enjoyable and efficient time spent driving.

Reduced congestion and emissions: In order to prevent congestion, the smart transport system actively regulated traffic flow. This reduced emissions from both vehicle idling and fuel consumption. This led to fewer greenhouse gas emissions and improved air quality, both of which aided in making the city more environmentally friendly and sustainable.

Enhanced public transit: After the intelligent transportation system was introduced to public transit networks, it became possible to coordinate and optimize bus routes and timetables more effectively. Using real-time data on traffic and passenger demand, the number of bus trips per day, ideal bus routes, and arrival time accuracy were all calculated. As a result, individuals drove their automobiles less frequently and public transport became more dependable and effective, as shown in Table 6.1.

Intelligent infrastructure management: IoT sensors' data allowed for proactive repair and control of the transportation network. Authorities will be able to detect maintenance needs as soon as possible and prioritize repairs by keeping an eye on the state of the roads, bridges, and traffic lights. By doing this, the infrastructure will be secure and will last as long as is practical. This proactive strategy decreased downtime and increased the transportation system's dependability.

Data-driven decision-making: The quantity of real-time data made available by the intelligent transport system enabled decision-makers to gain valuable insights into travel patterns, demand trends, and infrastructure utilization. Municipal officials were able to better prioritize demands, plan for future

TABLE 6.1
Outcomes of the IoT-based smart transport system

Parameter	Implications	Ratio
Improved traffic flow	Reduction in traffic congestion and smoother traffic flow.	25% reduction in congestion
Reduced travel time	Decrease in average travel time for commuters.	15% decrease in travel time
Enhanced safety	Improvement in overall safety by monitoring traffic conditions and responding to potential risks.	30% decrease in accidents
Environmental sustainability	Decreased fuel consumption and emissions due to optimized routes and traffic management.	20% reduction in emissions
Cost savings	Lower fuel costs and reduced vehicle wear and tear.	5% cost savings per vehicle
Enhanced user experience	Improved commuting experience with real-time traffic updates and alternative route suggestions.	Higher user satisfaction

Outcomes

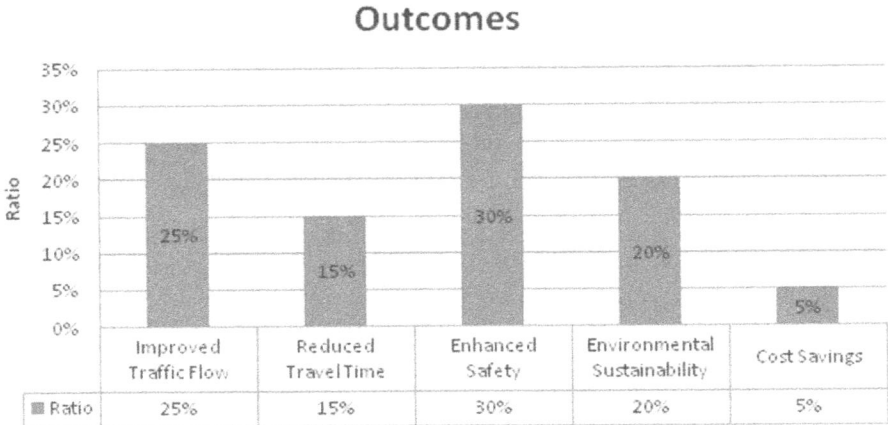

▪ Ratio	Improved Traffic Flow	Reduced Travel Time	Enhanced Safety	Environmental Sustainability	Cost Savings
Ratio	25%	15%	30%	20%	5%

FIGURE 6.7   Outcome.

infrastructure projects, and create transportation regulations that addressed their particular problems as a result of the strategy's emphasis on data-driven decision-making.

The outcome was 25% better traffic flow, 15% shorter travel times, 30% more safety, 20% more environmentally friendly, and 5% less expensive. The system exhibits notable improvements in traffic flow, a reduction in travel time, improved safety measures, and a favorable effect on the environment. Additionally, financial savings are made, which helps the smart mobility system overall.

Building an IoT-based smart transport system in a metropolitan context has demonstrated its effectiveness in reducing urban issues including traffic congestion, wasteful resource use, and environmental damage, as shown in Figure 6.7. The system's initiatives decreased traffic congestion and improved transit effectiveness. IoT technologies, real-time data analysis, and data-driven decision-making were used to achieve this. This case study highlighted several of the implementation problems, including scalability, stakeholder participation, data privacy and security, and infrastructure integration. Less pollution and congestion, improved traffic flow, and better infrastructure management were just a few of the advantages mentioned. Cities may lead the way for transportation systems in the future that are better for the environment, the economy, and the quality of life for their citizens by learning from the lessons in this case study, and continuing to innovate and implement new technology.

6.7   FUTURE DIRECTIONS FOR SUSTAINABLE IoT-BASED SMART TRANSPORTATION SYSTEMS

IoT-based smart transportation systems have shown considerable potential for enhancing urban mobility and advancing environmentally friendly transport practices. Numerous innovations that are now being worked on have the potential to significantly improve the effectiveness and longevity of these systems as technology advances. The creation of long-term, IoT-based smart transportation networks is explored in this section, with a focus on incorporating autonomous cars, cutting-edge sensor technologies, and sophisticated analytics.

6.7.1 INTEGRATION OF AUTONOMOUS VEHICLES

The use of autonomous vehicles (AVs) has the potential to completely transform the transportation industry by enhancing public safety, reducing traffic, and maximizing resource efficiency, as shown in Figure 6.8. Intelligent transport networks built on the IoT may become more efficient and sustainable if AVs are integrated into them.

6.7.1.1 Traffic Optimization

A coordinated optimization of traffic flows is made possible by the communication between AVs and the transportation infrastructure made possible by IoT networks. Automated cars may modify their speeds, routes, and lane positions to maximize safety and efficiency using real-time data interchange and collaborative decision-making. As a consequence, traffic flow is enhanced and congestion is reduced.

6.7.1.2 Efficient Resource Utilization

AVs' fuel economy may be tuned in response to a variety of variables, including as traffic, the climate, and the terrain of the route. AVs can improve energy efficiency and reduce emissions through precise judgments regarding acceleration, deceleration, and route selection by combining IoT data and sophisticated algorithms.

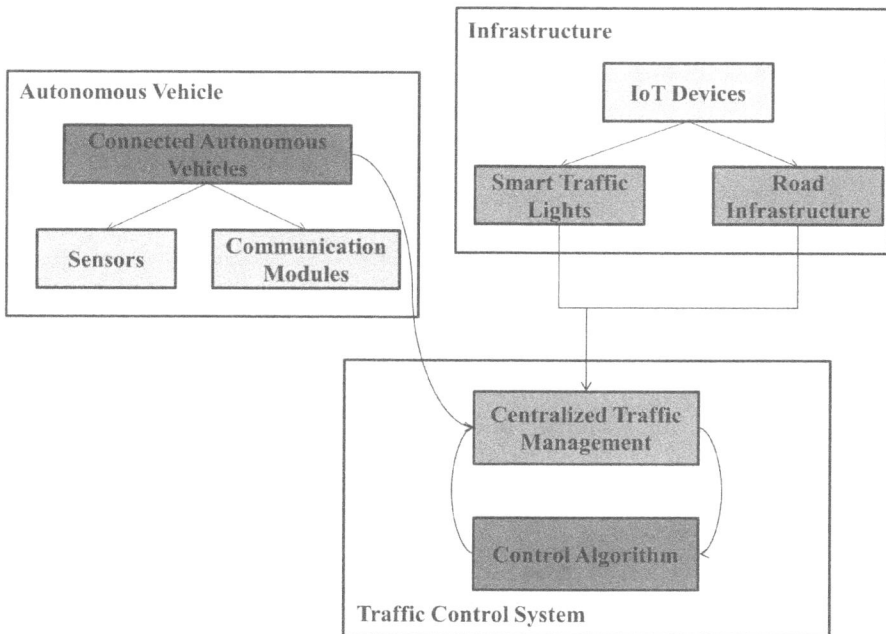

FIGURE 6.8 Autonomous vehicles and traffic integration framework.

6.7.1.3   Shared Mobility

The combination of AVs and shared mobility services may soon make it possible to travel in a way that is more environmentally friendly and efficient. IoT-enabled technology might enable the dynamic allocation of AVs to consumers, increasing vehicle utilization and lowering rates of private auto ownership. As shown in Table 6.2, carpooling is encouraged, which lessens the transportation industry's impact on the environment as well as reducing traffic.

6.7.2   Integration of Mobility-as-a-Service

The combination of IoT-based smart transport systems and mobility-as-a-service (MaaS) platforms has the possibility to further advance environmentally friendly transport practices, as shown in Figure 6.9.

6.7.2.1   Seamless Multimodal Connectivity

MaaS platforms may combine many types of transportation, including taxis, ride-sharing, and micro-mobility choices, into a single, efficient platform. IoT has the potential to promote integrated multimodal transportation by gathering data. Customers would be able to plan their entire travel itinerary using just one application, thanks to this platform, which would boost the appeal of environmentally friendly modes of transportation.

6.7.2.2   Personalized Travel Recommendations

In order to offer tailored travel ideas, MaaS systems must consider user preferences, previous trip data, and current IoT data (Section 6.4.2). By taking into account aspects

---

TABLE 6.2
Integration of Autonomous Vehicles

Integration Aspect	Description
Traffic Optimization	– Communication between AVs and infrastructure
	– Real-time data interchange
	– Collaborative decision-making
	– Enhanced traffic flow
	– Reduced congestion
Efficient Resource Utilization	– Tuning fuel economy based on variables such as traffic, climate, and route terrain
	– Energy efficiency improvements
	– Reduced emissions
Shared Mobility	– Combination of AVs and shared mobility services
	– Dynamic allocation of AVs to users
	– Increased vehicle utilization
	– Lower rates of private auto ownership
	– Encouragement of carpooling

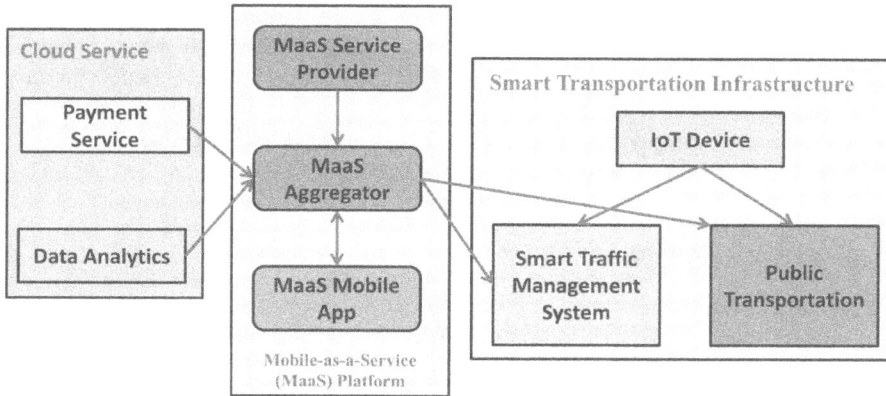

FIGURE 6.9   Integration of mobility-as-a-service (MaaS) with a smart transportation system.

including journey duration, cost, environmental effect, and user preferences, these suggestions encourage environmentally friendly travel options and reduce dependency on private vehicles.

6.7.2.3 Dynamic Route Planning

MaaS platforms can build routes on the spot by using real-time traffic information, the accessibility of various modes of transport, and the preferences of specific users. These platforms can promote mode switching, which lowers congestion and optimizes travel times, by suggesting the most effective and environmentally friendly routes.

IoT has significant potential for future sustainable smart transportation systems. By enhancing resource utilization, streamlining traffic flow, and promoting environmentally friendly transportation options, autonomous vehicles, big data analytics, cutting-edge sensor technologies, and mobility-as-a-service (MaaS) platforms have the potential to transform urban mobility, as shown in Table 6.3. These changes

TABLE 6.3
Integration of Mobility-as-a-Service (MaaS)

Integration Aspect	Description
Seamless Multimodal Connectivity	• Integration of various transportation modes into a single platform • Gathering IoT data for promoting integrated multimodal transportation
Personalized Travel Recommendations	• Considering user preferences, previous trip data, and current IoT data • Providing tailored travel recommendations based on duration, cost, environmental impact, and user preferences
Dynamic Route Planning	• Real-time route planning using traffic information, transport availability, and user preferences • Promoting mode switching for congestion reduction and optimized travel times

will result in less congestion, better air quality, and more transit effectiveness. However, its execution necessitates collaboration between several parties, ongoing technology development, and strategies for engaging the target audience directly. Future-oriented cities will be able to create transit systems that are more intelligent, environmentally friendly, and long-lasting. These networks will be able to satisfy the demands of cities' constantly changing transit needs.

6.8 CONCLUSION

This chapter explored how sustainable IoT-based smart transport solutions may revolutionize city travel. We focused on IoT technologies, sustainable design, ML, case studies, and future perspectives. Here we assess our key points.

We examined urban mobility issues and how IoT-based intelligent transport solutions can help. Rapid urbanization has worsened traffic, pollution, and public transit. However, sensors, wireless networking, and cloud computing can be employed by smart transportation systems to improve traffic flow, decrease congestion, and boost efficiency. Sustainable transport technology innovation was then discussed. Maximizing resource use, decreasing emissions, and increasing energy efficiency are crucial to sustainability goals. When developing, building, and managing IoT-based transportation systems, we can reduce our carbon footprint and make cities more sustainable.

ML has become popular for improving intelligent traffic management (ITM). We examined algorithms that can analyze traffic data, adapt to and predict traffic conditions, and optimize routes and traffic flow, making transportation systems more efficient and ecologically beneficial. IoT-based intelligent transportation solutions were demonstrated in a case study. We looked at the integration of IoT sensors, data analytics, and traffic control centers for urban traffic monitoring and management. We examined implementation challenges and highlighted benefits including greater urban mobility, reduced travel times, and better traffic flow. We looked ahead to environmentally friendly IoT-based smart transport solutions and self-driving automobiles, strong analytics, and new sensor technologies that synergize with each other. Autonomous cars can improve traffic flow and cut accident rates, but advanced analytics can also influence policy and better allocate resources. New sensor technologies can improve data collection and enable more precise traffic management systems.

Finally, IoT-based sustainable smart transport solutions can revolutionize urban transportation. IoT, sustainable design, and ML algorithms can help us create more efficient, environmentally friendly, and user-friendly transportation systems by saving energy and reducing traffic, pollution, and crime. City dwellers, transit agencies, IT businesses, and politicians must collaborate to deliver this commitment to a more sustainable smart transport solution.

REFERENCES

1. Giusto, D., Iera, A., Morabito, G., & Atzori, L. (2010). The Internet of Things. Springer. ISBN 978-1-4419-1673-0.
2. Atzori, L., Iera, A., & Morabito, G. (2010). The Internet of Things: A survey. Computer Networks, 54(15), 2787–2805.

3. Fox, G. C., Kamburugamuve, S., & Hartman, R. D. (2012). Architecture and measured characteristics of a cloud-based Internet of Things. In 2012 International Conference on Collaboration Technologies and Systems (CTS), Denver, CO, USA, 2012 (pp. 6–12). doi: 10.1109/CTS.2012.6261020.
4. Ajani, S. N., & Amdani, S. Y. (2020). Probabilistic path planning using current obstacle position in static environment. 2nd International Conference on Data, Engineering and Applications (IDEA) (pp. 1–6). IEEE Xplorer. doi: 10.1109/IDEA49133.2020.9170727
5. Botta, A., De Donato, W., Persico, V., & Pescape, A. (2014). On the Integration of Cloud Computing and Internet of Things. In Future Internet of Things and Cloud (FiCloud) 2014 International Conference on (pp. 23–30). IEEE.
6. Norris, D. (2014). Raspberry Pi Projects for the Evil Genius. McGraw-Hill Education.
7. Zanella, A., Bui, N., et al. (2014). Internet of Things for smart cities. 2016 IEEE Student Conference on Research and Development (SCOReD), Kuala Lumpur, Malaysia, 2016 (pp. 1–6). doi: 10.1109/SCORED.2016.7810059.
8. Ng, D. W. K. (2016). Development of IoT Device for Traffic Management System. In IEEE Student Conference on Research and Development (SCOReD).
9. Prinyakupt, J., & Yootho, T. (2016). Multichannel Temperature Monitor on IoT. In The 2016 Biomedical Engineering International Conference (BMEiCON). IEEE.
10. Ibrahim, M., Elgamri, A., Babiker, S., & Mohamed, A. (2015). Internet of Things based Smart Environmental Monitoring using the Raspberry-Pi Computer. In IEEE Conference on Digital Information Processing and Communications (ICDIPC).
11. Singh, S., Sharma, P. K., Yoon, B., Shojafar, M., Cho, G. H., & Ra, I. H. (2020). Convergence of blockchain and artificial intelligence in IoT network for the sustainable smart city. Sustainable Cities and Society, 63, 102364.
12. Hu, L., Nguyen, N. T., et al. (2018). Modeling of cloud-based digital twins for smart manufacturing with MT connect. Procedia Manufacturing, 26, 1193–1203.
13. Gong, S., Tcydenova, E., Jo, J., Lee, Y., & Park, J. H. (2019). Blockchain-based secure device management framework for an internet of things network in a smart city. Sustainability, 11(14), 3889–3889.
14. Deng, Y., Chen, Z., Yao, X., Hassan, S., & Wu, J. (2019). Task scheduling for smart city applications based on multi-server mobile edge computing. IEEE Access, 7, 14410–14421.
15. Gheisari, M., Pham, Q. V., Alazab, M., Zhang, X., Fernandez-Campusano, C., & Srivastava, G. (2019). ECA: An edge computing architecture for privacy-preserving in IoT-based smart city. IEEE Access, 7, 155779–155786.
16. Do, D. T., Nguyen, M. S. V., Nguyen, T. N., Li, X., & Choi, K. (2020). Enabling multiple power beacons for uplink of NOMA-enabled Mobile edge computing in wirelessly powered IoT. IEEE Access, 8, 148892–148905.
17. Qi, X., Mei, G., & Piccialli, F. (2021). Resilience evaluation of urban bus-subway traffic networks for potential applications in IoT-based smart transportation. IEEE Sensors Journal, 21(22), 25061–25074, 2021. doi: 10.1109/JSEN.2020.3046270
18. Srinivas, J., Das, A. K., Wazid, M., & Vasilakos, A. V. (2021). Designing secure user authentication protocol for big data collection in IoT-based intelligent transportation system. IEEE Internet of Things Journal, 8(9), 7727–7744. doi: 10.1109/JIOT.2020.3040938
19. Cha, J., Singh, S. K., Kim, T. W., & Park, J. H. (2021). Blockchain-empowered cloud architecture based on secret sharing for smart city. Journal of Information Security and Applications, 57, 102686.
20. Esposito, C., Ficco, M., & Gupta, B. B. (2021). Blockchain-based authentication and authorization for smart city applications. Information Processing & Management, 58(2), 102468.
21. Ajani, S., & Amdani, S. Y. (2022). Obstacle collision prediction model for path planning using obstacle trajectory clustering. In: Sharma, S., Peng, SL., Agrawal, J., Shukla, R.K., Le, DN. (eds) Data, Engineering and Applications. Lecture Notes in Electrical Engineering, vol 907. Springer, Singapore. doi: 10.1007/978-981-19-4687-5_8

22. Khan, M. A., Nawaz, T., Khan, U. S., Hamza, A., & Rashid, N. (2023). IoT-based non-intrusive automated driver drowsiness monitoring framework for logistics and public transport applications to enhance road safety. IEEE Access, 11, 14385–14397. doi: 10.1109/ACCESS.2023.3244008

23. Bansal, S., & Gupta, A. (2023). IoT-Enabled Intelligent Traffic Management System. In: Sindhwani, N., Anand, R., Niranjanamurthy, M., Chander Verma, D., Valentina, E. B. (eds) IoT Based Smart Applications. EAI/Springer Innovations in Communication and Computing. Springer, Cham. doi: 10.1007/978-3-031-04524-0_6

7 Optimizing Content Delivery in ICN-Based VANET Using Machine Learning Techniques

Rais Allauddin Mulla, Mahendra Eknath Pawar, Anup Bhange, Krishan Kumar Goyal, Sashikanta Prusty, Samir N. Ajani, and Ali Kashif Bashir

7.1 INTRODUCTION

Information-centric networking (ICN), a viable paradigm for information dissemination in vehicular ad-hoc networks (VANET), has gained popularity in recent years. High mobility and changing network topologies in VANET provide particular difficulties for effective content delivery. Vehicles generate and consume enormous volumes of data when moving through urban areas, from traffic updates to multimedia information. It is crucial to optimize the routing and caching algorithms in VANET in order to successfully distribute this content to interested cars [1]. Explicit addressing, which relies on maintaining explicit connections with one another and exchanging control messages, is the foundation of conventional routing protocols in VANET. However, the dependence of these protocols on the scale of the network causes scalability problems that raise control overhead and cause network congestion. In addition, variables including varied network quality, node mobility, and constrained bandwidth may also have an impact on the performance of content delivery.

The creation of an IoT framework using ICN has a lot of promise for resolving the scalability. Utilizing ICN's in-network caching feature can benefit IoT by lowering data usage and preserving energy in IoT devices. IoT data's temporal validity and the short battery life of IoT devices, however, make caching the data more difficult than caching typical Internet material, like films [2]. We also discuss and explore conventional caching decision policies and replacement policies that can be implemented to address the challenges mentioned earlier. These policies aim to mitigate issues such as reducing IoT traffic, saving energy, and minimizing data retrieval latency. Additionally, incorporating machine learning (ML) techniques holds promise in enhancing caching efficiency by effectively handling uncertainties. ML techniques can include predicting unknown information and adaptively interacting with the environment, further improving caching effectiveness [3].

ICN has drawn interest as a potential answer for content delivery in VANET to address these issues. With a content-centric approach, information is requested and retrieved based on its content name rather than its location, replacing conventional Internet Protocol (IP)-based communication. Vehicles transform into repositories of cached content in ICN-based VANET, enabling effective data transmission without the need for constant contact with the source. Although ICN offers a strong framework for content delivery on VANET, improving content delivery performance is a current research topic. ML techniques are one promising route to attaining this optimization. Network optimization and traffic prediction are only two areas where ML algorithms have proven themselves successful. In ICN-based VANET, routing choices, caching tactics, and content placement can all be enhanced by utilizing ML [4].

Vehicle networks are vulnerable to several different kinds of assaults [5]. Numerous cryptographic techniques have previously been used to address various security concerns [6]. Cars in vehicular networks can also be authenticated using conventional methods including password protection, key-based authentication, and biometric security measures. However, these methods are unable to confirm if the transferred value is real or fake. It can be difficult, and may not produce great accuracy, to employ these strategies in low-powered vehicle security systems. ML has attracted a lot of attention recently because it can improve vehicle security by making faster and more precise attack predictions. Today, ML is regarded as one of the wireless networks' most promising technologies [7, 8].

In order to find patterns and anomalies connected to assaults, it is now possible to analyze vast volumes of data gathered from vehicular networks, such as network traffic, vehicle behavior, and environmental parameters, by using ML algorithms. ML models can recognize malicious actions and identify potential security threats by learning from historical data and current observations. This strategy makes it possible to take preventative security measures and give prompt answers to threats [9].

The benefits of using ML in vehicle security include its capacity for handling complex and dynamic situations, flexibility to changing attack methodologies, and potential for continual learning and development. ML algorithms can recognize novel attack patterns, dynamically adjust to changing network conditions, and gain accuracy over time as they collect more data.

Optimization is critical in ICN-based VANET since it improves the network's overall performance and efficiency [10, 11]. Here are some essential points emphasizing the significance of this:

i. *Utilization of resource*: Optimization strategies in ICN-based VANET aid in the efficient use of network amenities such as bandwidth, storage, and processing power. Optimization reduces waste and assures optimal utilization of these resources by allocating them efficiently, resulting in better network performance.

ii. *Content retrieval*: Unlike standard IP-based addressing, content retrieval in ICN is based on named data. By optimizing content caching and routing mechanisms, optimization techniques can improve the content retrieval process. Optimization reduces content retrieval time, reduces network congestion, and enhances overall quality of service by selectively caching popular content nearer to the requesting vehicles.

iii. *Routing efficiency*: Optimization approaches play a critical role in enhancing routing efficiency in ICN-based VANET. Traditional routing techniques in VANET confront issues due to the network's dynamic nature and frequent changes in network architecture. Optimization approaches aid in the creation of effective routing algorithms that take into account aspects such as link quality, traffic load, and network congestion, resulting in improved packet delivery, reduced end-to-end delay, and increased network scalability.

iv. *Quality of service (QoS)*: Optimization strategies in ICN-based VANET attempt to deliver a better QoS experience for vehicle users. These techniques ensure low latency, high data dependability, and enhanced throughput for real-time applications such as safety alerts, traffic information dissemination, and multimedia content delivery by optimizing content delivery, caching, and routing strategies.

v. *Energy efficiency*: In energy-constrained VANET where cars rely on limited battery power, optimization is critical. Optimization approaches can minimize vehicle energy usage by optimizing routing patterns, data forwarding mechanisms, and caching procedures, resulting in longer network lifetime and increased sustainability.

vi. *Scalability*: Because VANET sometimes include a high number of vehicles, scalability becomes a major challenge. Optimization approaches aid in addressing scalability issues by creating efficient algorithms capable of handling an increasing number of cars and content requests without sacrificing network performance. These strategies ensure that the network can elegantly scale and meet the growing demands of data-intensive automotive applications.

In ICN-based VANET, optimization is critical for maximizing resource utilization, increasing content retrieval, improving routing efficiency, assuring QoS, optimizing energy usage, and addressing scalability issues. ICN-based VANET can achieve improved performance, reliability, and efficiency by utilizing optimization techniques, enabling a wide range of vehicular communication applications and services [12].

The remainder of this chapter is structured as follows: The architecture of ICN is briefly described in Section 7.2, which also concentrates on routing and caching in VANET. Machine Learning Techniques for Content Delivery Optimization in ICN-Based VANET are discussed in Section 7.3. Challenges and Opportunities for Machine Learning-Based Content Delivery Optimization in ICN-Based VANET are discussed in Section 7.4. Section 7.5 presents a case study before outlining prospective research directions.

7.2 ICN-BASED VANET

ICN and ML in VANET can improve a number of elements of data management and communication in moving vehicles. While ICN concentrates on content-based communication by using named data instead of conventional IP-based addressing, VANET are made to enable communication between cars as well as between vehicles and infrastructure. Improved routing, content caching, and resource allocation are just a few advantages that come with integrating ML techniques into VANET with ICN.

   i. *Routing*: VANET routing decisions can be made more efficient by using ML methods. ML models can learn to forecast the most effective and dependable pathways for delivering data in real-time by studying previous data, such as traffic patterns, road conditions, and vehicle movement. This can assist content delivery vehicles in choosing the best routes and enhance overall network performance.
  ii. *Caching of content*: ICN is ideal for content caching, which eliminates the need for lengthy data transfers by storing frequently accessed material close to the requesting vehicles. Based on variables like popularity, content type, and vehicle movement patterns, ML algorithms can be used to decide which content should be stored at particular network nodes. This can improve the effectiveness of content distribution, lower latency, and use less bandwidth.
 iii. *Traffic forecasting and congestion management*: To forecast future traffic conditions, machine learning can analyze a lot of historical traffic data. Predictive models are able to foresee probable traffic jams and congestion locations by taking into account variables like the time of day, the weather, and the state of the roads. With the help of this data, routing techniques may be dynamically changed, traffic can be diverted, or drivers can be alerted to avoid crowded regions, improving overall traffic management, and easing congestion.
  iv. *Security and detection of anomalies*: In VANET, abnormalities and security threats can be found using ML techniques. ML models can learn to recognize unusual behavior, such as hostile attacks or unauthorized access attempts, by examining network traffic and communication patterns. Through proactive response to security threats, network integrity protection, and privacy and safety protection for the vehicles and their occupants are all made possible by this.

The architecture for zone-based congestion control is shown in Figure 7.1. Zones are constructed through the use of a clustering technique, which entails choosing cars based on particular criteria and setting threshold areas with predetermined attributes for each individual zone. The architecture also includes roadside units (RSUs) and a base station/content manager to support data collecting through zone-wise vehicular adaptation allocation [13]. Pre-cached content is initially obtained from RSUs. When an RSU is unable to deliver the requested content, the RSU forwards the request to the base station/central manager. The RSUs within each zone's designated area must provide data, and each zone is in charge of doing so.

The trained AI model is then given the data gathered from the zones/clusters. To forecast crucial variables like cache hit ratio, productivity, and average latency, this model uses an ICN-based strategy. Making informed decisions about the content cache process, paying special attention to important and safe communications, is made possible by the ML method. Additionally, the suggested strategy incorporates both vehicle-to-infrastructure (V2I) and vehicle-to-vehicle (V2V) communication

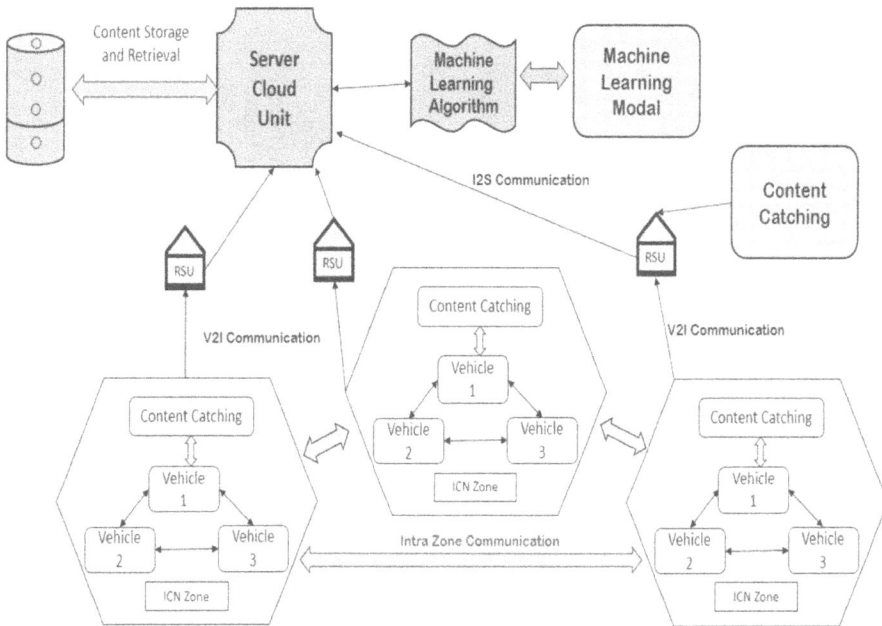

FIGURE 7.1 Architecture for zone-based congestion control in VANET.

techniques into the caching procedure. Table 7.1 gives the summary of different methods, its key finding, and drawbacks.

7.2.1 ICN-Based VANET Architecture

In VANET, ICN has become a potential paradigm for information dissemination. High levels of mobility, changeable network topologies, and the requirement for effective content delivery are characteristics of VANET. In this section, we will look at the specifics of the ICN-based VANET architecture, highlighting its essential elements and examining how they work together. Information is contained in named data objects (NDOs) in ICN-based VANET, as shown in Figure 7.2. A globally distinctive name and the data payload are both carried by an NDO, which is a standalone unit. Vehicles can communicate their interest in particular data items by using the name as the content's identifier. NDOs are adaptable for diverse VANET applications since they can represent a variety of data types, including text, photos, videos, or sensor measurements.

Figure 7.2 shows the basics of ICN Architecture used in VANET for establishing and operating communication. The architecture is responsible for establishing the connection through the network media, and other devices that are connected on an ad-hoc basis to the network. In the ICN-based VANET design, content is at the center of communication, and vehicles communicate their interest in certain data items rather than focusing on particular sources. Information is contained in NDOs,

TABLE 7.1

Summary of Different Methods and Common Pitfalls

Sr. No.	Paper	Methods	Attribute	Key Finding	Drawbacks
1	[14]	The algorithm for choosing the pre-caching nodes	Accuracy of the prediction and average delay	Pre-cache the content effectively	When loading content to the node, there is overhead
2	[15]	Cached content based on names	Performance and network delay	Manage the material using standard caching	Does not improve the accuracy of cache prediction
3	[16]	Game Stackelberg Method	Congestion and amount of content	Combat network sluggishness	Does not function properly in terms of network speed and performance metrics
4	[17]	Distributed content caching with two layers	Capacity for caching and transmission rates	Performance assessment	Throughput, handling of congestion, and network performance
5	[18]	A reward system for caching in video games	Caching of content	Better content caching outcomes	The price of caching is really expensive
6	[19]	Content pre-caching based on a hidden Markov model	Hop count, latency in the network values, delivery ratio, and ML approach	Boost the effectiveness of the caching system	ML method for effective content caching
7	[20]	RSU caching and pre-caching for vehicles	Pre-caching content, delivery speed, and load balancing	Performance of caching and delivery time	Ignores the cache hit ratio, average delay, or cache prediction accuracy
8	[21]	An integrated approach to networking	Approach to deep reinforcement learning	Caching of content and system improvements	There is no discussion of network performance or how to handle congestion
9	[22]	Joint content caching optimization strategy	Content caching based on ML, jointly implemented strategy, and policy	Efficiency in zones or clusters	Zones or clusters that are ineffective
10	[23]	A mobile-friendly approach to edge caching	Using network performance and analytics to handle	Cost estimating model for large-scale mobility	Does not offer post content caching strategies or policies for pre-caching material

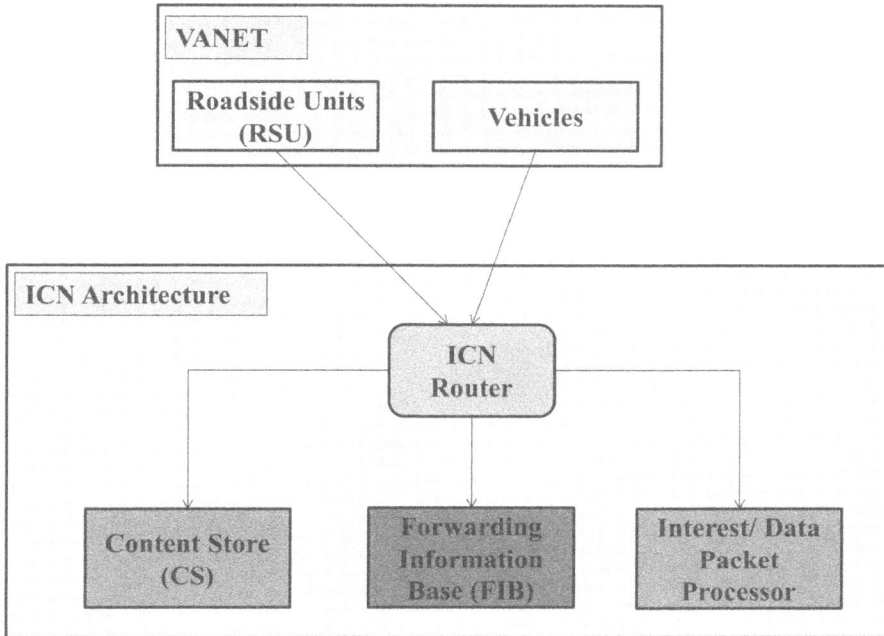

FIGURE 7.2   Basic ICN architecture in VANET.

which also contain names that are globally unique. Using an interest/data exchange approach, content routing provides effective distribution of material based on content names. By storing popular or often accessed data items in vehicles' caches that are carefully positioned throughout the network, content caching optimizes content delivery. By incorporating security controls into the material itself, content-based security ensures the authenticity and integrity of the content. Incorporating machine learning approaches can improve content delivery optimization in ICN-based VANET by adjusting to changing network conditions, vehicle movements, and trends in content popularity.

7.2.2 ICN COMMUNICATION MODEL

In contrast to conventional host-based communication, the ICN communication model places more emphasis on content. Instead of using a host's IP address, ICN names and requests data based on its content identifier (CID) as shown in Figure 7.3. A change in how information is requested, stored, and delivered inside a network is brought about by the ICN communication paradigm. Improved content delivery, effective caching, increased scalability, and support for mobile and dynamic situations are just a few benefits of the ICN communication model.

It also presents difficulties with regard to name conventions, content synchronization, caching techniques, and network scalability, as shown in Figure 7.3. In order to further explore and hone the ICN communication model, a number of ICN designs,

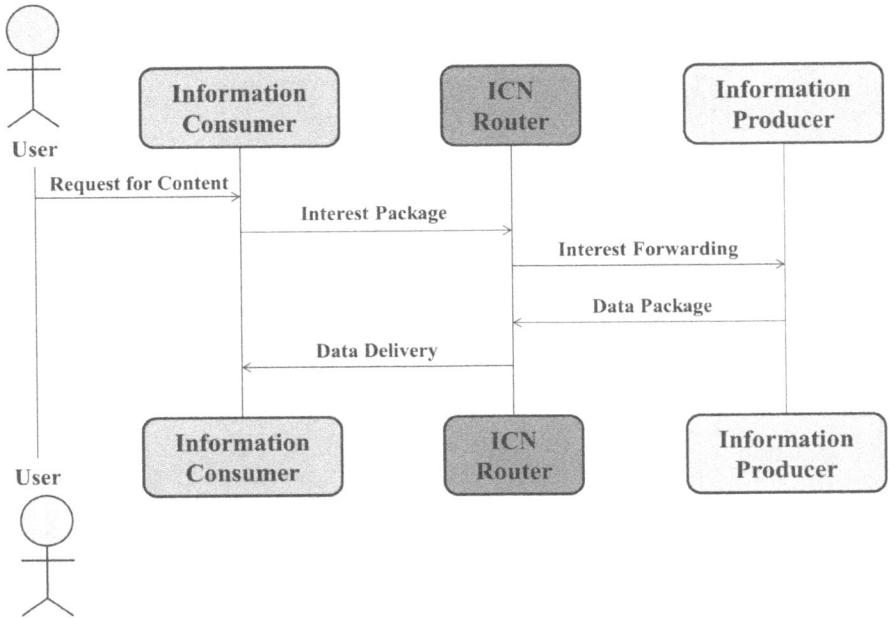

FIGURE 7.3   Communication model in information-centric networking (ICN).

including content-centric networking (CCN) and named data networking (NDN), have been developed [10].

7.2.3   ICN-BASED CONTENT DELIVERY IN VANET

The distribution and delivery of material in vehicle contexts are improved through ICN-based content delivery in VANET, which makes use of the ICN's guiding principles shown in Figure 7.4. VANET can more effectively distribute and retrieve material, address mobility issues, and enhance communication performance by adopting ICN ideas. The deployment of ICN requires the development of specific caching placement algorithms and replacement procedures in the context of vehicle contexts. In-network caching for VANET is the main topic of this paper's investigation. To assess the effectiveness of various tactics under various scenarios, we run simulations and do comparison assessments [1].

  i. *Efficient routing*: ICN-based content delivery in VANET optimizes routing decisions by considering content popularity, network conditions, and vehicle movement patterns. Routing protocols can utilize historical data and real-time observations to learn and predict the most efficient routes for content delivery. This improves routing efficiency, reduces latency, and enhances overall network performance.
 ii. *Multicast and content dissemination*: ICN-based VANET can effectively distribute content to numerous interested vehicles all at once by using

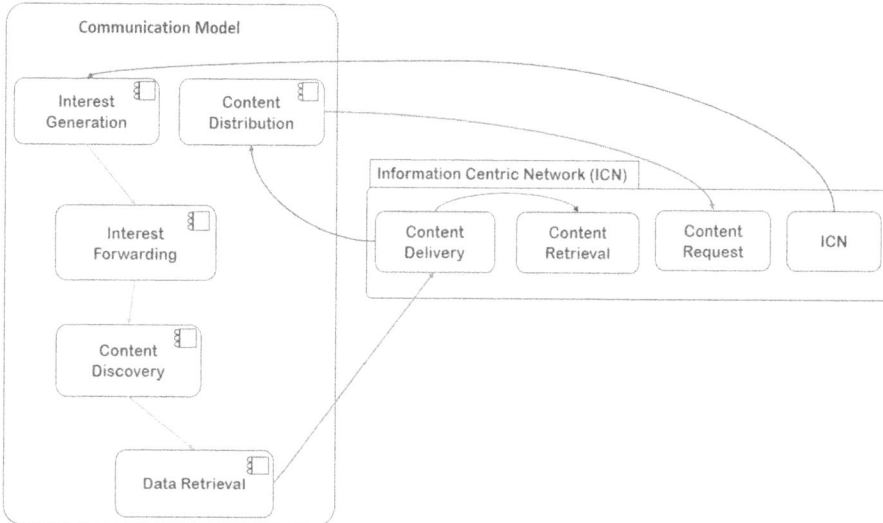

FIGURE 7.4 ICN-based content delivery.

multicast communication. Multiple nodes that have the needed content can satisfy content requests, allowing for effective data distribution to a fleet of cars. Reducing duplicate data transmissions and conserving network resources are two benefits of multicast-based multimedia distribution.

7.2.4 ICN-BASED ROUTING IN VANET

When discussing routing and data delivery in VANET, the term "ICN-based routing" refers to the use of ICN principles and protocols. By concentrating on content-centric communication and effective data dissemination, ICN-based routing seeks to meet the particular characteristics and difficulties of vehicular networks.

i. *Routing based on naming convention*: ICN-based routing in VANET uses name-based routing in place of conventional IP-based routing. Data packets are routed and forwarded by vehicles using content names. Based on the accessibility and closeness of the content, routing options are made, allowing cars to retrieve the content from nearby sources or caches.

ii. *Based on content name and its retrieval*: Instead of focusing on individual host addresses, ICN-based routing in VANET uses content names. Vehicles can request specific content by expressing interest in the associated CIDs, which allow content to be uniquely identifiable and uniquely identified. Effective content retrieval and dissemination are made possible by this content-centric strategy [3].

iii. *Dynamic mobility assistance*: Highly mobile and dynamic vehicles are used in VANET. The shifting positions and relationships of vehicles are taken into consideration via ICN-based routing in VANET. By separating material from specific host addresses, it makes it possible for cars to access content regardless of their position or mobility.

7.2.5   ICN-Based Caching in VANET

The ideas of ICN are combined with intelligent caching mechanisms in ICN-based caching in VANET using ML techniques. This is in order to optimize caching decisions taking into account a variety of parameters like content popularity, user requests, network circumstances, and vehicle movement. ML techniques are essential in VANET for streamlining caching choices and enhancing content delivery. In order to generate insightful forecasts about content popularity, user demand, network conditions, and vehicle mobility, these algorithms combine historical data with in-the-moment observations. VANET can improve content availability, lower data retrieval latency, and adapt caching tactics to changing network conditions by integrating ML into caching decision processes. Mobile-aware caching, user demand analysis, network status monitoring, content popularity prediction, and reinforcement learning for caching rules are all made possible by machine learning [10]. These methods work together to facilitate effective and intelligent caching in VANET, which enhances the performance of content delivery.

7.3   MACHINE LEARNING TECHNIQUES FOR CONTENT DELIVERY OPTIMIZATION IN ICN-BASED VANET

7.3.1   Introduction to Machine Learning Techniques

The content caching and placement optimization have both shown success when machine learning (ML) techniques have been used. On the basis of past performance data, present-day observations, and forecast models, caching decisions can be dynamically optimized. To make wise caching judgments, ML algorithms can assess elements including content popularity, user demand, network circumstances, and mobility patterns. The optimization of content distribution in a variety of systems, including VANET, IoT networks, and conventional internet infrastructures, has benefited greatly from the development of ML techniques [11]. ML algorithms can optimize content delivery, decrease latency, and enhance user experience by examining data trends and making intelligent predictions. These methods use historical information, present-day observations, and forecast models to improve the effectiveness of content distribution [5]. Here are some examples of often employed machine learning methods in this situation:

i. *Collaborative filtering*: A common strategy is collaborative filtering, which forecasts user preferences and content popularity using the past actions and shared interests of users. Collaborative filtering can be used in ICN-based VANET to identify content items that are likely to be well-liked by users and give them priority when it comes to caching.

ii. *Deep learning* (*DL*): For the purpose of optimizing content distribution, DL techniques, such as deep neural networks, can analyze huge volumes of data and uncover complicated patterns. DL algorithms are able to learn hierarchical data representations, making it possible to anticipate user preferences, network circumstances, and the popularity of certain material with accuracy.

iii. *Clustering and classification*: Users or vehicles can be grouped using clustering techniques like k-means clustering based on their shared preferences for particular types of information or mobility patterns. Based on numerous criteria and traits, classification algorithms such as decision trees or support vector machines can forecast content popularity or user desires.

iv. *Genetic algorithms*: Using genetic algorithms, caching placement and replacement tactics can be made more effective. These algorithms employ evolutionary concepts to repeatedly enhance fitness-based caching decisions, improving content availability and lowering latency.

v. *Time-series analysis*: It includes autoregressive integrated moving average (ARIMA) models or recurrent neural networks (RNNs), can be used to examine temporal trends in user preferences or the popularity of certain material. Future trends can be predicted using these techniques, and caching decisions can be optimized accordingly.

7.3.2 MACHINE LEARNING TECHNIQUES FOR CONTENT DELIVERY OPTIMIZATION IN ICN-BASED VANET

When it comes to optimizing content distribution in VANET built on ICN, ML techniques are crucial. These methods increase the effectiveness of content distribution, decrease latency, and enhance overall network performance by utilizing historical data, real-time observations, and prediction models. The conventional host-centric communication paradigm is replaced in ICN-based VANET by a content-centric strategy. This change calls for effective content delivery and distribution systems. The use of ML techniques offers useful tools for overcoming the difficulties faced by VANET and improving content delivery. The optimization of content distribution in ICN-based VANET can benefit from supervised learning, unsupervised learning, and reinforcement learning (RL) techniques. Each of these strategies has special benefits and can be used for various areas of content delivery optimization. Using these three learning paradigms can be summarized as follows:

7.3.2.1 Supervised Learning

To make predictions or judgments, a model is trained on labeled data through supervised learning. Supervised learning can be utilized in the context of ICN-based VANET for a variety of purposes.

The popularity of content items can be predicted using supervised learning algorithms that have been trained on previous data. These models can predict the chance of content being requested by looking at features including content attributes, user behavior, and network circumstances. Decisions regarding caching can be improved by using this knowledge.

Supervised learning can be used to forecast QoS metrics including latency, throughput, and reliability for various content delivery channels. Predictions can be generated to choose the best path for content delivery, ensuring specified QoS levels, by training models using labeled data that contains network measurements and QoS parameters.

7.3.2.2 Unsupervised Learning

Analyzing unlabeled data in order to find patterns, structures, or relationships is known as unsupervised learning. Unsupervised learning can be helpful in ICN-based VANET in the following ways:

Based on their patterns of content consumption, unsupervised learning techniques like clustering can discover groups of similar users. By adapting content delivery to the preferences and requirements of various user categories, this information can be used to personalize content delivery.

Algorithms for unsupervised learning can examine usage trends and content characteristics to find comparable content items. Similar material can be grouped together to create more effective caching and delivery techniques, eliminating content duplication and increasing cache utilization.

7.3.2.3 Reinforcement Learning

By maximizing cumulative rewards, reinforcement learning (RL) teaches agents to make decisions in a given environment. The following scenarios can use RL in ICN-based VANET:

By analyzing the rewards received in accordance with the network circumstances, the availability of the material, and the delivery efficiency, RL systems can learn the best routing policies. Agents can consider variables like traffic patterns, network congestion, and vehicle movement when making routing decisions to ensure effective and dependable content delivery.

The caching choices made by cars or RSUs can be optimized via RL. Agents can learn to cache the most pertinent and well-liked content pieces by structuring caching as a sequential decision-making process based on the observed rewards and penalties. This adaptive learning strategy boosts content availability and cache utilization.

7.3.3 Comparison of Machine Learning Techniques for Content Delivery Optimization in ICN-Based VANET

For optimizing content distribution in ICN-based VANET, ML techniques are a useful tool, shown in Figure 7.5. Here is a comparison of the three main machine learning approaches – supervised learning, unsupervised learning, and reinforcement learning – used in ICN-based VANET for content delivery optimization. A summary of the comparison is shown in Table 7.2.

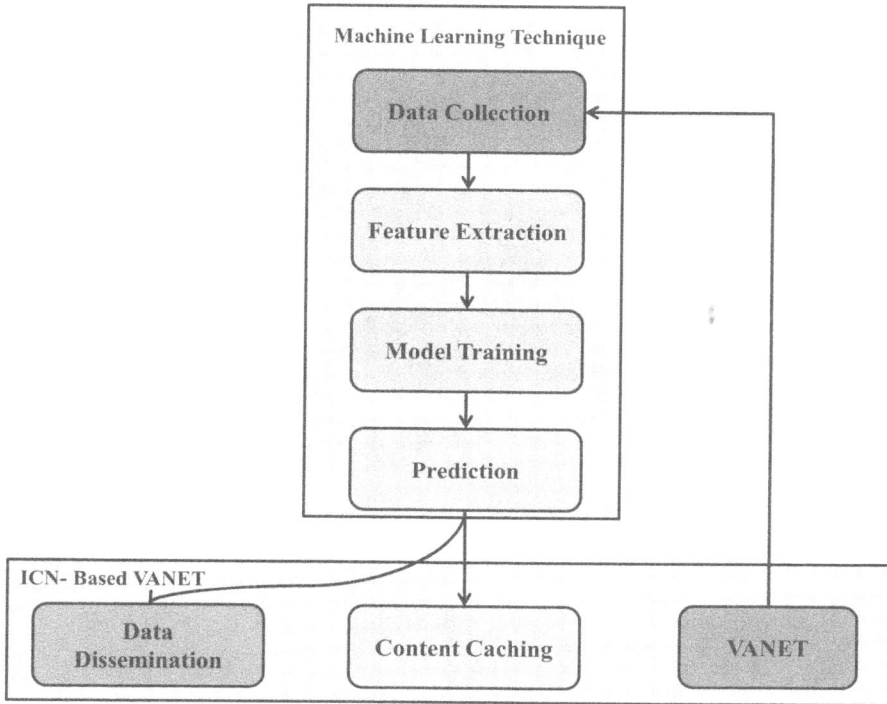

FIGURE 7.5 VANET machine learning model.

7.3.3.1 Deep Learning

A network forwarder for NDN using DL techniques. In order to create an effective network forwarder, the researchers investigated various ML strategies and conducted a survey of the available AI approaches in the networking industry as shown in Figure 7.6.

The goal was to use DL methods to improve the speed and scalability of NDN [31]. It was decided to use ICN and AI for Internet of Things (IoT) applications shown in Figure 7.6. The emphasis was on utilizing ICN's built-in caching capabilities and applying them to data-centric IoT situations. The plan also included the use of machine learning techniques to establish appropriate caching strategies for individual data items and to enhance node behavior predictions [32].

7.3.3.2 Collaborative Filtering

The use of collaborative filtering powered by AI in the edge cloud, this method attempted to manage cloud database traffic by intelligently caching material on edge nodes. The suggested technique attempted to optimize content caching choices based on user behavior and preferences by utilizing collaborative filtering, shown in Figure 7.7. A thorough analysis of the proposed strategy's performance in comparison to a number of benchmark strategies was carried out to determine its efficacy. The analytical outcomes shown that, in terms of cache hit ratio, content

TABLE 7.2

Comparison Emphasizing the Key Traits of Machine Learning Methods

Paper	Technology	Performance (%)	Advantage	Limitation	Future Scope Area	Category
[5]	ICN-CIoT-AI	93.45	Enhances the security of IoT devices	Metrics related cloud-based IoT intelligence	Hunting, threat intelligence, privacy protection, and intrusion detection	In the field of security
[6]	ICN-V-CIoT	92.02	Improves IoT network security and privacy	Does not focus on other privacy metrics related to ICN in IoT	A blockchain-based security framework for IoT network designs, ad-hoc networks for unmanned aircraft, and smart buildings	
[7]	ICN-IoT-AI	90.21	Enhances user privacy while retaining data utility	For security related to device not considered	Sparse coding to increase user privacy protection in the future without reducing data utility	
[24]	ICN-V-CIoT	93.23	Allows for incentive-based involvement in IoT networks while also improving security and privacy	Metric for data quality not included	The algorithm has to be improved to allow incentive-based participation while maintaining security and privacy with the least amount of cryptographic overhead	
[8]	ICN-V IoT	96.56	Improves caching and forwarding policies while dealing with increasingly complicated attacker scenarios	More scenarios of different attackers need to be considered	Policies for caching and forwarding, use of increasingly complicated attacker scenarios	
[25]	ICN-IoT	96.34	Improves performance in content-oriented ICN-IoT networks by improving naming, routing, security, and scalability	Need to focus on distributed device network	More realistic topologies will need to be tried with the proposed technique	
[9]	ICN-WSN (Wireless Sensor Networks)	91.12	ML techniques improve the performance of ICN-WSN models	Closed group model discuss. Security related algorithm not considered	The ICN-WSN models can perform better when using ML approaches	Performance using ML algorithm

(Continued)

TABLE 7.2 (Continued)
Comparison Emphasizing the Key Traits of Machine Learning Methods

Paper	Technology	Performance (%)	Advantage	Limitation	Future Scope Area	Category
[26]	ICN-IoT	93.22	Content-oriented ICN-IoT networks by improving naming, routing, security	Unmentioned ML algorithms or performance metrics	Performance is improved by the naming, routing, security, and scalability of content-oriented ICN-IoT networks	
[27]	NDN-V-CIoT	95.45	Increases the efficiency of NDN-V in cloud or mobile contexts by taking context-aware data ownership into account	Context-based device performance not considered for evaluation	Context-aware data ownership boosts NDN-V's efficiency in cloud- or mobile-based environments	
[4]	ICN-5G-CIoT	96.45	The impact of various parameters on latency and capacity in ICN-5G-CIoT systems is investigated	Poor security mechanism or security parameter not considered	Latency and capacity versus latency are impacted by cache size, effective cache replacement rules, and wireless channel parameters	Caching of network
[28]	ICN-CIoT	90.23	Uses the compressive big data analytics (CBDA) framework and DL-based replication to supply secure and environmentally friendly content in IoT systems	Mentioned framework applicable for connected device only	To provide secure and energy-efficient content, a CBDA platform and DL-based reduplication on index tables could be used	
[29]	ICN-V-IoT	94.56	More scalable method of transportation system in ICN network for privacy	Scalability increase but does not consider the metrics of performance	Communications and information pertaining to the entertainment system for automobile passengers	
[30]	ICN-IoT	93.70	The effect of data distribution in cache storage on transmission acquisition efficiency is investigated	Latency in communication	The emergency message and the other two messages can only be obtained from the original producer node, the allocation of three types of data in the cache storage that has not been fully investigated greatly lowers acquisition efficiency	

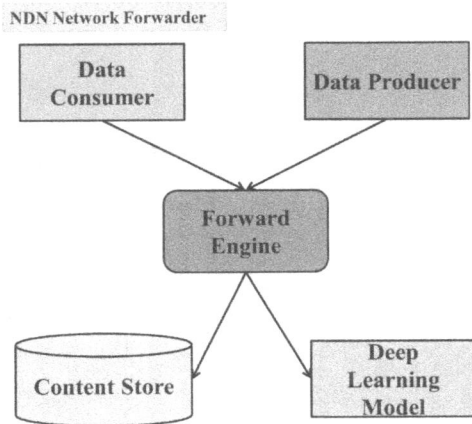

FIGURE 7.6   NDN using deep learning.

retrieval delay, and average hop count, the proposed method exceeded the best-performing benchmark strategies. These results indicate the possibility of improving the effectiveness and performance of content caching in edge cloud systems using a collaborative filtering strategy based on AI [33]. A well-designed caching strategy must consider a wide range of attributes related to the requested content

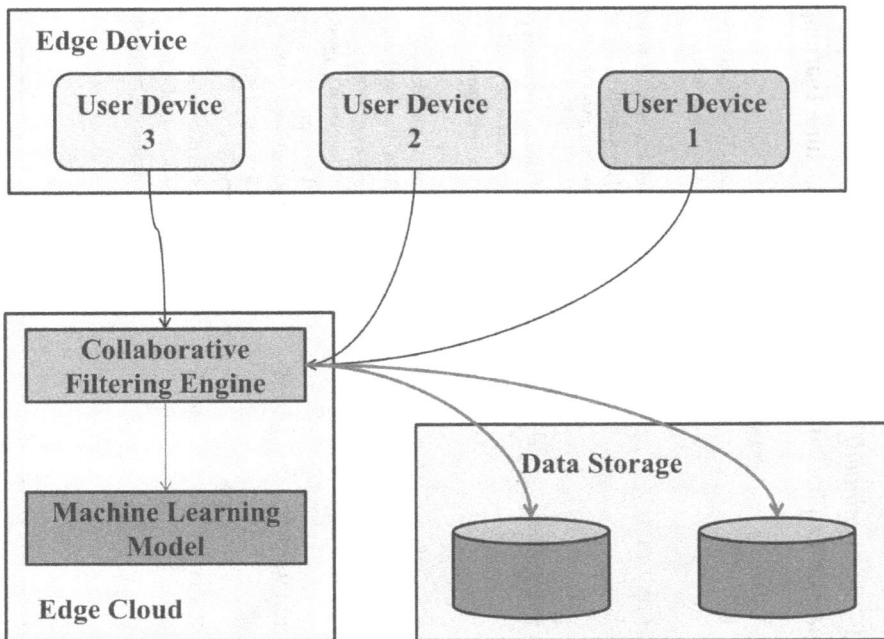

FIGURE 7.7   Collaborative filtering powered by AI in the edge cloud.

and take into account the various caching capacities of all the edge nodes in order to achieve a scalable network with low content retrieval latency. The summary of DL and ML analysis is discussed in Table 7.3. By doing this, the strategy can improve the network's overall performance by optimizing content placement and content delivery effectiveness.

7.3.3.3   Ensemble Learning

An effective and dependable solution to the problems of content delivery and network performance degradation in VANET brought on by dynamic topology and sporadic connectivity is the extension of the CCN architecture [34]. By integrating CCN into VANET, the network can give priority to content-centric communication, putting less emphasis on individual node-to-node communication and more on the delivery and accessibility of content. By using distinct names rather than exact node addresses, the CCN extension in VANET provides content-based routing and caching methods.

TABLE 7.3
Deep Learning/Machine Learning Analysis in Detail

AIML Technique	Model	Observation	Experiment/Simulation
Deep Neural Network (DNN) [35]	A two-layer neural network with a feed-forward architecture that uses linear units for the output layer and sigmoid units for the hidden layer	99% accuracy for this model	ndnSIM (Named Data Networking SIM)
Artificial Neural Network (ANN) [36]	Model, parameter, and inference-based decisions for caching and replacement	–	No simulation
Natural Language Programming (NLP) [37]	Sematic/synthesis based analysis	Shows exemplary performance in crisis situations	Cloud SIM
Machine Learning (ML) [38]	Extended learning classifier system under reinforcement learning (RL), deep belief network-restricted Boltzmann machines (DBN-RBM) for unsupervised learning, and multilayer perceptron (MLP) for supervised learning	The extended classifier system (XCS) algorithm using RL techniques outperforms the competition	MATLAB® used root mean square error
Machine Learning (ML) [11]	Frequent pattern (FP) growth algorithm for association rule mining	Increases the effectiveness of data collection and decreases the time needed to access data	Edge node

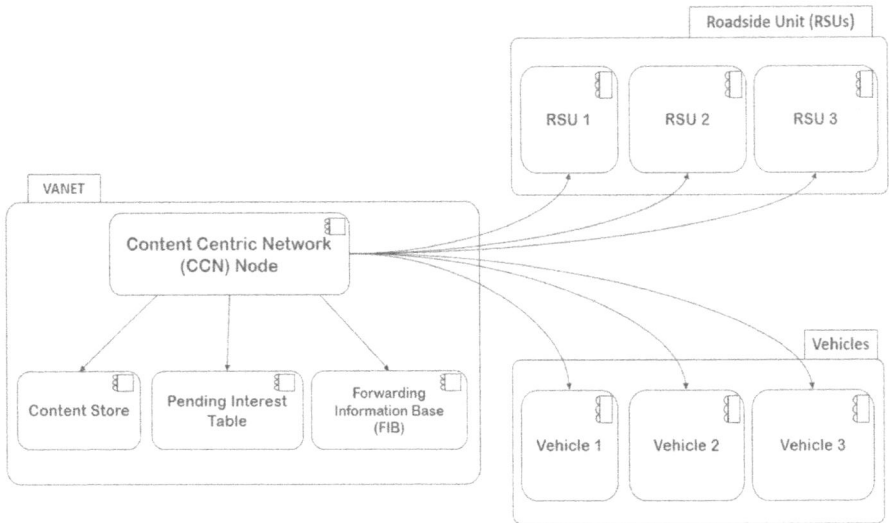

FIGURE 7.8   CCN architecture.

Despite their location or the level of connectivity of the content source, this enables cars to request and retrieve content based on their interests. CCN in VANET improves the effectiveness and dependability of content delivery by utilizing this content-centric strategy. The dependency on distant content sources can be reduced and the delays brought on by sporadic connectivity can be minimized by vehicles caching frequently sought content. Additionally, CCN's, shown in Figure 7.8, built-in multicast capabilities make it possible to effectively distribute material around the network to interested cars.

7.4   CHALLENGES AND OPPORTUNITIES FOR MACHINE LEARNING-BASED CONTENT DELIVERY OPTIMIZATION IN ICN-BASED VANET

The fast-developing topic of ICN-CIoT-AI has attracted a lot of attention from academics. It provides countless chances to investigate unresolved problems and deal with different difficulties. Security, in-network caching, heterogeneous networking, mobility, and automation are a few of these difficulties. Investigating these topics can help the development of ICN-CIoT-AI in significant ways.

7.4.1   CHALLENGES

The challenges for machine learning based content delivery optimization in ICN-based VANET are:

Limited data availability: Limited data availability is a major issue in the field of ML and AI. Large and varied training datasets are essential to the performance of machine learning algorithms. Nevertheless, it can be difficult to get enough labeled data in some fields or applications.

Heterogeneous data sources and imbalanced data: Due to a lack of data, data-sets may be biased or unbalanced, with some classes or cases being over- or underrepresented. As a result, the algorithms might not be sufficiently exposed to all potential scenarios, which might result in biased models and erroneous predictions. The dynamic nature of the network and the dearth of real-world labeled datasets make it difficult to obtain labeled training data for ML algorithms in VANET. It can be difficult to produce representative and diverse labeled data for training accurate models.

Security and privacy concerns: ML algorithms that depend on vehicle coop-eration and data sharing increase privacy and security issues. In VANET, it might be difficult to protect sensitive data while permitting efficient content delivery optimization.

Resource constraints and scalability: VANET are typically made up of sev-eral vehicles with modest processing capabilities. It is difficult to create ML algorithms that can scale to manage a large number of vehicles while taking into account the resource limitations of individual vehicles.

7.4.2 OPPORTUNITIES

Following are the opportunities for machine learning based content delivery optimi-zation in ICN-based VANET.

Automated content caching: Intelligent content caching decisions in VANET can be made thanks to ML techniques. ML models can forecast and opti-mize content caching solutions by examining user demand patterns and net-work circumstances, increasing content availability and lowering latency.

Adaptive routing and traffic control: ML algorithms can adaptively optimize routing decisions based on real-time network conditions. They can also optimize traffic control mechanisms to alleviate congestion and improve overall network performance.

User behavior analysis: ML models can analyze user behavior patterns, prefer-ences, and content demand to personalize content delivery in VANET. This can lead to improved user satisfaction and efficient resource utilization.

Network anomaly detection: ML algorithms can detect network anoma-lies, such as malicious activities or abnormal behavior, in VANET. This enhances the security and integrity of the network by enabling proactive measures against potential threats.

7.5 CONCLUSION AND FUTURE WORK

AI and ML are hampered by the lack of available data. When working with small amounts of data, researchers may encounter problems such as inadequate training samples, biased or unbalanced data, and difficulties in caching unusual events, and the danger of overfitting. However, a number of techniques, including collaborative data sharing, transfer learning, active learning, and data augmentation can help to mitigate these issues and enhance model performance. The scope of machine learn-ing techniques was discussed and it can be used to optimize content distribution in

VANET built on ICN. The above-mentioned future research directions can improve the functionality, security, and usability of ICN-based VANET further, making them more effective and dependable for a variety of vehicular communication applications.

Future research should focus on creating effective algorithms for making dynamic decisions. While utilizing ML approaches to optimize content delivery in ICN-based VANET has shown encouraging results in our study, there are still a number of areas that might use more investigation and advancement. With advances in ML and decision tree algorithms, it might be possible to enhance content delivery optimization.

Large-scale studies or the deployment of ICN-based VANETs in real-world testbeds may provide important information and draw attention to potential issues or restrictions in real implementation in the future.

The incorporation of additional cutting-edge technologies like edge computing, blockchain, and 5G/6G networks can be advantageous for ICN-based VANET. Future study should focus on examining the interactions and potential advantages of these technologies in streamlining information delivery and improving VANET's overall functionality.

REFERENCES

1. H. Khelifi, S. Luo, B. Nour, and H. Moungla, "In-Network Caching in ICN-based Vehicular Networks: Effectiveness & Performance Evaluation," ICC 2020 - 2020 IEEE International Conference on Communications (ICC), Dublin, Ireland, pp. 1–6, 2020. doi: 10.1109/ICC40277.2020.9148950
2. Z. Zhang, C.-H. Lung, X. Wei, M. Chen, S. Chatterjee, and Z. Zhang, "In-network caching for ICN-based IoT (ICN-IoT): a comprehensive survey," in IEEE Internet of Things Journal. 2023. doi: 10.1109/JIOT.2023.3274653
3. S. Ajani, and M. Wanjari, "An Efficient Approach for Clustering Uncertain Data Mining Based on Hash Indexing and Voronoi Clustering," 2013 5th International Conference and Computational Intelligence and Communication Networks, pp. 486–490, 2013. IEEE. doi: 10.1109/CICN.2013.106
4. M. Sakthivanitha, and S. Saradha, "Survey based on security aware caching scheme for IoT based information centric networking," EAI Endorsed Transactions on Energy Web, vol. 8, no. 32, p. e2, 2020.
5. N. Moustafa, "A new distributed architecture for evaluating AI-based security systems at the edge: network TON_IoT datasets," Sustainable Cities and Society, vol. 72, p. 102994, 2021.
6. K. Lei, J. Fang et al., "Blockchain-based cache poisoning security protection and privacy-aware access control in NDN vehicular edge computing networks," Journal of Grid Computing, vol. 18, no. 4, pp. 593–613, 2020.
7. S. N. Ajani, and S. Y. Amdani. Agent-Based Path Prediction Strategy (ABPP) for Navigation Over Dynamic Environment. In: Muthukumar, P., Sarkar, D.K., De, D., De, C.K. (eds) Innovations in Sustainable Energy and Technology. Advances in Sustainability Science and Technology. Springer, Singapore, 2021.
8. N. Magaia, and Z. Sheng, "ReFIoV: a novel reputation framework for information-centric vehicular applications," IEEE Transactions on Vehicular Technology, vol. 68, no. 2, pp. 1810–1823, 2018.
9. S. N. Ajani, P. K. Ingole, and A. V. Sakhare "Modality of multi-attribute decision making for network selection in heterogeneous wireless networks", Ambient Science - National Cave Research and Protection Organization, India, Vol. 9, no. 2, pp. 26–31, 2022, ISSN-2348 5191.

10. G. Deng, L. Wang et al., "Distributed Probabilistic Caching strategy in VANETs through Named Data Networking", IEEE Conference on Computer Communications Workshops (INFOCOM WKSHPS), pp. 314–319, 2016.
11. V. Kirilin, A. Sundarrajan, S. Gorinsky, and R. K. Sitaraman, "RL-cache: learning-based cache admission for content delivery," IEEE Journal on Selected Areas in Communications, vol. 38, no. 10, pp. 2372–2385, Oct. 2020. doi: 10.1109/JSAC.2020.3000415
12. I. U. Din, B. Ahmad, A. Almogren, H. Almajed, I. Mohiuddin, and J. J. Rodrigues, "Left-right-front caching strategy for vehicular networks in ICN-based internet of things," IEEE Access, vol. 9, pp. 595–605, 2021.
13. S. Kannan, G. Dhiman, Y. Natarajan, A. Sharma, S. N. Mohanty, M. Soni, U. Easwaran, H. Ghorbani, A. Asheralieva, and M. Gheisari, "Ubiquitous vehicular ad-hoc network computing using deep neural network with IoT-based bat agents for Traffic Management," Electronics, vol. 10, p. 785, 2021.
14. H. Guo, L. L. Rui, and Z.-P. Gao, "A zone-based content pre-caching strategy in vehicular edge networks," Future Generation Computer Systems, vol. 106, pp. 22–33, 2020.
15. L. C. Fourati, S. Ayed, and M. A. B. Rejeb, "ICN clustering-based approach for VANETs," Annals of Telecommunications, vol. 76, pp. 745–757, 2021.
16. A. Alioua, S. Simoud, S. Bourema, M. Khelifi, and S.-M. Senouci, "A Stackelberg Game Approach for Incentive V2V Caching Insoftware-Defined 5G-Enabled VANET," In Proceedings of the 2020 IEEE Symposium on Computers and Communications (ISCC), Rennes, France, pp. 1–6, 7–10 July 2020.
17. L. C. Liu, D. Xie, S. Wang, and Z. Zhang, "CCN-Based Cooperative Caching in VANET," In Proceedings of the 2015 International Conference on Connected Vehicles and Expo (ICCVE), Shenzhen, China, pp. 198–203, 19–23 October 2015. IEEE.
18. Y. Wang, Y. Lin, L. Chen, and J. Shi, "A stackelberg game-based caching incentive scheme for roadside units in VANETs," Sensors, vol. 20, p. 6625, 2020.
19. Z. Xue, Y. Liu, G. Han, F. Ayaz, Z. Sheng, and Y. Wang, "Two-layer distributed content caching for infotainment applications inVANETs," IEEE Internet of Things Journal, vol. 9, pp. 1696–1711, 2021.
20. R. S. Pereira, L. Guan, M. Ye, and Z. Zhang, "Faster Content Delivery using RSU Caching and Vehicular Pre-caching in Vehicular Networks," arXiv 2021, arXiv:2112.02692.
21. L. Yao, Z. Li, W. Peng, B. Wu, and W. A. Sun, "Pre-Caching Mechanism of Video Stream Based on Hidden Markov Model in Vehicular Content Centric Network," In Proceedings of the 2020 3rd International Conference on Hot Information-Centric Networking (HotICN), Hefei, China, pp. 53–58, 12–14 December 2020. IEEE.
22. M. Zhang, S. Wang, and Q. Gao, "A joint optimization scheme of content caching and resource allocation for internet of vehicles in mobile edge computing," Journal of Cloud Computing: Advances, Systems and Applications, vol. 9, p. 33, 2020.
23. Y. He, Z. Zhang, F. R. Yu, N. Zhao, H. Yin, V. C. M. Leung, and Y. Zhang, "Deep-reinforcement-learning-based optimization for cache-enabled opportunistic interference alignment wireless networks," IEEE Transactions on Vehicular Technology, vol. 66, pp. 10433–10445, 2017.
24. L. Nkenyereye, S. R. Islam, M. Bilal, M. Abdullah-Al-Wadud, A. Alamri, and A. Nayyar, "Secure crowd-sensing protocol for fog-based vehicular cloud," Future Generation Computer Systems, vol. 120, pp. 61–75, 2021.
25. T. Zhi, Y. Liu, and J. Wu, "A reputation value-based early detection mechanism against the consumer-provider collusive attack in information-centric IoT," IEEE Access, vol. 8, pp. 38262–38275, 2020.
26. R. Hussain, S. H. Bouk, N. Javaid, A. M. Khan, and J. Lee, "Realization of VANET-based cloud services through named data networking," IEEE Communications Magazine, vol. 56, no. 8, pp. 168–175, 2018.

27. R. Ullah, M. A. U. Rehman, M. A. Naeem, B.-S. Kim, and S. Mastorakis, "ICN with edge for 5G: exploiting in-network caching in ICN-based edge computing for 5G networks," Future Generation Computer Systems, vol. 111, pp. 159–174, 2020.

28. V. Khetani, Y. Gandhi, S. Bhattacharya, S. N. Ajani, and S. Limkar, "Cross-domain analysis of ML and DL: evaluating their impact in diverse domains," International Journal of Intelligent Systems and Applications in Engineering, vol. 11, no. 7s, pp. 253–262, 2023.

29. C. Chen, J. Jiang, R. Fu, L. Chen, C. Li, and S. Wan, "An intelligent caching strategy considering time-space characteristics in vehicular named data networks," IEEE Transactions on Intelligent Transportation Systems, vol. 23, no. 10, pp. 19655–19667, 2022. doi: 10.1109/TITS.2021.3128012

30. S. Anamalamudi, M. S. Alkatheiri, E. Al Solami, and A. R. Sangi, "Cooperative caching scheme for machine-to-machine information-centric IoT networks," Canadian Journal of Electrical and Computer Engineering, vol. 44, no. 2, pp. 228–237, 2021.

31. M. T. R. Khan, M. M. Saad, M. A. Tariq, J. Akram, and D. Kim, "SPICE-IT: smart COVID-19 pandemic controlled eradication over NDN-IoT," Information Fusion, vol. 74, pp. 50–64, 2021.

32. M.-O. Pahl, S. Liebald, and L. Wüstrich, "Machine-learning based IoT data caching," in 2019 IFIP/IEEE Symposium on Integrated Network and Service Management (IM), Arlington, VA, USA, pp. 9–12, April 2019.

33. D. Gupta, S. Rani, S. H. Ahmed, S. Verma, M. F. Ijaz, and J. Shafi, "Edge caching based on collaborative filtering for heterogeneous ICN-IoT applications," Sensors, vol. 21, no. 16, p. 5491, 2021.

34. Q. Zhang, J. Wu, M. Zanella, W. Yang, A. K. Bashir, and W. Fornaciari, "Sema-IIoVT: Emergent semantic-based trustworthy information-centric fog system and testbed for intelligent internet of vehicles," IEEE Consumer Electronics Magazine, vol. 14, p. 1, 2021.

35. Y. Tang, K. Guo, J. Ma, Y. Shen, and T. Chi, "A smart caching mechanism for mobile multimedia in information centric networking with edge computing," Future Generation Computer Systems, vol. 91, pp. 590–600, 2019.

36. C. Campolo, G. Lia, M. Amadeo, G. Ruggeri, A. Iera, and A. Molinaro, "Towards Named AI Networking: Unveiling the Potential of NDN for Edge AI," International Conference on Ad-Hoc Networks and Wireless, Springer, pp. 16–22, 2020.

37. Q. Zhang, J. Wu, M. Zanella, W. Yang, A. K. Bashir, and W. Fornaciari, "Sema-IIoVT: emergent semantic-based trustworthy information-centric fog system and testbed for intelligent Internet of vehicles," IEEE Consumer Electronics Magazine, vol. 14, p. 1, 2021.

38. C. Safitri, R. Mandala, Q. N. Nguyen, and T. Sato, "Artificial Intelligence Approach for Name Classification in Information-Centric Networking-Based Internet of Things," in 2020 IEEE International Conference on Sustainable Engineering and Creative Computing (ICSECC), Indonesia, pp. 158–163, December 2020.

8 The Role of IoMT Technologies in Revolutionizing Healthcare
A Comprehensive Overview

*Sushmita Sunil Jain, Saurabh Manoj Kothari,
and Satyam Kumar Agrawal*

8.1 INTRODUCTION TO IoMT

The Internet of Medical Things (IoMT) is an application of the Internet of Things (IoT) in healthcare [1, 2]. It involves the use of medical devices attached to patients to monitor their medical parameters and communicate the information to healthcare professionals [3, 4]. However, ensuring patient privacy is critical for IoMT security [5]. Creating a robust authentication method for IoMT is challenging due to limited resource capacity, diverse platforms and protocols used by different providers, and the inherent vulnerability of IoMT systems to security breaches [6].

Despite these challenges, the IoMT environment, consisting of clinical sensors, frameworks, and computer systems, has significantly improved healthcare services by collecting patient physiological data and providing access to authorized healthcare professionals. It holds the key to closing the gaps in diagnosis, care, and patient well-being while reducing costs and improving clinical efficacy. However, data leaks and breaches remain a significant concern in the IoMT environment. Personal information transmitted from clinical sensors over open Internet routes may be wiretapped or overheard by malicious actors, leading to data breaches [7].

Remarkable developments in the healthcare sector are being driven by notable breakthroughs in the miniaturization of devices and technology, wireless technology, processing power, and computing capacity. With the ability to collect, produce, analyze, and transfer data, connected medical devices have now become a reality as a result of this advancement. It is an interconnected ecosystem made up of software applications, medical devices, healthcare services, and computing systems. These devices and the data they produce collectively comprise its core. The IoMT can assist in monitoring, alerting, and informing not only carers but also healthcare professionals by providing real-time data to spot problems before they become life-threatening or by enabling early intervention [8].

DOI: 10.1201/9781003437079-8

To safeguard the security of IoMT systems and protect sensitive patient data, a set of 11 security necessities has been derived from the following considerations: confidentiality, integrity, availability, non-repudiation, and authentication (CIANA). To list them in full, the 11 security requirements are: confidentiality/privacy, integrity, availability, non-repudiation, authentication, authorization, anonymity, forward/backward secrecy, secure key exchange, key-escrow resilience, and session key agreement. These requirements aim to ensure that data is kept private, protected from tampering, and available at all times and that only authorized users can access it. Techniques such as data encryption, access control, digital signatures, and smart cards can be used to meet these requirements [9].

Traditional cryptographic methods, such as data encryption and access control, are commonly used to safeguard security during wireless communication. However, the computational resources needed challenge medical sensors' low memory, processing power, and energy capacity, which all require innovative solutions. The successful implementation of IoMT security requires a comprehensive understanding of the risks and challenges involved, along with a proactive approach to addressing them [10].

The role of IoMT in a smart healthcare system is an important application of IoT. Smart IoMT devices include a wide variety of gadgets, such as smartphones, wireless sensors, and wearables like smartwatches. These devices are equipped to gather a variety of health metrics in real-time, including blood pressure (BP), body temperature, heart rate (HR), respiration rate (RR), blood glucose (BG), as well as location and time information. These metrics are vital for healthcare professionals who need to obtain and monitor necessary and important health information digitally, allowing quick patient interventions and individualized care [10]. IoMT is discussed in more detail in the following sections.

8.2   HISTORY AND DEVELOPMENT OF IoMT

Significant turning points and shifts have been noticeable in IoMT evolution over time. Compared to its current state, the IoMT's initial landscape and functionality were very different. Its roots can be found in the early days of the IoT, when the idea of linking various devices first developed. From there, attention eventually switched to the implementation of IoT technology in the medical domain, giving rise to the IoMT. Since then, there have been notable improvements and advancements in the industry, allowing for the seamless integration of medical equipment, sensors, and healthcare systems. Table 8.1 shows the differences between IoT and IoMT.

A brief history of IoMT is as follows:

The first IoT gadget developed in 1990–1991: The first IoT device was a toaster that John Romkey created in 1990 and allowed for remote on/off control over the Internet. He added a crane system in 1991 and wholly mechanized the operation of putting the bread in the toaster [11].

Mark Weiser constructed a water fountain stock market in 1998: Ubiquitous computing pioneer Mark Weiser created a fountain outside his office that

TABLE 8.1
Differences between IoT and IoMT

S. No.	IoT	IoMT	Reference
1	It serves a specific purpose and is spread over a wide area.	Used in or near the human body, a medical facility, or another constrained or limited geographic region.	[17]
2	Both sun and wind energy are potential energy sources. If the nodes are stationary, they might be continuously powered.	The use of heat, stress, and motion by IoMT nodes allows them to obtain energy from the human body as well.	[18]
3	Node size varies depending on the environment and application.	The nodes are scaled down to minimize their presence.	[17]
4	Sensor deployment is very simple.	Deployment is challenging, especially when it comes to implants, which almost always require surgery.	[19]
5	It aims to preserve data integrity. Errors are offset by redundancy.	The data is sent and preserved with maximum integrity.	[19]

replicated real-time stock market volume and price trends using water flow and height. It would have been amazing to witness the 2008 financial catastrophe by this method [12].

The phrase "Internet of Things" established in 1999: Kevin Ashton, the founder of Auto-ID, named his presentation, "The Internet of Things." Radio frequency identification (RFID) and the Internet were merged by Ashton, who impressed the audience with his originality [13].

The first smarthome gadget built in 2005: The Nabaztag, a rabbit-shaped Wi-Fi enabled ambient electronic device that could alert its owner and provide information on the weather and stock market developments, was first introduced in June 2005. It was a primary prototype of smarthome gadgets similar to Alexa and Google Home. It communicated with Really Simple Syndication (RSS) and needed to be fed, for example. With a newly-installed Raspberry Pi brain, the recognizable rabbit made a brief comeback in 2019 [13].

IoT began in 2008–2009: The Cisco Internet Business Solutions Group (IBSG) announced the start of IoT as between 2008–2009, when, according to Cisco, there were more connected machines than people on Earth. Currently, there are almost three times as many connected devices (21.5 billion) as there are people in the world [14].

IoT augments the hype cycle for new technology in 2011: The "Internet of Things" was added to the list of new technological developments, by market research firm Gartner in 2011, which also developed the popular "hype cycle for emerging technologies." For the seventh year in a row, Vodafone was recognized, for its IoT connectivity services, by Gartner as a leader [13].

IoT gadgets begin employing sensors after 2013: For precision sensing of one's surroundings, sensors were introduced in 2013 into thermostats and home lights. Thermostats, garage doors, or home lighting, and many similar devices, could all be managed from an individuals' phone. IoT was now available to everyone [13].

The first "smart city" was constructed in 2014: The Smart Docklands in Dublin was presented as a "testbed" for smart cities, allowing developers to demonstrate advanced technological solutions to topical problems which include smart streetlights, flood-level monitors, and city noise monitoring devices [15].

IoT enters the healthcare and insurance sectors in 2018: One of the IoT market's flourishing segments is represented by healthcare tools. By 2026, it is expected that the market size of this industry, sometimes referred to as the IoMT, will be USD 176 billion. The quality of wearable medical equipment is improved by IoT technology, which also gives healthcare providers access to patient data. Pacemakers, fall detection, geo-fencing, location tracking, and the monitoring of blood glucose levels and heart rate are a few examples of medical IoT applications [13].

In response to the COVID-19 problem, IoMT intensifies in 2020: In order to gauge people's body temperatures as a way of checking who had COVID-19, thermal cameras that measured body heat using infrared technology began to emerge in public locations in 2020. However, this technology was not primarily aimed at medical applications, but was usually employed by the police to search for hidden suspects and to detect embers [16].

New advancements in IoMT continue to shape the healthcare landscape. An overview of the advancements so far is shown schematically in Figure 8.1. The revolutionary aspect of IoMT is the fusion of numerous technologies, including wearables, telemedicine platforms, data analytics, and artificial intelligence. To achieve better healthcare results, this dynamic ecosystem promotes communication between doctors, patients, and technology specialists. As long as IoMT continues to be researched and developed, we can anticipate more advancement, broader capabilities, and a more integrated healthcare system that empowers both patients and clinicians.

8.3 THE ARCHITECTURE OF IoMT

IoMT systems are composed of several architecture layers, similar to IoT. These layers are depicted in Figure 8.2.The underlying structure of IoMT or IoT deployments is composed of the following layers:

- *The connectivity layer*: It is responsible for shifting data, using connectivity technologies like network and gateways, from the perception layer to the cloud and vice versa. All of the medical equipment on the web is connected by this layer, which also transfers data [20].

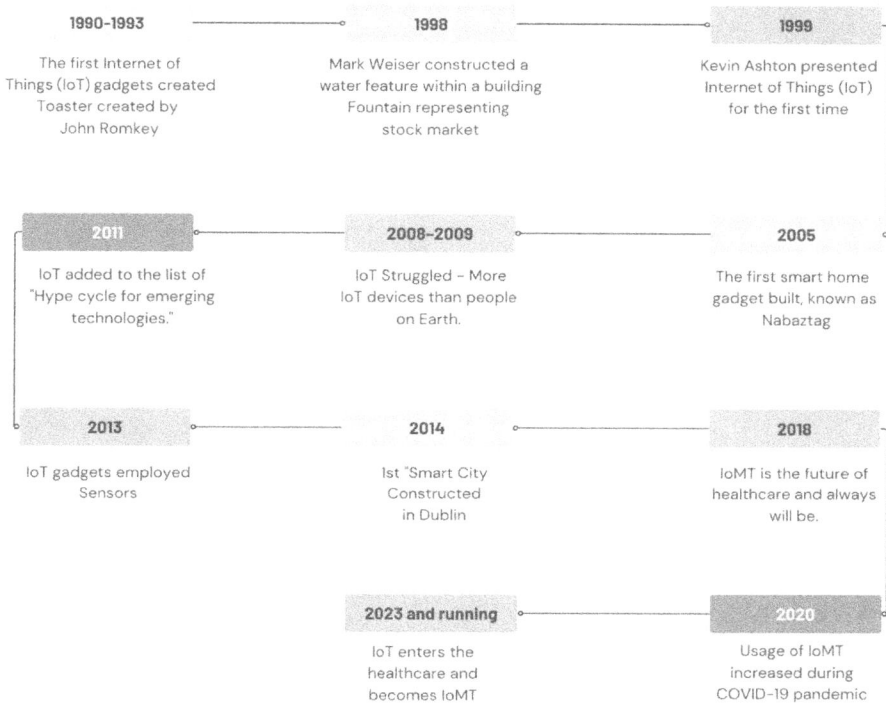

1990-1993	1998	1999
The first Internet of Things (IoT) gadgets created Toaster created by John Romkey	Mark Weiser constructed a water feature within a building Fountain representing stock market	Kevin Ashton presented Internet of Things (IoT) for the first time

2011	2008–2009	2005
IoT added to the list of "Hype cycle for emerging technologies."	IoT Struggled – More IoT devices than people on Earth.	The first smart home gadget built, known as Nabaztag

2013	2014	2018
IoT gadgets employed Sensors	1st "Smart City Constructed in Dublin	IoMT is the future of healthcare and always will be.

2023 and running	2020
IoT enters the healthcare and becomes IoMT	Usage of IoMT increased during COVID-19 pandemic

FIGURE 8.1   Development of IoMT.

- *The perception layer*: It is often referred to as the sensing layer. This is the level at which all the sensors in an IoT-enabled system can detect and gather user data [21]. The perception layer is represented by various smart medical devices that collect health data. Through data-sensing acquisition procedures, it collects patient medical characteristics from sensors, actuators, edge servers, handheld devices, and medical sensors [22].
- *The processing layer*: It is provided by cloud middleware or IoT platforms in order to store and manage data. This layer also offers other crucial functions including handling, analyzing, and storing the gathered medical data [23].
- *The application layer*: It offers end users opportunities for data analytics, reporting, and device management using software solutions. The user may connect, manage, and display healthcare devices and data here [24].

8.4   DIFFERENT APPROACHES IN IoMT

IoMT is divided into various clusters as different approaches so that we can have a thorough grasp of its process flow. We can more clearly understand the interconnection and intricacy of the IoMT by grouping it into these clusters, which helps

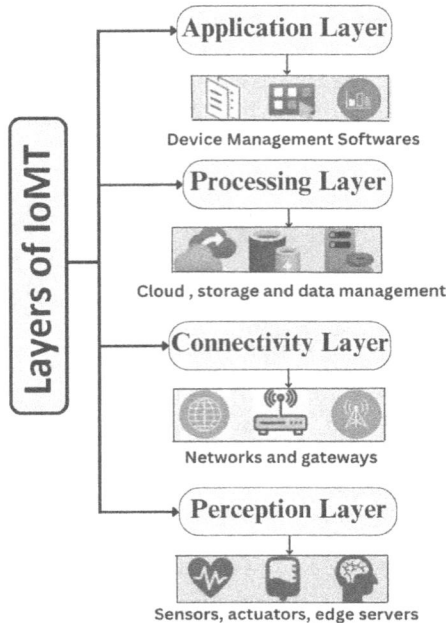

FIGURE 8.2   Architecture of IoMT.

us investigate its potential uses and consequences. A comprehensive understanding of the IoMT ecosystem is made possible by the distinct insights and considerations offered by each cluster, paving the path for improvements in patient outcomes and healthcare delivery.

- *Diagnostics*: The use of IoMT for patient monitoring and remote diagnosis enables clinicians to make quicker and more precise diagnoses. Blood glucose monitoring, sleep apnea diagnosis, and ECG monitoring are just a few of the uses for IoMT [25].
- *Chronic care*: Remote management of chronic conditions like diabetes, hypertension, and asthma can be done using IoMT. By utilizing IoMT, physicians can monitor patients in real-time, improving disease management and lowering readmission rates to the hospital [26].
- *Prophylactics*: IoMT can be used to provide preventative healthcare, such as advising patients on how to lead healthy lifestyles and engage in regular exercise. It is possible to track physical activity, sleep habits, and other health-related data with wearable devices like fitness trackers and smartwatches.
- *Recuperation*: IoMT can be utilized for post-recovery monitoring, including observing patients following surgery or hospital discharge. Through IoMT, doctors can keep an eye on patients from a distance, promoting quicker healing and lower medical expenses [24].

8.5 TYPES OF IoMT DEVICES

IoMT devices are categorized in different ways by different scientists, based on their usage and applications. Figure 8.3 shows the categorization which is purely based on proximity to the patient. They are as follows:

- *Implantable devices*: They are inserted into the patient's body to assist with medical procedures, such as a camera capsule or a cardiac sensor, and surgical procedures. The continuous monitoring of blood sugar levels eliminates the need for traditional glucose meters, and is one of the benefits offered for people with diabetes [27].
- *Wearable devices*: Necklaces, wristbands, and watches, for example, contain sensors that monitor the patient's vital signs. These devices contain various elements that include sophisticated tracking systems, state-of-the-art medical monitoring technology, and security alarms designed for personal safety, among others [26].
- *Bearable devices*: In order to satisfy the unique needs of patients, these are equipped with specialized computational capabilities. In daily life, these tools frequently help people access a variety of services. They include

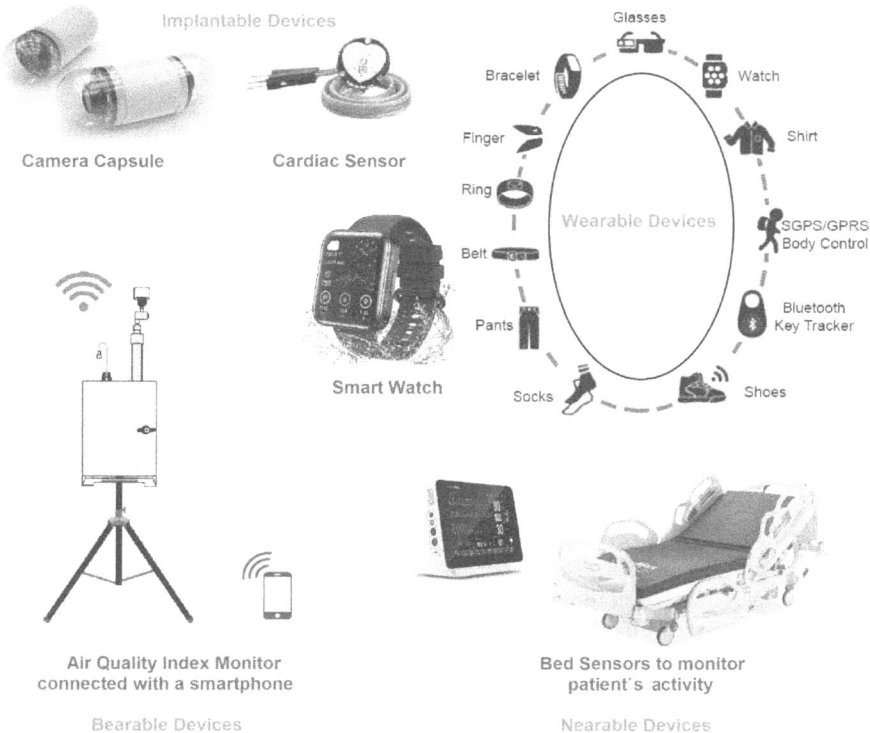

FIGURE 8.3 Types of IoMT devices.

computers that provide specific services to the patient, like an air quality monitor connected to a smartphone for asthma patients [24].

- *Nearable devices*: These are stationary and anchored to objects like doors or beds, use sensors to monitor the patient's activity and context, and can alert healthcare providers or family members in case of abnormal signs [28].

IoMT devices have also been categorized by D. Hemanth et al. [25] based on where they are used. Also, Amsaveni Avinashiappan and Bharathi Mayilsamy [29] added some other categories as follows:

- *On-body devices*: Public health wearables and surgical wearables are two broad categories of on-body devices. Professional fitness app developers, sports watch manufacturers, smart clothing companies, and wristband manufacturers are examples of public health wearables [30]. Although some of these products have not been given the green light by health authorities, they might be permitted based on objective scientific research and expert business analysis for specific safety applications. The use of wearables is anticipated to rise as they become more popular and are a "must-have" in the healthcare sector. Surgical wearables are normally certified for use by health agencies and involve approved items and network support. Any comparable tools are used as per professional advice or a doctor's prescription.
- *In-home devices*: The in-home gadgets include virtual patient management systems (VPMS), private emergency management systems (PEMS), and virtual telehealth services (VTS). PEMS helps older people who are housebound or have limited mobility become more independent by combining a wearable device/communication network with a live emergency contact center. With the aid of this equipment, users can easily connect and request emergency medical assistance.
- *Community devices*: The five different categories belonging to these devices are:
  - Mobility technologies enable passenger automobiles to monitor physiological variables while moving.
  - Medical staff in the first aid, ambulance, and emergency departments are supported by emergency management technology.
  - With the help of mobile touchscreens, kiosks in senior centers can offer the elderly products or services like communication, interactive puzzles, and health tips.
  - Medical care gadgets are tools that traditional healthcare facilities employ outside the house, for instance in a medical camp for underprivileged communities.
  - Logistics refers to the large movement and distribution of drugs, medical and surgical supplies, equipment for performing surgeries, and other supplies needed by healthcare experts.
- *In-clinic devices*: Devices used in clinics are instruments with both functional and therapeutic applications. The fundamental distinction between medical care devices in clinics and community devices is that when a

device in a clinic is operated by qualified personnel, the operator may have to use a centralized control system, rather than manually accessing the system.
- *In-hospital devices*: In-hospital IoMT-compatible devices fall into four different management and monitoring categories:
  - Product management keeps track of high-priced machinery and portable assets like wheelchairs and infusion pumps.
  - Personnel management keeps an eye on the productivity and success of the workers. By removing bottlenecks and enhancing patient care, patient flow management increases hospital efficiency.
  - Resource management makes it easier to buy, store, and use hospital supplies, consumables, medicines, and medical equipment to reduce production costs and improve worker productivity.
  - Energy and environmental monitoring, including temperature and humidity, keep an eye on electricity use and make sure that patient care areas and storage spaces, where medical supplies are kept, are kept in the best possible shape.

8.6 APPLICATIONS OF IoMT

We have already looked at a variety of typical applications for IoMT that are an excellent example of the important contribution that IoMT technology has made to healthcare. These applications have paved the way for revolutionary improvements in patient care, remote monitoring, telemedicine, and personalized healthcare services by using the power of networked medical devices, sensors, and data analytics. More specific primary applications that have emerged as a result of IoMT integration are briefly described in the following section, and are shown in Figure 8.4, with an emphasis on how they have improved patient satisfaction and healthcare results.

- A monitoring system for elderly people has been developed by Avik Ghose et al. His study offers a technique for the entire medical healthcare system to track the patient. The IoMT approach, which is a back-end platform, is used by the system [31].
- Cancer detection and diagnosis can be done by using IoMT. For example, an IoMT-based diagnostic system has been developed to identify early-stage breast cancer using artificial neural networks (ANN) and convolutional neural networks (CNN) with hyper-parameter optimization. Hyperparameters directly affect the training algorithms and model performance [32].
- Skin lesion detection and classification can be conducted using optimal segmentation and restricted Boltzmann machines (OS-RBM). The OS-RBM incorporates several phases: image acquisition, GF-based preprocessing, segmentation, feature extraction, and classification [33].
- For automatic seizure identification and management, a unified drug delivery system (iDDS) based on the IoMT has been proposed. Seizure detection and drug delivery units make up two components of iDDS. Once the target area has been identified, the drug is injected by a piezo-electrically

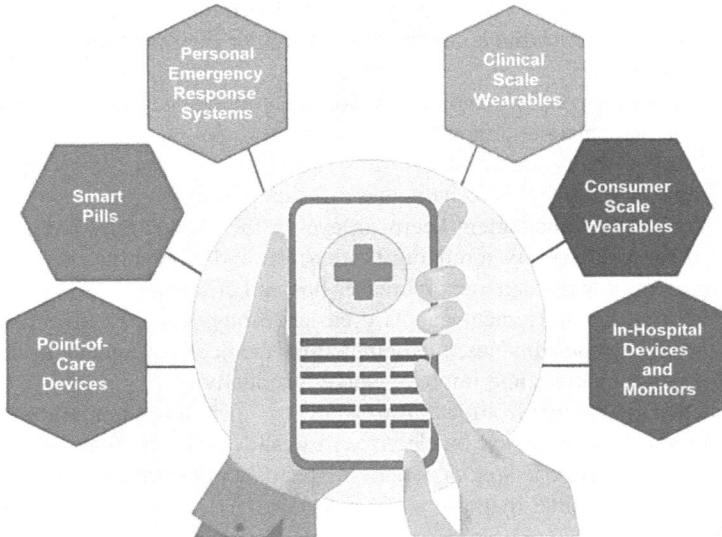

FIGURE 8.4 Different applications of IoMT.

activated valveless double reservoir micropump, which is also used for fault tolerance and improved medication control [34].
* Dr. Salah S. Al-Majeed and colleagues suggested developing a device for monitoring patients' physiological parameters that is essentially a medical sensing device, affordable, and IoMT based. Dr. Al-Majeed's study focuses mostly on synchronization and message transmission [35].

8.7   ADVANTAGES OF IoMT

IoMT offers a wide range of advantages that have completely changed the healthcare industry. IoMT technology has created new opportunities for better patient care, enhancing healthcare outcomes, and revolutionizing healthcare delivery by integrating medical devices, sensors, and data analytics seamlessly. In the present section, we'll discuss the wide range of benefits IoMT provides, including remote patient monitoring, real-time data analysis, improved efficiency, personalized treatment, and the potential for proactive interventions. These advantages demonstrate how IoMT has the power to fundamentally alter healthcare by creating a system that is more connected, effective, and patient-focused. IoMT has the following benefits:

Systems for monitoring patients: Patients with chronic diseases can have their health continuously monitored thanks to IoMT. It also allows doctors to have a better idea of the patient's home environment, which might affect the treatment [17].

Accessibility: Patients have easier access to healthcare resources and education due to IoMT. Using telehealth applications, patients can access more healthcare services whenever needed [17].

Cost management: The costs relating to an in-person visit to a medical facility are reduced in part via remote patient monitoring and telehealth. Faster processing of health data also saves medical professionals' time and money, allowing them to refocus their efforts on the more urgent problems.

Enhanced patient experience: IoMT makes it feasible to employ new-age technology that can ease pressure on outpatient self-service devices in healthcare facilities and can even decrease the necessity for in-person visits [18].

Accuracy: IoMT offers additional information, providing medical professionals with more precise insights into their patients' medical situations.

Logistics: In hospitals, IoMT devices are utilized to monitor the equipment and trigger alarms when maintenance or other problems need to be addressed. They are also employed as trackers to follow patients and medications across the campuses of medical facilities, resulting in fewer mix-ups and errors [18].

8.8 CURRENT STATUS OF IoMT

IoMT is a fast-evolving field of IoT applications in healthcare. IoMT is positioned to transform patient care, improve remote monitoring capabilities, and boost healthcare outcomes as a result of technological improvements and rising acceptance. In this section, we discuss the current status of IoMT in India, as well as in other countries.

8.8.1 INDIA

India has begun integrating IoMT into its hospitals and medical equipment thanks to the adaptability of the technology which enables it to be used in clinical and non-clinical situations.

The adoption of IoMT devices is anticipated to play an important role in determining the future of Indian healthcare, and the Indian government is actively supporting public–private partnerships to upgrade the nation's healthcare infrastructure. The potential benefits of IoMT for India's healthcare system are numerous and include improving healthcare access, especially in rural situations, quality, efficiency, and affordability. However, implementing IoMT technology poses challenges, including ensuring secure data transmission, accessibility for all individuals, and the high cost of infrastructure requirements [36].

8.8.2 GLOBAL

To enable healthcare organizations world-wide to achieve better results, increase efficiency, and provide patients with cutting-edge care, a crucial interaction is needed between IoMT and global MedTech businesses who are the main stakeholders in developing and supplying the IoMT devices after through research and development [37].

According to a McKinsey study, the healthcare sector is anticipated to experience the fastest growth in IoT deployments across all industries. The research also predicts that by 2025, IoMT will likely have a USD 1.6 trillion global economic impact [38]. This should accelerate the market expansion for Internet-connected medical devices. By 2025, the market for IoMT is anticipated to reach USD 322.2 billion, expanding by 29.9% from 2019 to 2020 [39].

Value-based healthcare is being introduced in numerous nations. Through many funding initiatives, the European Union promotes the transition to digital healthcare and a patient-centric strategy. Hospitalization at home and digital healthcare are two significant aspects of future healthcare. There are various reasons why this would be beneficial for all [40]. First, hospital beds are expensive and hard to come by, which often results in shortages. Second, the global population is living much longer – there will be twice as many elderly people on the planet in 2025 as there were in 2010. Third, it can be difficult for those who live in remote areas to receive healthcare routinely. According to the present demographic, there is an increasing number of people battling chronic disease conditions [41]. Between 2012 and 2024, it is anticipated that the United States national health expenditures (NHE) will have nearly doubled. The nation's economy may soon feel the impact of these circumstances, and a significant proportion of the population might not have access to healthcare [42, 43].

8.9   FUTURE TRENDS OF IoMT

The combination of IoT and industrial wireless sensor networks (WSNs) offers exciting prospects for industrial automation, revolutionizing procedures and activities across a variety of industries [44]. Similarly, the future for the subset of IoT, i.e., IoMT, looks promising, with the potential to transform the way healthcare is delivered. Dogra, Roopali & Rani (2020) suggested ways to identify forest fires in real time by integrating and using Wireless Sensor Network (WSN). Similarly, real-time remote data collection is a potential future application in smart healthcare, along with IoMT devices and technology [45].

IoMT with augmented reality (AR), virtual reality (VR), artificial intelligence (AI), blockchain, and sensors will be the next future of the medical world. IoMT can enable doctors and healthcare providers to remotely monitor and diagnose patients in real-time, leading to faster diagnosis and treatment. The use of IoMT can also help in reducing healthcare costs, by preventing hospital readmissions and optimizing healthcare services. Furthermore, integrating AI and machine learning (ML) into IoMT can enable more accurate diagnoses and personalized treatments for patients. IoMT devices can automate some medical processes and potentially help with early disease identification using AI-powered analysis. With the aid of these technologies, healthcare professionals can make data-driven decisions that will improve patient outcomes and raise the standard of care as a whole. Additionally, the IoMT ecosystem may evolve and adapt over time, boosting its diagnostic precision and treatment suggestions thanks to the continuous learning capabilities of AI and ML algorithms. Healthcare systems may open up new horizons in personalized medicine by merging the connectivity and data-gathering powers of IoMT devices with the computing power and intelligence of AI and ML. All these advancements will increase effectiveness, patient happiness, and ultimately save lives.

8.10   LIMITATIONS OF IoMT

IoMT has issues that must be resolved for it to be implemented successfully, just like any other technology. This section examines IoMT's drawbacks, such as issues with data security and privacy, interoperability problems between various devices and systems, potential rifts in patient-physician relationships, regulatory and compliance issues, and the digital divide that might prevent some groups from taking advantage of IoMT's benefits. Understanding these restrictions is essential for creating risk-reduction measures and ensuring that IoMT implementations prioritize patient safety, privacy, and equal access to healthcare technologies.

One major challenge that faces the implementation of IoMT is the lack of a robust security system to protect the sensitive health-related data collected and transmitted by these devices. This can leave the devices vulnerable to cyber-attacks and data breaches, compromising patient privacy and leading to serious consequences. In addition, there are concerns around privacy protection, with patients worried about who has access to their health data and how it is being used [19]. Another challenge is how to ensure adequate training and understanding of IoMT among healthcare professionals. While IoMT has the potential to improve patient outcomes and streamline healthcare delivery, it requires specialized knowledge and expertise to operate effectively. This can be a significant barrier, especially for healthcare professionals who may need more training or experience to work with IoMT devices and applications [46].

Regulations governing IoMT are constantly changing and differ by continent and nation. For example, the Food and Drug Administration (FDA) in the USA specifies the requirements for the businesses that create IoMT devices and services, as well as for pharmaceutical firms and other stakeholders, in terms of efficacy and safety. Challenges may arise when legacy systems and outdated devices, such as patient ventilators and infusion pumps, are connected to IoMT systems and new appliances. The IoMT infrastructure is distributed, therefore software platforms and hardware must be able to connect securely. However, security guidelines and protocols for these kinds of integration are constantly evolving [47].

Overall, addressing these challenges is crucial to ensure the successful implementation and adoption of IoMT in healthcare. This requires investment in developing secure and privacy-protective systems, providing adequate training and education to healthcare professionals, and raising patient awareness and understanding of the benefits and risks of IoMT. By doing so, we can unlock the full potential of IoMT to improve healthcare delivery and patient outcomes.

8.11   CONCLUSION

In conclusion, IoMT has the potential to transform completely the way that healthcare is provided. Further developments in IoMT hardware, communication, and data analytics are to be expected as technology develops. Health outcomes will be enhanced, thanks to remote patient monitoring, quicker and more precise diagnoses, individualized treatment plans, and IoMT integration with healthcare systems. The IoMT ecosystem will continue developing, resulting in a seamless and integrated healthcare environment as wearable technology, smart sensors, and AI are widely adopted. To guarantee confidence in, and confidentiality of, patient data, it is crucial to solve the security and privacy issues related to IoMT.

REFERENCES

1. Abawajy, J. H., & Hassan, M. M. (2017). Federated Internet of Things and cloud computing pervasive patient health monitoring system. IEEE Communications Magazine, 55(1), 48–53.
2. Sangaiah, A. K., Dhanaraj, J. S. A., Mohandas, P., & Castiglione, A. (2020). Cognitive IoT system with intelligence techniques in a sustainable computing environment. Computer Communications, 154, 347.
3. Rafque, W., Qi, L., Yaqoob, I., Imran, M., Rasool, R. U., & Dou, W. (2020). Complementing IoT services through software-defined networking and edge computing: A comprehensive survey. IEEE Communications Surveys Tutorials, 22, 1–45.
4. Abomhara, M., & Koien, G. M. (2014). Security and privacy in the Internet of Things: Current status and open issues. International Conference on Privacy and Security in Mobile Systems, IEEE, 1–8.
5. Badotra, S., Nagpal, D., Panda, S. N., Tanwar, S., & Bajaj, S. (2020). IoT-Enabled Healthcare Network with SDN. 2020 8th International Conference on Reliability, Infocom Technologies and Optimization (Trends and Future Directions) (ICRITO), IEEE, 38–42.
6. Chen, S., Xu, D., Liu, B., & Wang, H. (2014). A vision of IoT: Applications, challenges, and opportunities with China perspective. IEEE Internet of Things Journal, 1(4), 349–359.
7. Bharathi, K. S., & Venkateswari, R. (2019). Security challenges and solutions for wireless body area networks. In Computing, Communication, and Signal Processing (pp. 275–283). Springer: Singapore.
8. Syed, L., Jabeen, S., Manimala, S., & Alsaeedi, A. (2019). Smart healthcare framework for an ambient assisted living using IOMT and big data analytics techniques. Future Generation Computer Systems, 101, 136–151.
9. Xiong, N., Vasilakos, A. V., Yang, L. T., Song, L., Pan, Y., & Kannan, R. (2009). Comparative analysis of quality of service and memory usage for adaptive failure detectors in healthcare systems. IEEE Journal on Selected Areas in Communications, 27(4), 495–509.
10. Mamdouh, M., Awad, A. I., Khalaf, A. A., & Hamed, H. F. (2021). Authentication and identity management of IoHT devices: Achievements, challenges, and future directions. Computers & Security, 111, 102491.
11. Gupta, J., & Singh, R. (2018). Internet of Things (IoT) and academic libraries a user-friendly facilitator for patrons. 2018 5th International Symposium on Emerging Trends and Technologies in Libraries and Information Services (ETTLIS), presented at the 2018 5th International Symposium on Emerging Trends and Technologies in Libraries and Information Services (ETTLIS), IEEE, Noida, pp. 71–74.
12. Alsubaei, F., Abuhussein, A., Shandilya, V., & Shiva, S. (2019). IoMT-SAF: Internet of medical things security assessment framework. Internet Things, 8, 100123.
13. Timeline of Internet of Medical Things. (2023) Available online: https://medicaliomt.com/timeline-of-internet-of-medical-things/ (as accessed on 05 May 2023).
14. Castañeda, L., Jain, P., & Müller, H. A. (2013). The Future of Internet Applications: A Survey of Future Internet Projects. University of Victoria: Victoria, British Columbia, Canada.

15. Novotný, R., Kuchta, R., & Kadlec, J. (2014). Smart City concept, applications and services. Journal of Telecommunications System and Management, 3, 117.
16. Barnawi, A., Chhikara, P., Tekchandani, R., Kumar, N., & Alzahrani, B. (2021). Artificial intelligence-enabled Internet of Things-based system for COVID-19 screening using aerial thermal imaging. Future Generation Computer Systems, 124, 119–132.
17. Lutkevich, B., & DelVecchio, A. (2023). Internet of Medical Things (IoMT) or health-care IoT. Online Available: https://www.techtarget.com/iotagenda/definition/IoMT-Internet-of-Medical-Things (accessed on 05 May 2023).
18. The Internet of Medical Things (IoMT): the connected future of Healthcare (January 2022). Online Available: https://blog.richardvanhooijdonk.com/en/the-internet-of-medical-things-iomt-the-connected-future-of-healthcare (accessed on 05 May 2023).
19. Alaba, F. A., Othman, M., Hashem, I., & Alotaibi, A. T. (2017). Internet of Things security: A survey. Journal of Network and Computer Applications, 88, 10–28.
20. Belkhouja, T., Sorour, S., & Hefeida, M. S. (2019). Role-based hierarchical medical data encryption for implantable medical devices. in 2019 IEEE Global Communications Conference (GLOBECOM), IEEE, pp. 1–6.
21. Parmar, M., Gupta, L., & Dutta, B. R. (2022). Storage optimized secured data transaction of IoT sensors. 2022 IEEE 2nd International Conference on Mobile Networks and Wireless Communications (ICMNWC), Tumkur, Karnataka, India, IEEE, pp. 1–6, doi: 10.1109/ICMNWC56175.2022.10031866
22. Timeline of Internet of Medical Things and Changes It Brings to Healthcare. (2021). Available online: https://www.altexsoft.com/blog/internet-of-medical-things/ (accessed on 05 May 2023).
23. Alsaeed, N., & Nadeem, F. (2022). Authentication in the internet of medical things: Taxonomy, review, and open issues. Applied Science, 12(15), 7487.
24. Alsaeed, N. I., & Aldahwan, N. S. (2020). Ubiquitous health care monitoring services (UHCMS): Review of opportunities and challenges. International Journal of Computer Application, 975, 8887.
25. Hemanth, J. A. D. J., & George, A. (2021). Internet of Medical Things: Remote Healthcare Systems and Applications; Springer: Berlin/Heidelberg, Germany.
26. Liyanage, M., Braeken, A., Kumar, P., & Ylianttila, M. (2020). IoT Security: Advances in Authentication; John Wiley and Sons: Hoboken, NJ, USA.
27. Bhatia, H., Panda, S. N., & Nagpal, D. (2020). Internet of Things and its applications in healthcare-a survey. 2020 8th International Conference on Reliability, Infocom Technologies and Optimization (Trends and Future Directions) (ICRITO), IEEE, pp. 305–310.
28. Alsubaei, F., Abuhussein, A., & Shiva, S. (2017). Security and privacy in the internet of medical things: taxonomy and risk assessment. In Proceedings of the 2017 IEEE 42nd Conference on Local Computer Networks Workshops (LCN Workshops), Singapore, IEEE, pp. 112–120.
29. Bharathi, M., & Amsaveni, A. (2021). Machine learning with IoMT: Opportunities and research challenges. In Internet of Medical Things: Remote Healthcare Systems and Applications (pp. 235–252). doi: 10.1007/978-3-030-63937-2_13.
30. Yaacoub, J. P. A., Noura, H. N., Noura, O., Salman, E., Yaacoub, R., Couturier, & Chehab, A. (2020). Securing the Internet of Medical Things systems: Limitations, issues, and recommendations. Future Generation Computer Systems, 105, 581–606.
31. Ghose, A., Sinha, P., Bhaumik, C., Sinha, A., Agrawal, A., & Choudhury, A. D. (2013). UbiHeld: Ubiquitous healthcare monitoring system for elderly and chronic patients. In Proceedings of the ACM Conference on Pervasive and Ubiquitous Computing Adjunct Publication, Zurich, Switzerland, pp. 1255–1264.

32. *Breast Cancer Histopathological Database (BreakHis)*. (2020) Available online: https://www.kaggle.com/ambarish/breakhis (accessed on 05 May 2023).

33. Peter Soosai Anandaraj, A., Gomathy, V., Amali Angel Punitha, A., Abitha Kumari, D., Sheeba Rani, S., Sureshkumar, S., et. al. (2020). Internet of medical things (IoMT) enabled skin lesion detection and classification using optimal segmentation and restricted Boltzmann machines. Cognitive Internet of Medical Things for Smart Healthcare, SSDC, vol. 311, Springer: Cham.

34. Sayeed, M. A., Mohanty, S. P., Kougianos, E., & Zaveri, H. (2020). iDDS: An edge-device in IoMT for automatic seizure control using on-time drug delivery. IEEE International Conference on Consumer Electronics (ICCE), IEEE, pp. 1–6..

35. Al-Majeed, S. S., Al-Mejibli, I. S., & Karam, J. (2015). Home Telehealth by Internet of Things (IoT). In Proceedings of the Canadian Conference on Electrical and Computers. Online Available: www.fortunebusinessinsights.com/industry-reports/internet-of-medical-things-iomt-market-101844 (accessed on 05 May 2023).

36. Singh, R. K. (February 2023). The potential of IoMT in Indian healthcare. Times of India, Online Available: https://timesofindia.indiatimes.com/blogs/voices/the-potential-of-iomt-in-indian-healthcare/ (accessed on 05 May 2023).

37. Internet of Things (IoT) in Healthcare Market is Expected to Grow at a CAGR of 29.9% to Reach $322.2 billion by 2025: Meticulous Research. (2020). Available online: https://www.globenewswire.com/news-release/2020/03/19/2003195/0/en/Internet-of-Things-IoT-in-Healthcare-Market-is-Expected-to-Grow-at-a-CAGR-of-29-9-to-Reach-322-2-billion-by-2025-Meticulous-Research.html (accessed on 05 May 2023).

38. Remuzzi, A., & Remuzzi, G. (2020). COVID-19 and Italy: What next?. Lancet, 395, 1225–1228. doi: 10.1016/s0140-6736(20)30627-9

39. Megari, K. (2013). Quality of life in chronic disease patients. Health Psychology Research, 1, 27. doi: 10.4081/hpr.2013.e27

40. Jost, T. S. (2006). Our broken health care system and how to fix it: An essay on health law and policy. Wake For. L. Rev., 41, 537. Available online: https://heinonline.org/HOL/Page?handle=hein.journals/wflr41&id=547&div=&collection= (accessed on 05 May 2023).

41. Gouda, K., & Okamoto, R. (2012). Current status of and factors associated with social isolation in the elderly living in a rapidly aging housing estate community. Environmental Health and Preventive Medicine, 17, 500–511. doi: 10.1007/s12199-012-0282-x

42. Keehan, S. P., Cuckler, G. A., Sisko, A. M., Madison, A. J., Smith, S. D., Stone, D. A., Poisal, J. A., Wolfe, C. J., & Lizonitz, J. M. (2015). National health expenditure projections, 2014–2024: Spending growth faster than recent trends. Health Affairs, 34, 1407–1417. doi: 10.1377/hlthaff.2015.0600

43. European Commission (2017). Uptake of Digital Solutions in the Healthcare Industry. Available online: https://monitor-industrial-ecosystems.ec.europa.eu/industrial-eco systems/health (accessed on 05 May 2023).

44. Zhang, X., Rane, K. P., Kakaravada, I., & Shabaz, M. (2021). Research on vibration monitoring and fault diagnosis of rotating machinery based on internet of things technology. Nonlinear Engineering, 10(1), 245–254.

45. Dogra, R., Rani, S., & Sharma, B. (2020). A Review to Forest Fires and Its Detection Techniques Using Wireless Sensor Network. doi: 10.1007/978-981-15-5341-7_101

46. Zaldivar, D., Tawalbeh, L., & Muheidat, F. (6 January 2020). Investigating the security threats on networked medical devices. In Proceedings of the 2020 10th Annual Computing and Communication Workshop and Conference (CCWC), Las Vegas, NV, USA, IEEE, pp. 0488–0493.

47. The Internet of Medical Things (IoMT): The connected future of healthcare (2022). Available online: https://blog.richardvanhooijdonk.com/en/the-internet-of-medical-things-iomt-the-connected-future-of-healthcare/ (accessed on 05 May 2023).

9 Deep Learning and Its Applications in Healthcare

Shubham Gupta, Pooja Sharma, and Thippa Reddy

9.1 INTRODUCTION

Every global citizen has a right to a good quality of life, and good health is a crucial component of that quality. Globally, doctors are working hard to keep up with the most recent research and refresh their knowledge base. When treating patients, they may have the necessary training and expertise but because of the nature of the human brain, individuals can't memorize all the information needed for every circumstance, and therefore a diagnosis is probably not always accurate. Additionally, some subject areas are lacking in medical specialists. However, it is a social responsibility to give medical care to patients effectively while taking into account the limited healthcare resources [1].

Currently, the growing accessibility of data is transforming the healthcare sector. A good example is the electronic health record (EHR) consisting of data collected during patient visits to a doctor or to hospital throughout their lifetime [2]. These are patient-centered, real-time records that make data instantly and securely accessible to authorized users. In recent years, these have been applied to academic research domains. As with the financial and retail services sectors in the previous ten years, the development of such a sizable volume of EHR data has made it a valued source for medical data analysis. Due to improvements in automated data processing and machine learning(ML) techniques, researchers are now able to carefully examine the massive digital footprints that were once overlooked.

Health is a multifaceted phenomenon. Economic research has shown that there are internal as well as external elements that affect health, including the community and the environment in which a person lives. The health of the community is directly influenced by a number of variables, including socioeconomic level, education, income, healthcare infrastructure, per capita health spending, environment, way of life, housing, sanitation, water supply, relationship status, prevalence of poverty, and religion. Just as socioeconomic factors affect health status, so do political and social problems in society, so it is vital that all of these elements work together to improve a population's health status [3].

A great deal of empirical study has been undertaken to examine how a health system is connected. First, early literature has noted the significance of health, with regard to human capital, in promoting economic progress. Income and health are

DOI: 10.1201/9781003437079-9

directly correlated. It has been made clear by Bloom [4] that health capital has a favorable effect on total economic output. Bloom came to the conclusion that the accumulation of health capital accounted for around one-quarter of growth in the economy, and that an increase in health is connected with an increase in annual economic growth of up to 4%. According to Kamiya [5], the gross domestic product (GDP) and having access to better sanitation are statistically significant in having a beneficial influence on lowering child mortality.

Second, public well-being and reading have an unbreakable connection, with higher levels of literacy resulting in better public health. Longevity and decreased mortality are more tied to public health than to individual disease therapy. Ghosh [6] looked at how education and standards of living play a vital role in accessing maternal health facilities in rural and urban regions. On the other hand, religion, caste and work status of a mother have a lesser impact on securing maternal health facilities.

Third, Wagstaff and Cleason [7] confirmed that good governance of public expenditure contributes to the reduction in child mortality rate. Enhanced public health spending increases reach out to the poor, yet does not rule out the use of private facilities. It reflects the child mortality rate, the favorable impact of public health expenditures on outpatient care utilization for the poorest 2 quintiles. Various impact assessments revealed a negative relationship between lower government health expenditure and poor health outcomes as a result of insufficient health care delivery (Global Monitoring Report).

Enhancing health status is currently one of the primary goals and the foundation for sustaining and stimulating a country's optimal level of financial efficiency and growth.

9.2 HEALTHCARE MONITORING

The need for more advanced healthcare monitoring is driven by the rising need for clinicians and family members to remotely track patients' health problems. Healthcare monitoring systems make it possible to track patients' vital body processes in real-time while they are living their daily lives (at home, at work, participating in sports, etc.) or while they are receiving medical treatment, all without interfering with their regular routines. According to a research, healthcare monitoring is intended to oversee the care of patients with a range of problems, such as cardiac disorders, hypo- and hypertension, diabetes, hyper- and hypothermia, etc.

The overall structure of the healthcare system, which uses a wireless on-body sensor network to monitor the patient, is shown in Figure 9.1. Any time there is an emergency, the information is sent via the network to the person's family, doctor, and emergency services (e.g., an ambulance).

The sensors built into the wearable device are used in the patient monitoring (data telemetry) healthcare system to track physiological data (temperature, blood pressure, glucose level), as well as vital signs (including breathing, heartbeat, etc.). An antenna operating at a specific frequency (often in the MHz range) provides wireless power to the wearable device. High-frequency (in the GHz range) wireless data transmission between the wearable and the external medical devices has been carried out. The wireless transmission device operates generally in a low-power standby

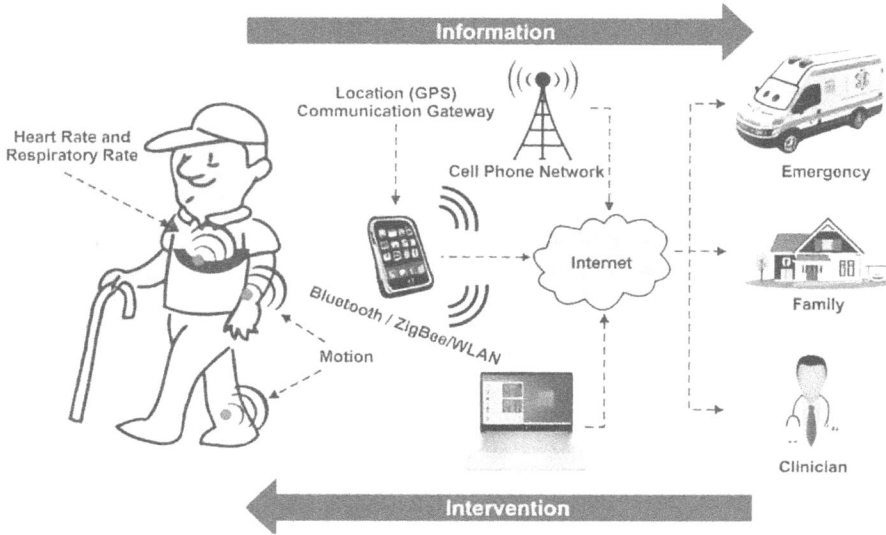

FIGURE 9.1 General structure of a healthcare network [8].

mode, but it can be activated by an outside signal to operate at full power at the frequency band approved for medical devices and transfer data. The information is sent to the healthcare professional at the medical center so they can assess the patient's condition. When abnormalities are found, the doctor is notified right away so that the appropriate steps may be taken as needed.

With international organizations investing heavily in research and development (R&D), the electronic and radio frequency (RF) components of healthcare monitoring systems that are inserted into a patient's body to support them on a temporary or ongoing basis are also witnessing substantial development. The RF antenna must communicate with all of the device's numerous parts.

Configuring minimal and low-profile antennas for a healthcare system is a huge task. There are two types of antennas used in this type of system: in-building or mobile antennas. The healthcare monitoring system, with the different types of antennas, is presented in Figure 9.2.

The majority of hospitals have extremely dense wall construction, which makes them difficult places for wireless transmissions. Additionally, the staff make use of a range of clinical apps for smartphones, which all demand a strong Wi-Fi network. Hospital designers have realized that deploying a multi-carrier DAS (distributed antenna system) is the only practical option to provide the wireless environment needed by today's medical staff. Figure 9.2 depicts patient monitoring in a hospital setting. Local practitioners continually track the patients through an in-building antenna system.

Real-time monitoring is necessary for elderly or patient care and can be done without interfering with their daily routines. Therefore, medical equipment that includes antenna and sensor components should be integrated into their daily lives.

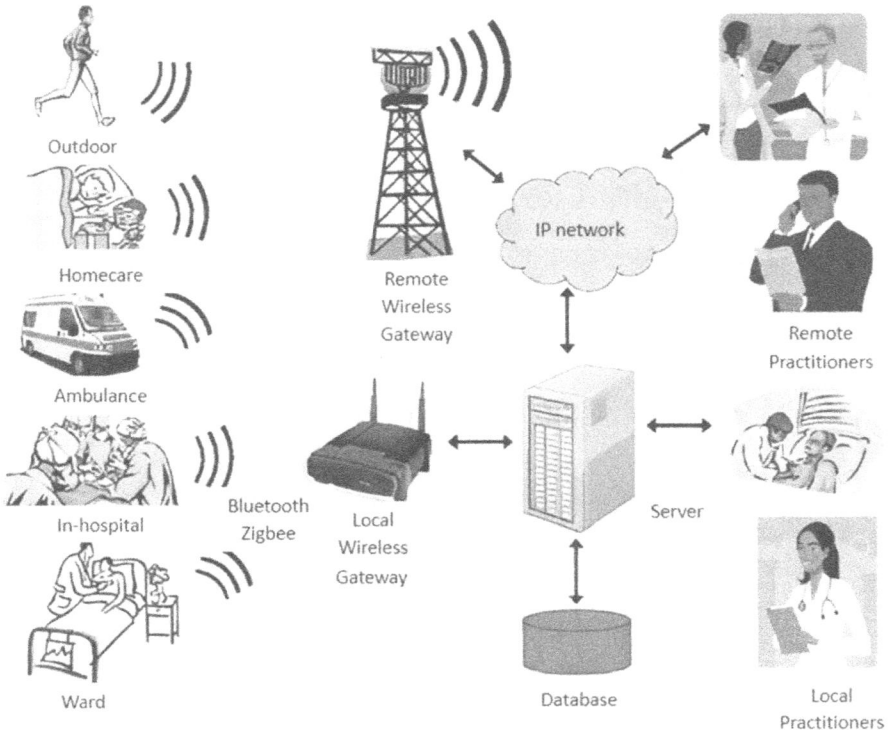

FIGURE 9.2   Healthcare monitoring system (using both in-building and mobile antennas) [9].

A practical approach that makes use of wearable antennas can be incorporated into clothing and utilized to remotely broadcast and receive sensor data from both outside and within the home or other setting, as shown in Figure 9.2. In this way, the mobile antenna system makes it simple for the remote practitioners (linked to a hospital computer) to keep an eye on the patients.

9.2.1   Scope of Indoor Patient Monitoring

In a hospital intensive care unit (ICU), health indicators need to be monitored continuously, coupled with one-on-one nursing care, to guarantee patients' safety. In conjunction with the attending physician, the ICU team will intervene right away in an emergency and report on the diagnosis and any further treatment needed.

After being moved from the ICU to a private or general ward, there is no option for continuous surveillance of the patient [10]. Since nurses examine patients at irregular intervals, some evident issues can be overlooked. The failure to continuously monitor patients is mostly to blame for 50% of patient mortality in hospitals. Patients who are weak and unable to receive close monitoring outside of the ICU are at a high risk of dying.

Recently, hospitals have seen a significant increase in the number of patients with complex health conditions. When there are two or more concurrent medical conditions, this might lead to unidentified complications and extremely volatile risk factors. Complex medications contra-indicated with other drugs also present significant hurdles. Therefore, health decline can easily go unreported where patient monitoring cannot be done continuously.

In the past, nurses did the ward rounds and manually monitored vital signs to keep records up to date. Currently, patients are checked on average every four to eight hours. In an Indian setting, the time gap between each visit by a nurse might be significantly longer. Individuals exhibit indicators of deterioration within 6 to 12 hours after any unfavorable incident. Non-continuous tracking leads to symptoms being missed, and treatment may only begin after vital signs have reached dangerous levels.

The medical industry acknowledges flaws in the current patient monitoring system and would ideally like to implement a "practical and affordable" real-time monitoring solution. This technology will soon be a major element of how patients are cared for, and it holds immense promise.

The difference between a patient living or dying might come down to the difference between late diagnosis or prompt life-saving treatment, which all depends on the efficacy of the patient monitoring system in place [11]. A hospital ward cannot replicate an ICU setup. Instead, a low-cost, accurate monitoring device can become a vital component of a hospital's non-ICU setup. Installing a patient monitoring system may offer the hospital a good return on their investment as the technology is expensive to implement. It is obvious that any hospital's reputation depends on its ability to save lives by putting in place a continuous monitoring network. Consequently, systems for healthcare monitoring are in high demand in the hospital setting.

9.2.2 HEALTHCARE MONITORING SYSTEM IN THE HOSPITAL SETTING

Utilizing an indoor tracking and location system, full vital-sign monitoring of a patient may be ensured without losing information and extra setup delays. As a result, there is more patient and worker satisfaction, as well as an improvement in safety. A monitoring system ensures the highest level of security in hospitals and allows severely ill patients to roam around without restriction. Telemetry gadgets are placed on a patient's clothing using RF system components so that the precise location of each device and so the patient's position may be determined immediately by a wall-mounted tracking system [12].

Furthermore, a patient's vital statistics (heart rate, blood pressure, and SpO2) may be read on a device via an application or directly. When a patient's blood pressure or saturation is too low, an alert will notify medical staff. If the patient is not in their original ward, the nearest nurse will be notified. This allows hospital workers to access information and determine patient location and vital signs.

Figure 9.3 depicts a patient monitoring and tracking system. The telemetry systems worn by the patients are recorded by a wall-mounted antenna system that is evenly dispersed throughout the hospital. Bluetooth, Wi-Fi, Zigbee, and ultra-wide band (UWB) are wireless systems that are used for short-range or interior interactions.

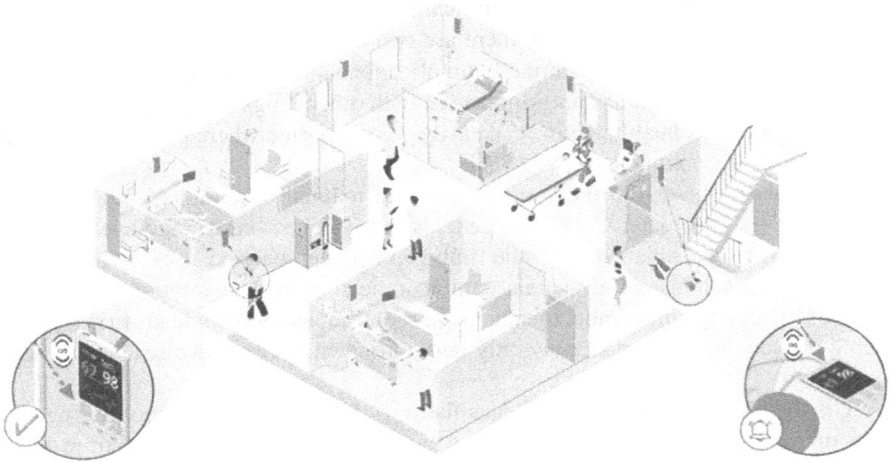

FIGURE 9.3   Patient monitoring and tracking in a hospital ward [12].

Out of all these technologies, UWB is preferred for applications in healthcare due to its unique characteristics, such as a higher bandwidth, lower power consumption, and lower design complexity. Low electromagnetic radiation is caused by the low radio power pulse of less than 41.3 dBm in the indoor environment. Low irradiation is appropriate for hospital uses and has no environmental impact. Furthermore, even at a close range, the low radiation is safer for the human body.

The scope of the wall-mounted monitoring equipment is determined by the antenna length. As a result, the layout of the antennas for an interior healthcare monitoring system has significant implications for the installation of the system.

The Recommender System (RS) in Figure 9.4 focuses mostly on the customized interactions of a single user. It is in charge of dynamically inferring user/person needs from the interaction between users and products. Also, installed in the proper platform it can identify user requirements using a list of keywords. However, the RS may also serve as a recommendation engines rather than relying on particular requirements based on its input capabilities. The intermediate stage of the RS idea began growing in the field in the 1990s, when the framework rating was emphasized for researchers. Numerous academics have examined the origins of RS, which determine predicting ideas, cognitive science, and information recovery.

9.3   APPLICATIONS OF RECOMMENDER SYSTEMS

a. *Healthcare services*: The Recommender System (RS) (Figure 9.4) may be used to engage with patients or expert health opinions to deliver health recommendations when the opinion is supported by consultants or medical professionals. The RS is able to provide the best option to doctors and patients who have expressed a desire for an automatic referral based on past consultant history.

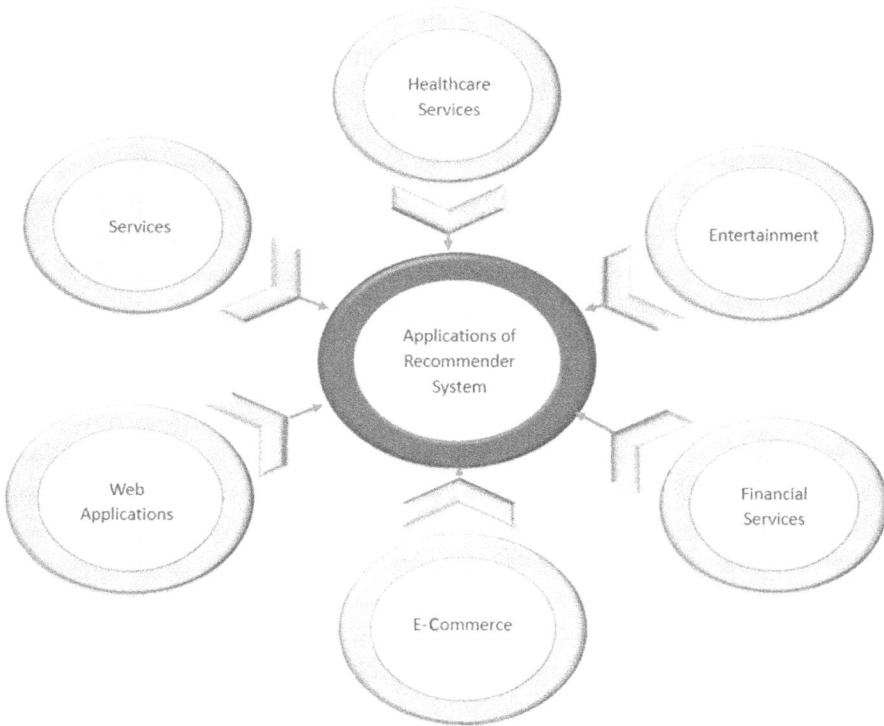

FIGURE 9.4   Applications of a Recommender System [13].

   b. *E-commerce*: According to the user's or client's prior health problem history or user profile record, e-commerce makes a recommendation to them. For example, on Amazon, the correlation among the often bought or liked products is going to be shown as a preferred choice for recommendation.

   c. *Online services/Web applications*: Only the unique user is given meaningful advice via online services. Additionally, viewers are offered recommendations based on trending material or historical data. The most popular items are email filters, media, and newspapers.

   d. *Entertainment and financial services*: Only the unique user is given meaningful advice via online services. Additionally, viewers of this project are offered recommendations based on trending material or historical data. The most popular items are email filters, media, and newspapers.

9.4   ARCHITECTURE OF THE HEALTH RECOMMENDATION SYSTEM

Health recommender systems (HRSs) offer the potential to motivate and engage users to change their behavior by sharing better choices and actionable knowledge based on observed user behavior. Data mining and analytics have increased due to big data's quick expansion in a number of industries, including healthcare,

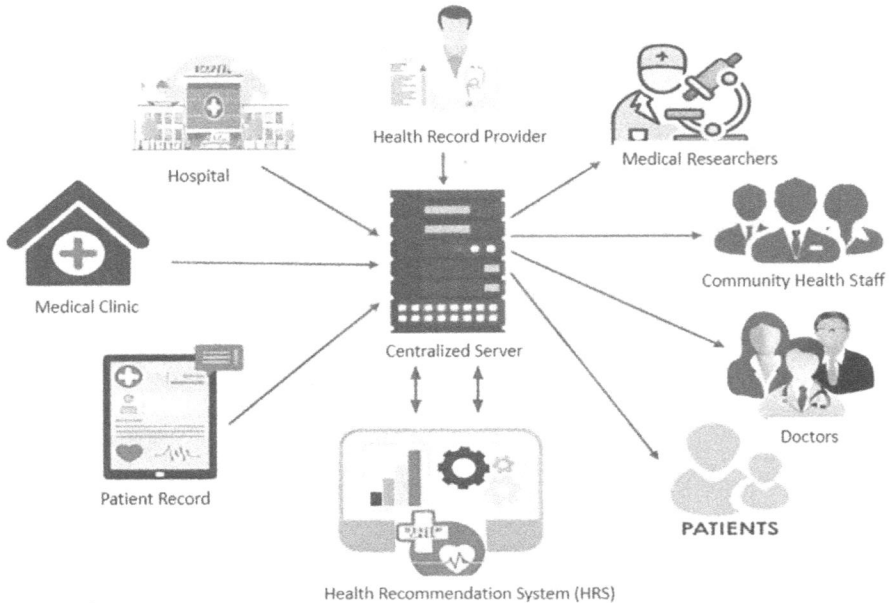

FIGURE 9.5 Overall structure of the Health Recommendation System (HRS) [14].

e-commerce, trading, and others. Big data analytics is a growing industry that has become recognized as an integral diagnostic service in medical facilities. Big data in the healthcare industry may be classified according to three characteristics: diversity, volume, and velocity [14]. Diversity refers to the fact that data may come from a wide range of sources, both internal and external, and that the data may take diverse forms. Figure 9.5 depicts the overall layout of the Health Recommendation System (HRS).

9.4.1 Categories of e-Health in HRS

In a patient care setting, patients can view and transmit their health information via an electronic program known as e-health. The three categories of e-health are described as follows:

   a. *Electronic health record (EHR)*: This is a type of electronic record that includes data about a single person's health. It is accountable for keeping track of medical professionals' and clinicians' data according to the healthcare system they use. Additionally, it keeps track of medical records and distributes electronic access to data. The following are some of the fundamental EHR characteristics and capabilities:
      i. Facilitate interaction by using the following three steps—identify, support, and create. Initial safe electronic communication among patients and clinicians is supported by the providers. Then EHR can locate

pertinent resources for patients, families, and carers. Finally, it can generate and record precise instructions.

ii. Organize and collect patient data, such as medical history, demographics, external clinical papers, clinical notes, issue lists, and test and laboratory results.

iii. Manage or gather patient care plans, offer appropriate guidelines and protocols, and support clinical decision-making.

b. *Personal health record* (*PHR*): A PHR is a database of health-related data from specific patients. It is in charge of providing patients and customers with legitimate, secure, and private data manipulation, access, and transmission. According to the International Organization for Standardization (ISO), there are four different types of PHR:

i. Web service providers handle and look after individual EHRs.

ii. Individually composed EHR, maintained and used by users/patients.

iii. Integrated element with EHR, solely managed by the patient.

iv. Integrated component with EHR, administered by the health provider [15].

c. *Mobile health* (*MHealth*): MHealth is a new mobile communication and network topology intended specifically for medical facilities. Rapid patient assistance with health information is crucial for reducing extra costs by preventing needless hospital visits. With the aid of specialists or medical professionals, the patient uses a monitoring gadget to track their symptoms in real-time for speedy diagnosis. Figure 9.6 illustrates the flow of information in the MHealth system.

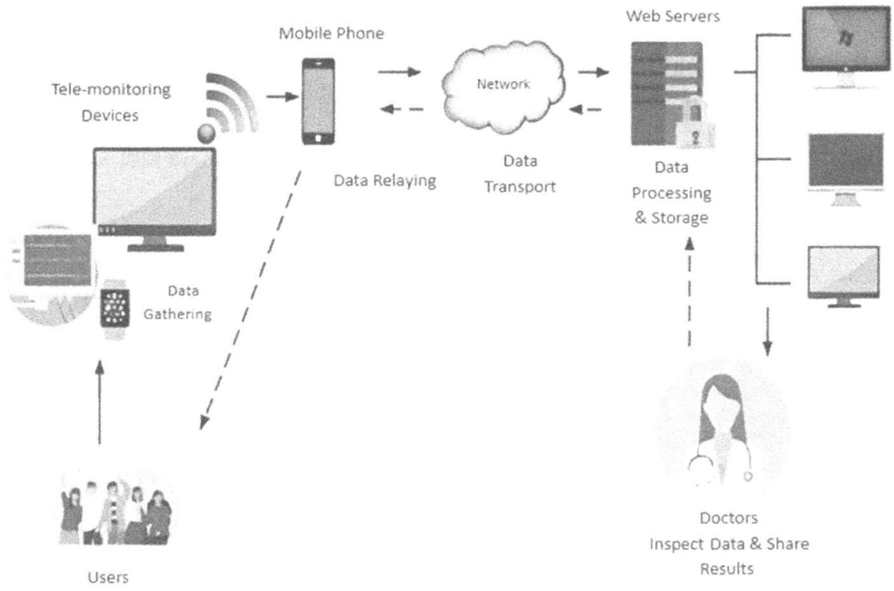

FIGURE 9.6   The flow of information in MHealth [15].

9.5 DEEP LEARNING FRAMEWORK

A general-purpose artificial intelligence (AI) technique called machine learning (ML) can infer associations from data without first defining them [16]. The key selling point is the capacity to generate models of prediction without making firm beliefs about the underlying methods, which are frequently unidentified or inadequately characterized. Data harmonization, representation learning, model fitting, and assessment make up the usual ML pipeline. Conventional procedures are constrained in their ability to deal with information from nature in their raw form since they only involve one, frequently linear, alteration of the input space [17].

In contrast to conventional ML, DL learns and models from the raw information. In reality, DL makes it possible for computational models with several neural-network-based processing phases to educate specifications of information with various degrees of abstraction [18]. The amount of human learning (HL), their connections, as well as DL's capacity to educate important abstractions of the inputs are the main distinctions between DL and conventional artificial neural networks (ANNs).

However, conventional ANNs are often limited to three stages and are trained to provide supervised models that are tailored to a certain task and are not transferable to other contexts. Each stage of a DL device uses the information it obtains as signals from the stage below to generate a representation of the observed trends by optimizing local unsupervised criteria [19]. The main feature of DL is that, rather than being designed through human engineers, it features stages that are learned automatically from data. Figure 9.7 provides a high-level illustration of these variations.

The architecture of DL is made up of multiple layers of neural networks, as opposed to ANNs, which typically consist of three stages and one conversion toward the final outcomes. Layer-wise unsupervised pre-training makes it possible to effectively tune deep networks and extract complex structures from inputs to utilize as higher-level functions to get suitable estimations [16].

Object identification in images, audio recognition, language translation, and natural language comprehension have all seen remarkable improvements because of DL's ability to effectively find complex patterns in high-dimensional data [20]. Clinically

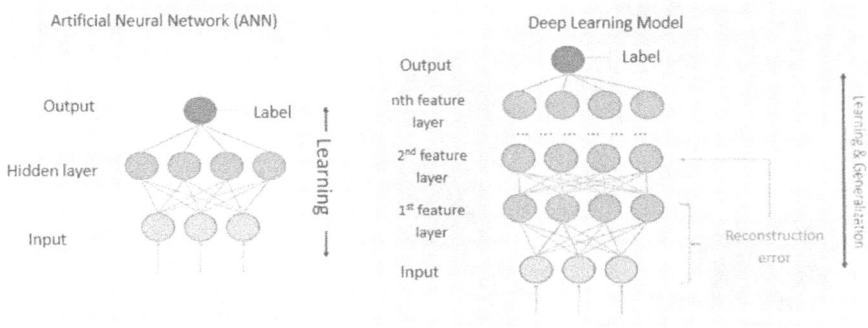

FIGURE 9.7 Artificial neural network (ANN) and deep learning (DL) topologies are compared.

relevant successes have also been achieved in healthcare (e.g., skin cancer classification, prediction of the sequence specificity of DNA- and RNA-binding proteins, and identification of diabetic retinopathy in retinal fundus photographs [21]), which may pave the way for a new creation of smart origins, depend on DL for application in biological practice.

After Bengio reviewed and compiled the well-known algorithms for DL in 2006, DL has become a topic of interest in ML. Several applications, including image retrieval, web search, natural language processing (NLP), healthcare, and many others, have successfully used DL [22].

9.5.1 ROLE OF DEEP LEARNING IN HEALTHCARE

During the COVID-19 pandemic, practical applications of a variety of deep learning (DL) techniques were needed. DL has demonstrated its value to governments all around the world in the field of healthcare. DL algorithms can classify or cluster data, perform image processing, or forecast when things will return to normal.

Based at the Coronavirus Research Center at the Johns Hopkins University [23], more than 213 million patients were confirmed to have COVID-19 in 2023, with 4.4 million dying as a result. This global pandemic was and still is a grave danger to human civilization. DL was not recognized as a viable method for health informatics prior to COVID-19 because of the vast amount of training information and computer resources required, in comparison to other methods that require similar efforts and adjustments. DL approaches were consistently criticized for their lack of accessibility. COVID-19, on the other hand, has formed a demand for rigorous study in ways to identify the optimal classification tools. To track the development of ML and DL in forecasting, identifying, and diagnosing COVID-19, search code techniques were used. Convolutional neural network (CNN), deep neural network (DNN), and support vector machine (SVM) algorithms have demonstrated detection accuracies of up to 99%.

DL was utilized in the medical area or healthcare to identify and differentiate between COVID-19 or healthy chest X-rays in patients utilizing image-processing abilities. For detecting COVID-19 in chest X-ray (CX-R) pictures, the COVID-DeepNet approach has been developed [24]. This method assists radiologists with experience in quickly and accurately understanding pictures. The outcomes of two distinct strategies rely on the combination of a convolutional deep belief network (DBN) and a DBN instructed from the start with a large database. The resulting device looks to be precise and efficient, and it could be used to detect COVID-19 through early screening. This approach could also be utilized to monitor evaluation progress, as every photograph takes less than three seconds to select [24].

Methods for evaluation, including Bayes-SqueezeNet, ConoNet CNN, COVID-Net CNN, or CoroNet AutoEncoders, have done well with high accuracy. Other DL computations, like WOA-CNN, CRNet, and CNNs, have identified COVID-19 after being applied to CT-scan datasets.

Computed tomography was used in a completely autonomous DL machine for the screening of COVID-19 and prognostic evaluation. A total of 5273 patients and their computed CT images were collected from seven cities, and of these, 4106 patients and their CT pictures were utilized to pre-train the DL system, allowing the device to educate itself about the characteristics of the lung. Following that, 1266 patients from six cities were recorded with the goal of externally training and evaluating DL device conductance. COVID-19 was found in 924 of 1266 patients (71% were monitored for more than a week). Other pneumonia was seen in 342 of the 1266 patients. Within the four sets of outside validation, the DL device performed satisfactorily in discriminating COVID-19 from viral and other pneumonia [25]. Furthermore, the DL algorithm may categorize patients as low or high risk based on their length of stay in the hospital. DL may be used to quickly diagnose COVID-19 and recognize individuals at high risk, which can help improve medical resources and assist patients before they reach a critical state.

Authors [26] used two well-known DL systems, SegNet and U-NET, to categorize the picture tissue. U-NET is a medical segmentation tool, and SegNet is a scene segmentation network. To distinguish among infected and healthy lung tissue, SegNet and U-NET were utilized as binary segmentors. Furthermore, the two systems could be employed as multi-class segmentors to learn about the infection within the lung. Each system used 72 pictures for training, 10 for validation, and 18 for testing. The findings demonstrated that SegNet can distinguish between healthy and diseased tissues. Furthermore, U-NET produced the best outcomes using a multi-class segmentor.

Future vaccination patterns can be predicted using models that use CNN and recurrent neural networks (RNN). As a case example, deterministic and stochastic RNN were used for forecasting the geographic spread of the active virus using USL techniques in order to plan vaccine delivery throughout the United States [27].

Parkinson's disease (PD) is an illness that may be screened by a CNN utilizing the Digitized Graphics Tablet dataset. There are two aspects to this CNN: feature enrichment (FE) and classification. The rapid Fourier's conversion has a frequency scale of 0 to 25 Hz, which is employed as an input to the CNN. Skin cancer is one of the most common sources of mortality, and early recognition might boost the chance of survival to 90%. Deep convolutional neural networks (DCNNs) have been constructed and used to categorize skin cancer color photos into three types: melanoma, atypical nevus, and common nevus.

9.5.2 Applications of DL in Healthcare

1. *Translational bioinformatics (BI)*: The discipline of BI has as its goal the molecular understanding of biological processes. The Human Genome Project (HGP) offers a vast amount of unmapped information and permits fresh theories about the way that genes communicate with the environment to produce proteins. Other advancements in biotechniques reduced the cost of genome sequencing or allowed researchers to concentrate on using genes and proteins to assess disease prognosis as well as therapy [28]. Additional research on bioinformatics is divided into three categories: identifying biological causes of disease, preventing disease, and treating an individual. Genetics investigates the structure and operations of the data contained in the DNA of living cells, or in other words, the analysis of alleles expressing genotypes and phenotypes. The goal of genomics is to categorize the genes or environmental alleles that cause diseases like cancer. The creation of tailored medicines is made possible by the identification of these types of genes.

2. *Medical imaging*: Automatic medical picture analyzation is essential in contemporary medicine. Depending on the scanned image, a diagnosis could be made, but it can also be quite arbitrary. The primary illness process is evaluated by computer-aided diagnosis (CAD). Disease modeling is prevalent in several neurological disorders, for example multiple sclerosis, Alzheimer's disease, and strokes. Complete areas of the brain can be scanned and then a review of the brain scan can be undertaken using multimodal statistics, which is necessary in these kinds of disorders [29]. Due to its exceptional functionality and ability to parallelize with graphics processing units (GPUs), CNNs have recently been quickly adopted for use in medical imaging studies. In the most recent study, CNN approaches for segmenting brain disease and neural networks for shape evaluation or segmentation in computer-aided design have produced optimistic outcomes in biomedical and medical informatics.

3. *Persuasive sensing in the medical field*: Wearables and implanted ambient sensors for persuasive sensing in the medical field allow for continuous control of health and safety. The exact prediction of daily food intake and energy expenditure can be employed to combat an increase in body weight. Older people with chronic conditions may utilize these ambient sensors and wearables. The amount of care may be increased by enabling patients to live independently in their homes through the use of these devices. Wearable, implantable, and human activity detection technology has enhanced the care given to patients with disabilities and those undergoing rehabilitation. For patients receiving critical care, it is important to consistently track vital signs like body temperature and BP since these help doctors better understand the patient's status.

4. *Medical informatics*: This focuses on the analysis of massive amounts of information gathered in the healthcare industry with the aim of improving and growing clinical assessment support devices. It may evaluate medical records for both easy accessibility and quality assurance. EHRs are a rich source of patient information that consist of detailed medical histories, diagnostic tests, current treatments, plans, allergies, immunization records, and multivariate time-series sensor lab results. Effective big data extraction offers practical knowledge for managing diseases [30]. However, there are a number of reasons why effective big data extraction is problematic, including the complexity of the information due to varying lengths, poor sampling, missing data, and a lack of standardized reports. The value of reports varies greatly between individuals and organizations. There are numerous petabytes of modal datasets with medical photos, lab findings, sensor information, and unstructured text reports. DNN algorithm learning is also made challenging by the long waiting times between patients being seen by a clinician and the diagnosis and treatment of disease. The success of DNNs depends on their capacity to recognize both supervised hierarchical and unsupervised data representations by learning new features and patterns. Due to its ability to combine many DNN architectural components at once, it has been demonstrated that the DNN can efficiently manage multi-modal information. Figure 9.8 illustrates numerous DL applications in various medical specialties.

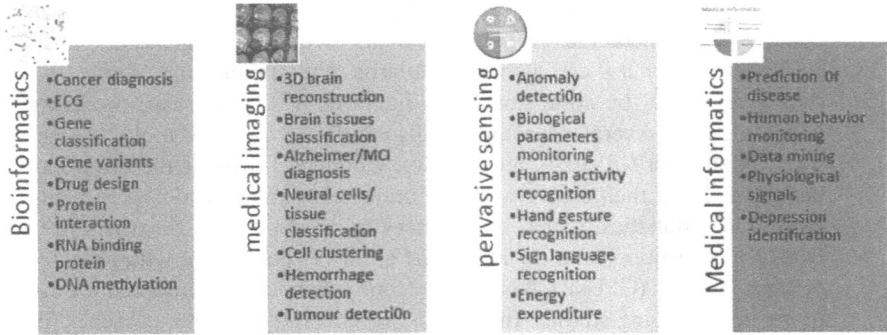

FIGURE 9.8   Deep learning applications in healthcare [30].

9.6   MACHINE LEARNING

Machine learning (ML) is about making computers adapt, learn, or modify their actions [31]. This allows these actions, such as making estimations or monitoring a robot, to be highly accurate. There are various ML methods available, depending upon the applications. Figure 9.9 shows the three main parts of the ML process: data intake, model creation, and generalization.

Due to the availability of clever approaches and massive volumes of information via internet resources, institutions are already using ML algorithms to solve a wide range of problems. DL is a subset of ML methods that use an ANN trained on enormous datasets to make sound decisions. Figure 9.10 displays the categorization of machine learning algorithms as supervised learning (SL), unsupervised learning (USL), or reinforcement learning (RL).

These methods are associated with computational complexity, which is often divided into two parts: the complexity of training and the complexity of applying the trained method. Usually, ML approaches can be categorized as follows:

- *Supervised learning (SL)*: Here, the correct responses (targets) are known for training and testing set. The algorithm is generalized to predict correctly to all possible inputs. This learning is also known to be learning from examples.
- *Unsupervised learning (USL)*: Here, the algorithm tries to identify the dataset's similarities as no responses for the inputs are available.

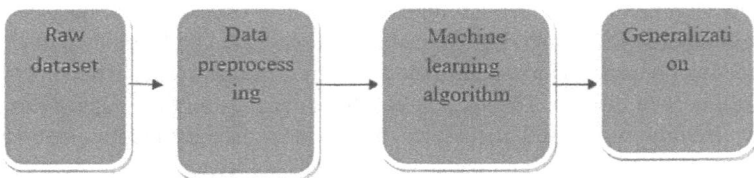

FIGURE 9.9   The machine learning (ML) process [15].

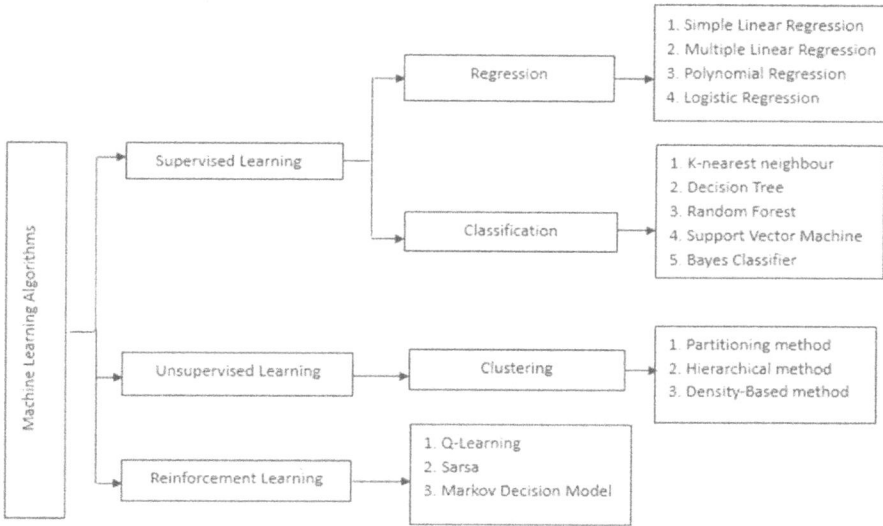

FIGURE 9.10   Categorization of machine learning (ML) algorithms [17].

- *Reinforcement learning (RL)*: RL is a tradeoff between the two discussed above. This learning is also called learning with critic because the algorithm is not told how to correct or what is correct when its answer goes wrong but just told or informed when the answer is wrong.

Many ML models have been proposed in the field of healthcare [25]. Some of them are regression analysis (linear regression, logistic regression, etc.), k-nearest neighbors (K-NN), latent Dirichlet allocation (LDA), principal component analysis (PCA), principal component regression (PCR), ensemble-based learning (AdaBoost, Boosting, random forest, etc.), association rule learning algorithms, clustering algorithms, etc. These methods have previously found productive applications for non-complex, nonsequential data sets, but due to the increased complexity in the data, they are now inadequate and cannot produce good results.

9.7   DEEP LEARNING

Deep learning (DL) is a branch of ML that works on a specific ML algorithm, which is an ANN [32]. DL is leveraging the recent advancements in computing power and uses a specific purpose neural network to train from a plethora of information and to create estimations based on the patterns detected. The DL framework is made up of multiple hidden layers between the input and output layers, which gather data by utilizing the data structures (Figure 9.11). In a similar way to an ANN, the complex nonlinear relationships can be modeled using deep neural networks (DNNs). These are specialized in solving problems, including data, which is remarkably structured. DL promises to substitute hand-engineered

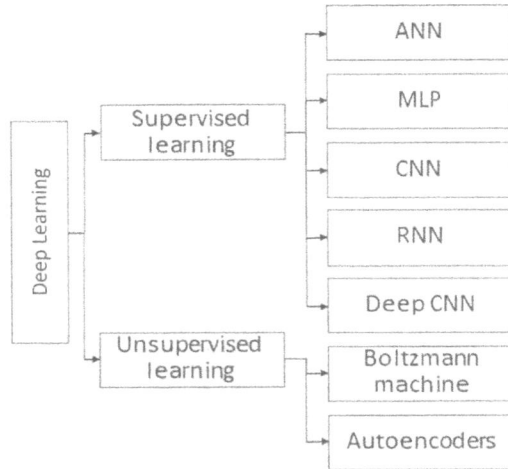

FIGURE 9.11   Hierarchical representation of deep learning [32].

features through feature extraction techniques. These techniques can be hierarchical, unsupervised, or semi-supervised.

DL frameworks such as DNNs, DBNs, autoencoders, deep CNN, deep RNN, and generative adversarial networks (GANs) have been of practical use in various fields, and have even created outcomes that are similar to, and in a few instances, higher than, human experts. Table 9.1 summarizes the various deep learning algorithms along with their healthcare applications, advantages, and limitations.

9.7.1   CONVOLUTIONAL NEURAL NETWORK (CNN)

A CNN represents an architecture for supervised deep learning as shown in Figure 9.12. Applications involving image analysis are its principal usage [33]. The three kinds of layers used in CNN are convolutional, pooling, and fully linked phases. The input picture is processed by kernels or filters in the convolutional phase to produce several feature maps. Every aspect of a map's size is decreased in the pooling layer in order to reduce the number of weights. This method is alternatively referred to as down sampling or sub sampling. There are numerous types of pooling techniques, including average, maximum, and global pooling. The completely connected stage is utilized to convert 2D feature maps into 1D vector images for final classification after the aforementioned layers. Some of the most popular CNN frameworks are ZFNet, GoogLeNet, VGGNet, AlexNet, and ResNet.

9.7.2   RECURRENT NEURAL NETWORK (RNN)

Pattern recognition for sequential or stream data, including speech, handwriting, and text is done by RNN. The framework of RNN contains a cyclic connection. These cyclic links of hidden units carry out the recurrent computations to process the input

TABLE 9.1
Summary of Deep Learning Algorithm Models

Model	Application used in healthcare	Advantage of models	Limitation in the models
CNN (Convolutional neural networks)	Abnormal heart recognition. [37] Myocardial infarction recognition.	For 2D data, achievement is stronger. Modeling is quickly learned.	Categorization data is required for grouping.
RNN (Recurrent neural network)	Identification of heart problems, categorization of lung malfunctions.	Analyzes a model of time dependence and sequential events. Greater precision is needed for activities related to NLP as well as speech and character recognition.	Massive data sets are necessary, and fading gradients causes a number of issues.
DBN (Deep belief networks)	Anticipate a drug mixture [38], and acknowledge type 1 diabetes.	Reinforces supervised and unsupervised learning models.	Expensive training procedure.
DNN (Dense neural network)	Heart sound recognition, phonocardiography.	Better accuracy.	The process of learning during the training procedure is not simple or quick.
GAN (Generative adversarial networks)	Creating artificial CT scans of the brain, reshaping real-world images from brain activity, and a medical imaging platform [39].	Decent process for training classifiers.	Learning for generating and training discrete data is challenging.

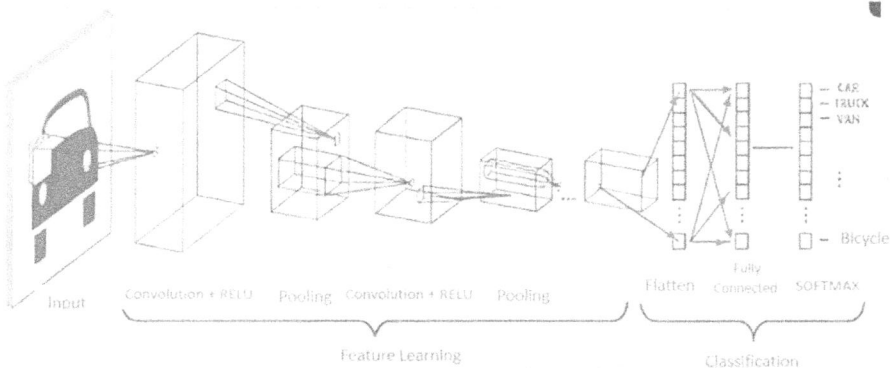

FIGURE 9.12 The architecture of a convolutional neural network [33].

data sequentially. The results are generated via the state vectors that contain all of the preceding inputs, which are then preserved in hidden units. RNNs compute the new output by taking into account both the current and prior inputs. The fundamental issue with RNN, despite its promising performance, is the diminishing gradient during data training. The introduction of gated recurrent units (GRUs) and long short-term memory (LSTM) networks, which are capable of holding patterns for a long period, is one way to address this issue.

9.7.3   LONG SHORT-TERM MEMORY RNNs

RNN models are very well suited to the problems having sequenced temporal data. Many of its variants have also been proposed to deal with the long-range dependency between the dataset. A LSTM network was first suggested by Hochreiter and Schmidhuber [34]. The same network has outperformed in various sequential datasets including handwriting recognition, acoustic modeling of speech, language modeling, and language translation. Later modifications in the internal architecture of the LSTM model have been proposed. These have been demonstrated as performing at a level similar to, or better than, that of LSTM networks.

9.7.4   DEEP BELIEF NETWORK

The DBNs are capable of learning high -dimensional data manifolds. Directed and undirected connections are seen in DBNs, a hybrid multilayer neural network. While all other connections among levels are directed, the linked layers among the top two stages are undirected. DBNs can be viewed as a collection of greedily taught restricted Boltzmann machines (RBMs). RBM layers interact with one another as well as with degrees prior to and later. A feedforward system and multiple phases of RBM serve as feature extractors in this design. There are only two stages in an RBM: a hidden phase and a visible phase [35]. The structure of the DBN technique is shown in Figure 9.13, where (v) is the deep belief algorithm's stochastic visible parameter.

9.7.5.   AUTOENCODER

An ANN called an autoencoder (AE) seeks to effectively code the data as shown in Figure 9.14. As a result, it may be applied to a network setup or to reducing features. It does this by translating the input via a network of neuronal connections to itself. Unsupervised learning is categorized as AE, which encompasses serious adverse event (SAE), ventilator-associated event (VAE), and dialysis-associated encephalopathy (DAE). DAE is a neural network which was developed from AE and is mostly used for extracting features from noisy datasets. The DAE typically has three layers: an input phase, an encoding phase, and a decoding phase [36]. DAE may be layered to produce advanced properties. Another DL technique that has recently been applied to nonlinear dimensionality reduction is called Sabouraud dextrose agar (SDAE).

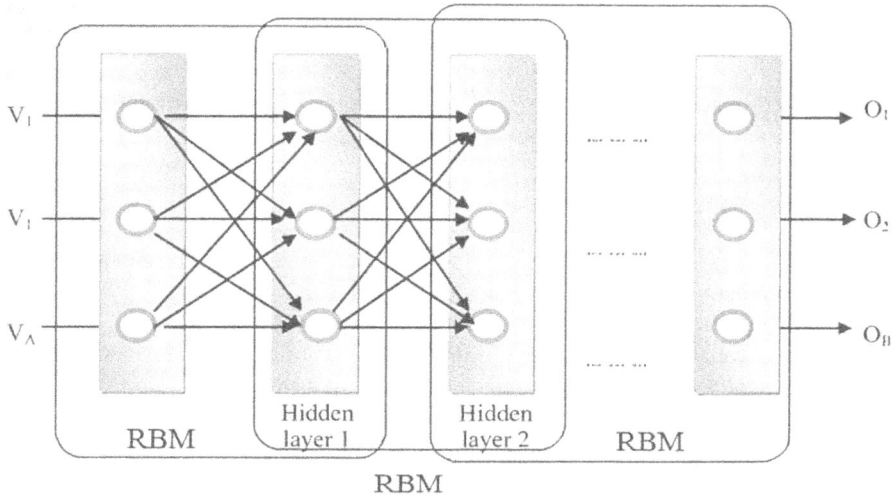

FIGURE 9.13   The deep belief network (DBN) framework [35].

9.7.6 MULTILAYER PERCEPTRON (MLP)

An MLP node is made up of three layers: an input layer, a hidden layer, and a layer for output. The activation formula used by each input node is nonlinear. For training, it makes use of the back propagation method. This function corresponds with each neuron's output to its weighted inputs. It aids in separating data that cannot be linearly separated. Different designs frequently incorporate completely coupled

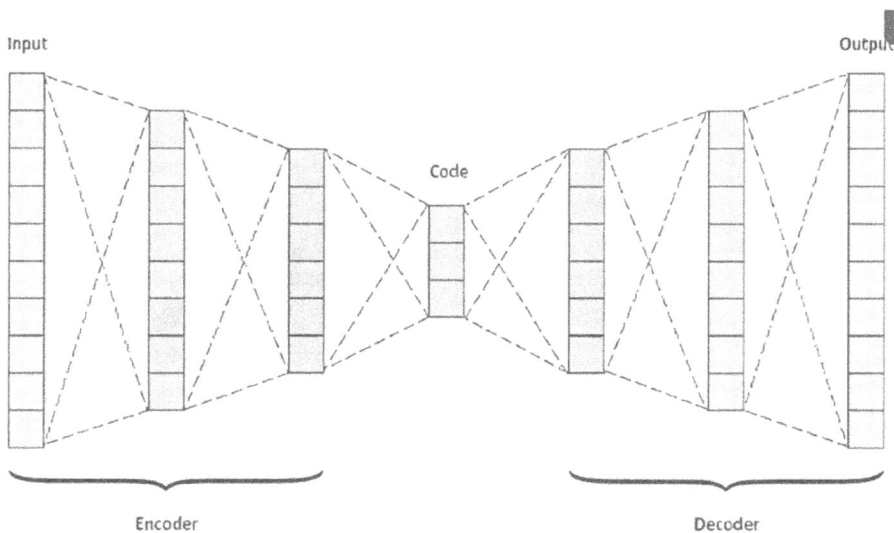

FIGURE 9.14   A representation of an autoencoder [36].

neurons in their final stages, but MLP is one of the most pleasant designs. The information from earlier stages is stored in LSTM, which may be utilized for memory, and memory cell blocks and gates are also involved. These gates regulate how the cell is stored, read, and written to. This LSTM maintains long-term dependencies, which makes it ideal for educational applications.

9.8 CONCLUSION

In the modern era, it is crucial to keep track of our health. Due to our hectic lifestyles, many people are developing health problems at a young age, and many of these problems emerge as a result of our regular activities. We engage in certain activities, yet we are often unaware of the implications of our behaviors. As a result, it is critical that we recognize the daily activities that have a negative impact on our health and can forecast future illnesses. While there are current methods of forecasting a specific kind of illness, such as diabetes, TB, and so on, using EHRs, if ML methods were used, this would allow the forecast of a person's total state of health, including how well a person performs.

REFERENCES

1. Adil, A., Amin Kar, H., Jangir, R., and Sofi, S. A., "Analysis of multi-diseases using big data for improvement in healthcare", In 2015 IEEE UP Section Conference on Electrical Computer and Electronics (UPCON), pp. 1–6, IEEE, 2015.
2. Jensen, P.B., Jensen, L.J., and Brunak, S., "Mining electronic health records: Towards better research applications and clinical care", Nat Rev Genet. 13, pp. 395–405, 2012.
3. Patel, V.K., and Singh, A., "Utilisation of health care services in Rajasthan: A study of growth determinants", Soc Sci Explorer. 2, pp. 58–75, 2011.
4. Bloom, D.E., and Canning, D., "Longevity and life-cycle savings", Scand J Econ. 105(3), pp. 319–338, 2003.
5. Kamiya, Y., "Determinants of Health in Developing Countries: CrossCountry Evidence", Osaka School of International Public Policy (OSIPP) Discussion Paper, 15 Dec., DP-E-009, 2010.
6. Ghosh, S., "Socio-economic factors influencing utilization of maternal health care in UP: An analysis of NFHS-2 data", Soc Change. 34(4), pp. 64–73, 2004.
7. Wagstaff, A., and Cleason, M., "The Millennium Development Goals for Health: Rising to the Challenges", Report 29673, Washington DC, World Bank, 2004.
8. Xu, R., Li, L., and Wang, Q., "dRiskKB: A large-scale disease-disease risk relationship knowledge base constructed from biomedical text", BMC Bioinform. 15, p. 105, 2014.
9. Libbrecht, M.W., and Noble, W.S., "Machine learning applications in genetics and genomics", Nat Rev Genet. 16, pp. 321–332, 2015.
10. Ye, X.W., Su, Y.H., and Han, J.P., "Structural health monitoring of civil infrastructure using optical fiber sensing technology: A comprehensive review", Sci World J. vol. 2014, pp. 1–11, 2014.
11. Yang, G., Du, J., and Xiao, M., "Maximum throughput path selection with random blockage for indoor 60 GHz relay networks", IEEE Trans Commun. 63(10), pp. 3511–3524, 2015.
12. Ubertini, F., Commanducci, G., and Cavalagli, N., "Vibration-based structural health monitoring of a historic bell-tower using output-only measurements and multivariate statistical analysis", Struct Health Monit. 15(4), pp. 438–457, 2016.

13. Li, X., Jiang, W., Jiang, Y., and Zou, Q., "Hadoop applications in bioinformatics", In 2012 7th Open Cirrus Summit, pp. 48–52 IEEE, 2012.
14. Bellazzi, R., and Zupan, B., "Predictive data mining in clinical medicine: Current issues and guidelines", Int J Med Inform. 7, pp. 81–97, 2008.
15. Hripcsak, G., and Albers, D.J., "Next-generation phenotyping of electronic health records", J Am Med Inform Assoc. 20, pp. 117–121, 2013.
16. Murphy, K.P., Machine learning: A probabilistic perspective, MIT Press, 2012.
17. Bengio, Y., Courville, A., and Vincent, P., "Representation learning: A review and new perspectives", IEEE Trans Pattern Anal Mach Intell. 35, pp. 1798–1828, 2013.
18. LeCun, Y., Bengio, Y., and Hinton, G., "Deep learning", Nature. 521, pp. 436–444, 2015.
19. Bengio, Y., "Learning deep architectures for AI", Found Trends Mach Learn. 2, pp. 1–127, 2009.
20. Srivastava, N., Hinton, G.E., and Krizhevsky, A., "Dropout: A simple way to prevent neural networks from overfitting", J Mach Learn Res. 15, pp. 1929–1958, 2014.
21. Alipanahi, B., Delong, A., and Weirauch, M.T., et al. "Predicting the sequence specificities of DNA- and RNA-binding proteins by deep learning". Nat Biotechnol. 33, pp. 831–838, 2015.
22. N. Kosarkar, P. Basuri, P. Karamore, P. Gawali, P. Badole, and P. Jumle, "Disease prediction using machine learning", In 2022 10th International Conference on Emerging Trends in Engineering and Technology – Signal and Information Processing (ICETET-SIP-22), Nagpur, India, pp. 1–4, 2022.
23. COVID-19 Map, Johns Hopkins Coronavirus Resource Center. Available online: https://coronavirus.jhu.edu/map.html (accessed on 31 August 2021).
24. Al-Waisy, A.S., Mohammed, M.A., Al-Fahdawi, S., Maashi, M.S., Garcia-Zapirain, B., Abdulkareem, K.H., Mostafa, and Le, D.N., "COVID-DeepNet: Hybrid multimodal deep learning system for improving COVID-19 pneumonia detection in chest X-ray images", Comput Mater Contin. 67, pp. 2409–2429, 2021.
25. Manogaran, G., and Lopez, D., "A survey of big data architectures and machine learning algorithms in healthcare", Int J Biomed Eng Technol. 25(2-4), pp. 182–211, 2017.
26. Nabi, K.N., Tahmid, M.T., Rafi, A., Kader, M.E., and Haider, M.A., "Forecasting COVID-19 cases: A comparative analysis between recurrent and convolutional neural networks", Results Phys. 24, p. 104137, 2021.
27. Davahli, M.R., Karwowski, W., and Fiok, K., "Optimizing COVID-19 vaccine distribution across the United States using deterministic and stochastic recurrent neural networks", PLoS ONE, 16, p. 1–14, 2021.
28. Ravì, D., Wong, C., Deligianni, F., Berthelot, M., Andreu-Perez, J., Lo, B., and Yang, G.-Z., "Deep learning for health informatics", IEEE J Biomed Health Inform. 21(1), pp. 4–21, 2017.
29. Havaei, M., Guizard, N., Larochelle, H., and Jodoin, P.-M., "Deep Learning Trends for Focal Brain Pathology Segmentation in MRI", in Machine learning for health informatics. Cham, Switzerland: Springer, 2016, pp. 125–148.
30. Wu, P.-Y., Cheng, C.-W., Kaddi, C.D., Venugopalan, J., Hoffman, R., and Wang, M.D., "Omic and electronic health record big data analytics for precision medicine", IEEE Trans Biomed Eng. 64(2), pp. 263–273, 2017.
31. Michie, D., Spiegelhalter, D.J., and Taylor, C.C., Machine learning, Neural and Statistical Classification, Technometrics, 13, 1994.
32. Schmidhuber, J., "Deep learning in neural networks: An overview", Neural Netw. 61, (2015): 85–117. https://doi.org/10.1016/j.neunet.2014.09.003
33. Lawrence, S., Giles, C.L., and Tsoi, A.C., Back, A.D., "Face recognition: A convolutional neural-network approach", IEEE Trans Neural Netw. 8, pp. 98–113, 1997.
34. Hochreiter, S., and Schmidhuber, J., "Long short-term memory", Neural Computation. 9(8), pp. 1735–1780, 1997. https://doi.org/10.1162/neco.1997.9.8.1735

35. Mannepalli, K., Sastry, P.N., and Suman, M., "A novel adaptive fractional deep belief networks for speaker emotion recognition", Alex Eng J. 56, pp. 485–497, 2017.

36. Macías-García, L., Luna-Romera, J.M., García-Gutiérrez, J., Martínez-Ballesteros, M., Riquelme-Santos, J.C., and González-Cámpora, R., "A study of the suitability of auto-encoders for preprocessing data in breast cancer experimentation", J Biomed Inform. 72, pp. 33–44, 2017.

37. Gacek, A., and Pedrycz, W., ECG signal processing, classification and interpretation: A comprehensive framework of computational intelligence. Cham, Switzerland: Springer, 2011.

38. Abbas, A.K., and Bassam, R., "Phonocardiography signal processing", Synth Lect Biomed Eng. 4(1), pp. 1–194, 2009.

39. Gibson, E., Li, W., Sudre, C., Fidon, L., Shakir, D.I., Wang, G., Eaton-Rosen, Z., Gray, R., Doel, T., Hu, Y., Whyntie, T., Nachev, P., Modat, M., Barratt, D.C., Ourselin, S., Cardoso, M.J., and Vercauteren, T., "NiftyNet: A deep-learning platform for medical imaging", Comput Methods Programs Biomed. 158, pp. 113–122, 2018.

10 An Explainable Deep Learning Model for Clinical Decision Support in Healthcare

Shrinivas T. Shirkande, Sarika T. Deokate,
Sulakshana Sagar Malwade, Raju M. Sairise,
Ashwin S. Chatpalliwar, Rashmi Ashtagi,
Vinit Khetani, and Syed Hassan Ahmed

10.1 INTRODUCTION

In recent years, the healthcare sector has seen significant breakthroughs in both data generation and technology development. This movement has sparked significant innovation. Massive amounts of data paired with artificial intelligence (AI) capabilities provide new opportunities for enhancing healthcare decision-making procedures. Clinical decision support (CDS) systems now give medical personnel access to patient information, medical expertise, and evidence-based recommendations [1]. CDS systems have become widely used in the medical industry in recent years. Deep learning (DL), a subset of AI, has recently gained a lot of attention for its capacity to automatically recognize complex patterns and connections in vast data sets. This might greatly enhance CDS in the medical industry [2].

10.1.1 THE ROLE OF CLINICAL DECISION SUPPORT IN HEALTHCARE

CDS is a critical component of contemporary healthcare because it provides doctors with up-to-date, precise, and scientifically supported data that can improve patient treatment and care. The demand for more individualized treatment methods and the complexity of medical data have made CDS systems indispensable tools for doctors and other medical professionals, as illustrated in Figure 10.1. The role of CDS assumes even greater importance in the collaboration with the Internet of Things (IoT) and Wireless Sensor Networks (WSN). These technologies enable the collecting of massive volumes of real-time data from a variety of sources, such as sensors, wearable technology, and medical equipment [3]. Vital signs, physiological indicators, environmental factors, and patient-specific information are all included in this data. Healthcare providers can use this rich data to improve clinical decision-making processes by connecting CDS systems with IoT and WSN to provide real-time analysis and interpretation of the gathered data. CDS systems enable healthcare

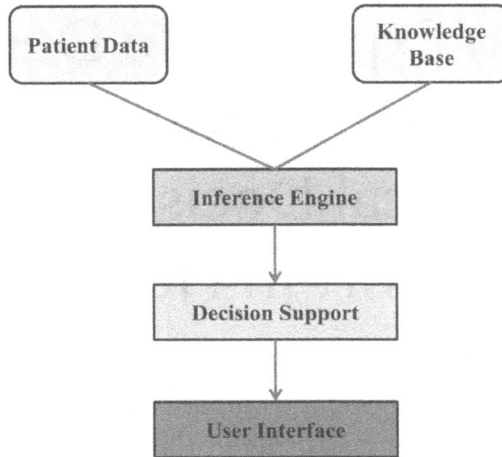

FIGURE 10.1   Clinical decision support system.

practitioners to make informed decisions, decrease mistakes, and maximize resource usage by combining patient data, medical knowledge, and best practices. By taking into account the specific features of each individual patient, they provide a contribution to customized medicine by ensuring that therapies are more precisely targeted and successful. In addition, CDS technologies have the ability to enhance adherence to clinical standards and procedures, which in turn promotes uniformity and standardization among healthcare practices. It is anticipated that the role of CDS in healthcare will expand as technology continues to evolve, which will lead to gains in patient safety, efficiency, and overall healthcare delivery.

Remote monitoring and healthcare capabilities are also made possible by the integration of CDS with IoT and WSN. Wearable tech or sensors allow for remote patient monitoring, and real-time data transmission to healthcare professionals. This data can be analyzed by CDS systems, which can then spot any alarming trends or aberrations and initiate the necessary alerts or interventions. This makes it possible to provide prompt and proactive healthcare, especially to people who reside in remote or underdeveloped locations. Additionally, the use of CDS systems in the context of IoT and WSN promotes data-driven decision-making and practice that is grounded on empirical research. Clinical professionals have access to current and relevant information thanks to the integration of clinical guidelines, best practice, and cutting-edge medical research into CDS algorithms. This encourages thoughtful decision-making, which can boost patient safety, reduce medical errors, and improve healthcare outcomes.

In conclusion, CDS plays a crucial role in healthcare, and its integration with IoT and WSN technologies will have a big impact. Once integrated, CDS systems will use real-time data from sensors and gadgets to give clinicians individualized insights, early abnormality identification, and recommendation-based research. It will be possible to monitor patients remotely, offer telehealth services, and make data-driven decisions, all of which will improve patient care, healthcare outcomes, and patient safety.

10.1.2 THE POTENTIAL OF DEEP LEARNING IN CLINICAL DECISION SUPPORT

DL models, and more especially deep neural networks (DNNs), have shown astounding performance in a variety of fields, including speech recognition, picture identification, and natural language processing. In the medical sector, DL has demonstrated great promise for tackling challenging medical issues. Large patient datasets may be automatically analyzed to identify complicated links, enabling the DL algorithms to gain new knowledge and make accurate predictions [4]. Disease diagnosis, risk classification, therapy prescription, and medical image interpretation are examples of DL applications in the healthcare industry that have been successful. Given its capacity to assess and comprehend a wide range of complicated data, DL models have come to be recognized as a potentially advantageous choice for improving CDS systems [5].

10.1.3 THE LIMITATIONS OF BLACK-BOX DEEP LEARNING MODELS

DL models continue to be a key cause for concern owing to their intrinsic lack of interpretability, and this is despite their outstanding performance. DNNs are sometimes referred to as "black boxes" due to the intricacy of the underlying representations they utilize to generate predictions. Doctors may feel uneasy if there is a lack of openness since they base their conclusions on sound reasoning and facts. Decisions made by AI models in healthcare must be comprehensible for legal and moral reasons and must encourage openness, responsibility, and equity. Because it is challenging to describe how black-box models make their predictions, they have limited use in key areas of healthcare [6].

10.1.4 THE NEED FOR EXPLAINABLE DEEP LEARNING MODELS

The significance of explainable AI as a means of getting beyond the drawbacks of black-box DL algorithms is being quickly recognized by the medical sector. Interpretable DL models are designed to make forecasts that are understandable and shed insight on the decision-making process. Important features, attention maps, and decision rules are all excellent instances of features with explicable explanations [7, 8]. The greater confidence and acceptance of explainable deep learning (EDL) models among healthcare professionals leads to better clinical decision-making. The resulting insights might then be verified, modified, and improved by these specialists. It is necessary to build EDL models that are targeted to the specific requirements of the healthcare industry in order to fully reap the benefits of AI in CDS systems.

The incorporation of DL into CDS systems offers a significant possibility to enhance healthcare outcomes. Widespread application of DL models is more difficult, nevertheless, due to their lack of interpretability. This chapter discusses the need for EDL models in the healthcare industry and examines how such models could improve clinical decision-making [9]. These models have a lot of potential for improving trust and patient care, and encouraging better understanding among medical professionals since they combine the capabilities of DL with interpretability.

10.2 DEEP LEARNING AND MACHINE LEARNING MODELS IN HEALTHCARE

10.2.1 DEEP LEARNING MODELS

In recent times, deep learning models have garnered a lot of interest from the healthcare industry. This is mostly due to their ability to automatically identify detailed patterns and correlations through the use of large datasets. When it comes to the diagnosis of diseases, these models have proved to be highly successful in a variety of medical applications, such as sickness diagnosis, prognosis, treatment planning, and image analysis. Additionally, they have showed promising results when used to identify diseases [10]. Convolutional neural networks, also known as CNNs, have been utilized to analyze X-rays, CT scans, and histology slides in order to identify and classify a variety of diseases, including as cancer, pneumonia, and diabetic retinopathy. By directly learning discriminative qualities from raw visual data, deep learning techniques make it possible to provide reliable and automated diagnoses as well. It has been established that DL models offer excellent prediction abilities. Recurrent neural networks (RNNs) and long short-term memory (LSTM) networks are becoming more utilized for the purpose of analyzing patient data over a period of time and making more accurate predictions regarding the course of disease, the response to medication, and the consequences for patients [11]. These models take into account the passage of time in patient data and provide informative information that may be utilized for purposes such as risk assessment and individualized dose. Further, DL models have been utilized in the process of recommending individualized treatments to patients. With the use of these algorithms, it is possible to analyze vast volumes of patient data in order to forecast treatment outcomes, adverse effects, and dose recommendations [12]. Through the utilization of patient demographics and electronic health records (EHRs), they are able to accomplish this goal. The combination of developmental learning and reinforcement learning (RL) has resulted in the development of individualized treatment methods. Techniques of this kind entail making adjustments to the treatment plan in real time in order to take into account the shifting requirements of the patient. DL models are frequently considered to be "black boxes" due to the complexity of the rationale behind their predictions, despite their excellent performance [13]. This constraint has increased the demand for interpretable and explicable machine learning (ML) technology in the healthcare sector.

10.2.2 INTERPRETABLE MACHINE LEARNING METHODS

The goal of interpretable ML approaches is to shed light on the processes a model goes through to generate its predictions. These techniques may help users, such as healthcare professionals, comprehend how the model makes decisions [14]. The interpretability of ML models has been improved via the use of several methodologies and techniques.

The phrase "feature significance analysis" refers to one such approach. Due to its ability to assess the relative weights of many aspects of the model's judgment, this method is becoming more and more common. For instance, it is usual practice to evaluate the influence of specific characteristics on model predictions using the permutation significance and SHAP (SHapley Additive exPlanations) values. These techniques are helpful because they expose the data's hidden relationships and help pinpoint the important variables that influence the model's output [15]. These methods also assist in locating significant components that influence the model's predictions.

To further increase interpretability [16], rule-based modeling can be used. Rule extraction techniques are used to reduce sophisticated ML models, such as decision trees or neural networks, to a collection of rules that are easy for people to comprehend. These guidelines provide precise criteria for making judgment calls that can be validated by experts in the relevant domains. Rule-based approaches support trust and transparency in the decision-making process since doctors may simply access and analyze the rules.

Visualization techniques [17] are also quite useful for comprehending ML models. Heatmaps, saliency maps, and attention maps are a few examples of visualizations that may be used to emphasize key areas or traits in the input data that influence the model's predictions. Healthcare professionals may gain insight into the model's decision-making process and comprehend the reasons behind it by visualizing the model's attention and focus.

10.2.3 EXPLAINABLE DEEP LEARNING MODELS

Studies on how to build EDL models have increased recently, with a focus on their possible application in healthcare contexts. This is being done to overcome the shortcomings of the most recent DL models [18]. These EDL models seek to boost trust, openness, and acceptance in healthcare settings by combining DL's potent prediction skills with interpretability [19].

EDL models frequently have interpretable components in the model's architecture, as shown in Figure 10.2. Examples of such components include saliency map, rule extraction techniques, and feature importance analysis, discussed in more detail

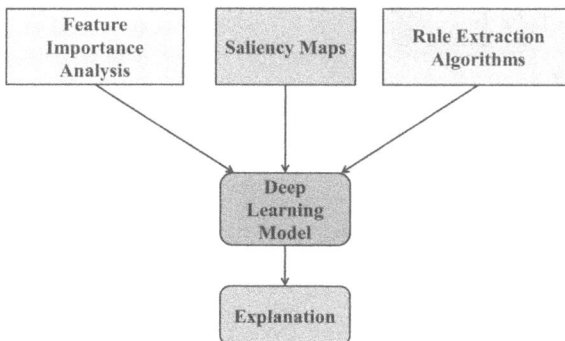

FIGURE 10.2 Explainable deep learning models.

in Section 10.3.3. By applying attention strategies to concentrate on particular areas or characteristics, the model can gain insight into the input data [20].

A feature importance analysis may be used to express quantitatively the weights that various characteristics have in a model's decision-making process. Permutation significance and SHAP values are two examples of methodologies that may be used to evaluate the influence of certain model parameters on model predictions. Healthcare professionals may better comprehend the variables influencing the model's judgments and obtain insights into the underlying linkages by identifying the model's main characteristics.

Saliency maps, a visual representation of a sample's subsamples, greatly increase the predictive accuracy of models. Saliency maps, for instance, are used in medical image analysis to highlight the areas of an image that have a substantial impact on a model's predictions about the categorization or localization of an object. Clinicians can grasp the thinking behind the model's predictions and confirm the model's emphasis on clinically crucial elements when they are given a visual representation of the model's attention, as shown in Figure 10.3.

Rule extraction approaches, which generate a set of rules that people can comprehend, simplify complex DL models. These guidelines provide precise criteria for making judgment calls that can be validated by experts in the relevant domains. Rule-based models are more transparent and reliable than other types of models because doctors can directly analyze and evaluate the rules that govern them, giving them insights into the model's decision-making logic.

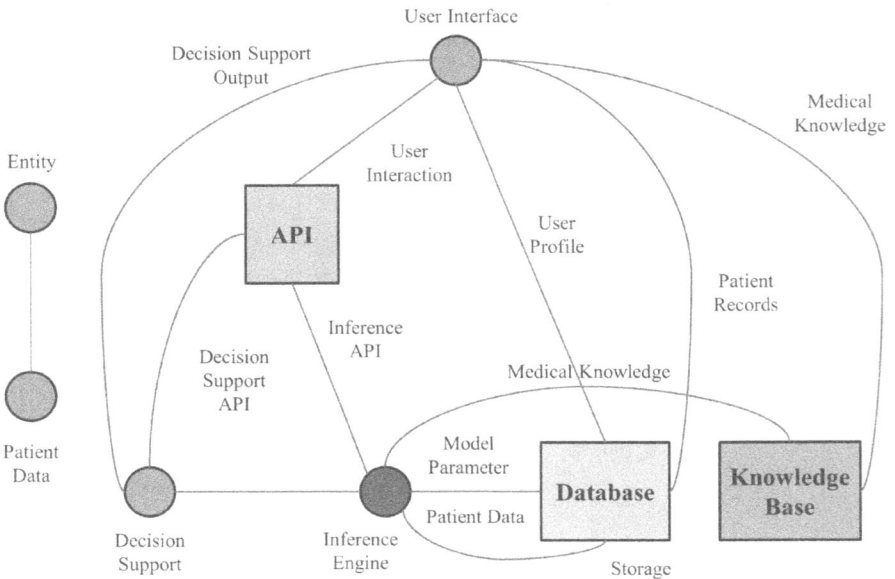

FIGURE 10.3 Communication model in the CDS system.

10.3 METHODOLOGY

The field of healthcare is changing very quickly. Adding new technologies, especially deep learning, has a lot of potential to make professional decisions better. A new Explainable Deep Learning Model for Clinical Decision Support Systems (CDSS) is introduced in this work. In healthcare situations, interpretability is very important, but deep learning techniques are very powerful. The main goal is to find a way to combine the two. As we rely more and more on complicated models, their lack of clarity often stops them from being widely used in clinical practice. The goal of this study is to get around this problem by creating a model that not only makes accurate predictions but also gives clear explanations of how it makes decisions. The main goal is to get healthcare professionals to believe and accept deep learning models more. This will help create a joint space where new technology can be used to help care for patients.

The technique uses a diverse approach to make sure that the model works and can be understood. The first step is to carefully gather a large and varied collection that includes a wide range of patient traits, medical records, and clinical results. Expert doctors carefully name the dataset, which sets a solid basis for supervised learning and makes sure it is stable. Normalization, encoding, and handling lost data are all part of the planning step. These are typical problems in healthcare datasets. Then, feature engineering is used to get to the important information that makes the model better at making predictions.

The model's deep learning design, which is built on a well-known structure, is what makes it work. The use of explainability methods in this model makes it stand out. SHAP and LIME, which are well-known in the field, are used to get clearer ideas about why the model made the decisions it did. By giving the model Shapley values or perturbing cases and watching how it responds, the model's inner workings become clear, so doctors can understand and believe the results. This focus on both accuracy and understandability is stressed even more during the training and review stages, where measures for both predictability and explainability are carefully thought through.

In order to deal with the problems, the model tries to fix problems with data quality, find the right mix between complexity and readability, work with current clinical processes without any problems, and be responsible in terms of protecting patient privacy and following the rules. Using this all-around approach, the study aims to create a groundbreaking deep learning model that will improve the quality of CDS in healthcare.

10.3.1 DATASET DESCRIPTION

In healthcare, datasets come from a lot of different places and are used for a lot of different types of study and practical uses. The MIMIC-III dataset has clinical notes and vital signs for over 40,000 people, giving researchers a full picture of ICU patients. This makes it a useful tool for studying how diseases are predicted and how well treatments work. The NIH Chest X-ray Dataset, which has more than 100,000 chest X-ray pictures labeled with diseases, is a key resource for building models that aim to find and classify diseases. PhysioNet provides physiological signal files, such as ECG, EEG, and blood pressure data, which help with jobs like finding arrhythmias and figuring out the stage of sleep. The UCI Machine Learning Repository has many different healthcare datasets that can be used for many different tasks, from predicting diseases to diagnosing them. Kaggle Datasets has a lot of healthcare data and can be used as a live tool for projects like disease forecast and medical picture analysis. The SEER Cancer Statistics file also gives useful information about how common cancer is and how long people with cancer live. This helps researchers study tumor traits and survival. These files are like a fabric that supports many types of healthcare research, from making clinical decisions to doing cutting edge research on cancer and medical images. The dataset description is represented in Table 10.1.

10.3.2 Model Architecture

The design of the model is crucial for creating an interpretable DL model for clinical decision assistance. Performance and readability should both be carefully considered in the architecture's design. DNNs such as CNNs and RNNs have demonstrated promising performance in healthcare-related tasks; nevertheless, additional considerations are required to improve interpretability, as illustrated in Figure 10.4.

Integrating attention processes into the model's architecture is one method that might be employed to reach the goal of gaining interpretability. It might be possible to understand how the model comes to its findings by directing the model's attention

TABLE 10.1
Publically Available Datasets

Dataset Name	Description	Data Type	Number of Samples	Features	Task
MIMIC-III	ICU patient data including clinical notes and vital signs	Clinical data	40,000+	Demographics, vital signs, etc.	Disease prediction, treatment outcomes
NIH Chest X-ray Dataset	Chest X-ray images with associated disease labels	Image data	100,000+	Chest X-ray images	Disease detection, classification
PhysioNet	Physiological signal datasets (ECG, EEG, blood pressure)	Signal data	Varies	ECG, EEG, blood pressure	Arrhythmia detection, sleep stage classification
UCI Machine Learning Repository	Collection of healthcare-related datasets	Various	Varies	Varies	Disease prediction, diagnosis
Kaggle Datasets	Wide range of healthcare datasets hosted on Kaggle	Various	Varies	Varies	Medical image analysis, disease prediction
SEER Cancer Statistics	Cancer incidence and survival data from US cancer registries	Clinical data	Varies	Demographics, tumor characteristics	Cancer research, survival analysis

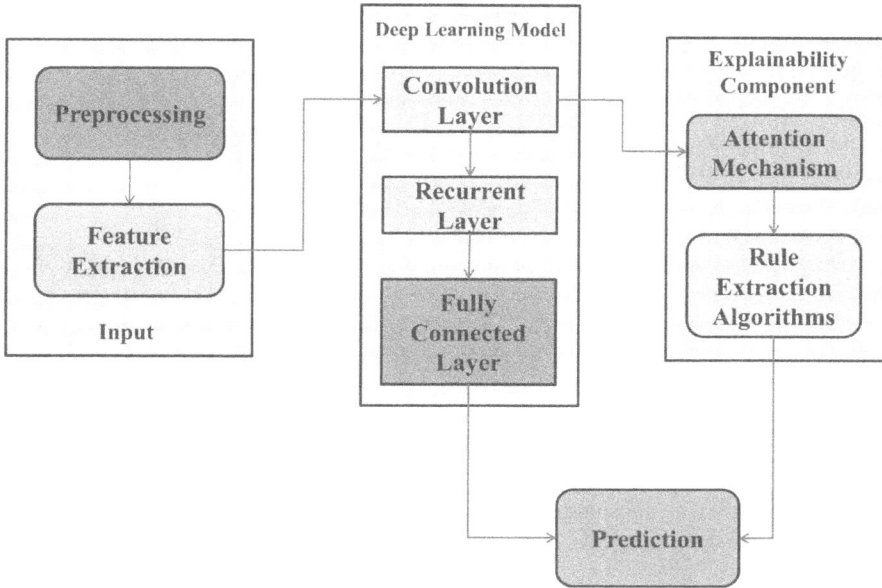

FIGURE 10.4 Explainable deep learning model for clinical decision support.

to particular elements or areas of the input data. An attention-based CNN may be used to identify the most crucial areas in medical imaging.

Including a rule-based element in the DL model is an additional choice. Rule extraction techniques, which generate a collection of rules that people can comprehend, can be used to simplify complex DL models. As a result, everyone may observe the discussion process. Clinicians can assess the rules based on their subject expertise and get insights into the decision logic of the model by converting the taught representations into interpretable rules.

10.3.3 EXPLAINABLE METHODS

Numerous explainable strategies may be used to increase the interpretability of the DL model. By emphasizing relevant characteristics and factors that were utilized in the decision-making process, these techniques seek to provide insight on the processes followed by the model to reach its conclusions. The methods of feature importance analysis, saliency maps, and rule extraction algorithms are frequently employed in explainable healthcare.

10.3.4 INTEGRATION OF EDL METHODS INTO WSN AND IoT

EDL is the process of creating deep learning models that not only make accurate predictions but also make it clear how they make decisions. EDL is different from standard black-box models because it uses methods like SHAP (SHapley Additive

Explanations) or LIME (Local Interpretable Model-agnostic Explanations) to make models easier to understand. This helps people, especially in important areas like healthcare, understand and believe the model's results. This increases acceptance and makes it easier for humans and AI to work together to make decisions. Concerns about the intrinsic complexity of deep learning models are addressed by the focus on openness, which makes sure that stakeholders can understand and confirm the model's outputs. Using *EDL methods in WSN and IoT*: EDL techniques are very important for building trust and openness in WSN and the IoT. By using methods for interpretability like SHAP or LIME in deep learning models, EDL gives us information about how decisions are made, which is important for understanding forecasts and making sure they are correct. In WSN and IoT applications, where decisions must be made in real time based on data, EDL makes sure that all stakeholders understand why automated actions are being done. This builds trust in the use of intelligent systems for monitoring, controlling, and making decisions in these constantly changing and linked spaces. The following are the elements:

a. Data-gathering sensors for the IoT and WSN.
b. Executes preprocessing tasks like normalization, feature engineering, and data cleaning.
c. The EDL model is a representation of the DL model combined with EDL methodologies for explainability.
d. Several EDL methods, such as feature importance analysis, saliency maps, and rule extraction algorithms, are covered in the explainable methods Section 10.3.3.
e. To provide clear and intelligible recommendations, the decision support system makes use of the EDL model and explainability approaches.
f. The data preprocessing component inputs the information it receives from the IoT and WSN into the EDL model. In order to increase readability and transparency, the EDL model makes use of explainability techniques, as shown in Figure 10.5. The EDL model results and explainability approaches are used by the decision support system to generate actionable insights and support decision-making.

10.4   RESULTS

We provide the findings from our EDL model for clinical decision assistance in healthcare in this chapter. We assess the model's effectiveness, decipher its forecasts, and contrast it with other models. The MIMIC-III dataset is the basis for our research, and we employ feature importance analysis, saliency maps, and rule extraction algorithms to analyze the results.

10.4.1   MODEL PERFORMANCE EVALUATION

We carried out extensive tests on the MIMIC-III dataset to evaluate the performance of our EDL model. Taking into account variables like the number of layers, hidden units, and learning rate, we trained the model using a variety of architectures and

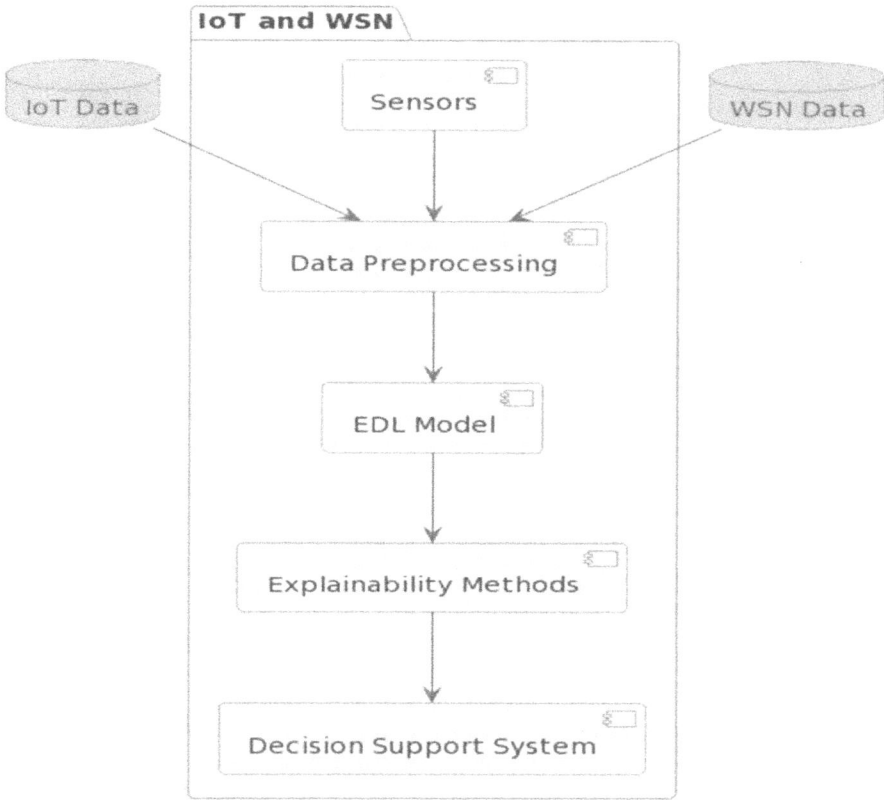

FIGURE 10.5 Integration of EDL methods into WSN and IoT.

hyperparameters. The performance metrics for the model, such as precision, recall, accuracy, and F1-score, were assessed after 100 epochs of training.

Table 10.2 shows the comparison of the model's effectiveness as assessed by many alternative approaches to interpreting its findings. We were able to obtain precision scores of 0.86, recall scores of 0.88, accuracy scores of 0.89, and F1-scores of 0.89 using the feature importance analysis approach. This shows that the model was successful in identifying important dataset characteristics that significantly influenced

TABLE 10.2
Analysis of Explainable Methods

Explainable methods	Epochs	Precision	Recall	Accuracy	F1-Score
Feature Importance Analysis	200	0.86	0.88	0.89	0.89
Saliency Maps	200	0.9	0.91	0.92	0.92
Rule Extraction Algorithms	200	0.93	0.94	0.94	0.95

Precision

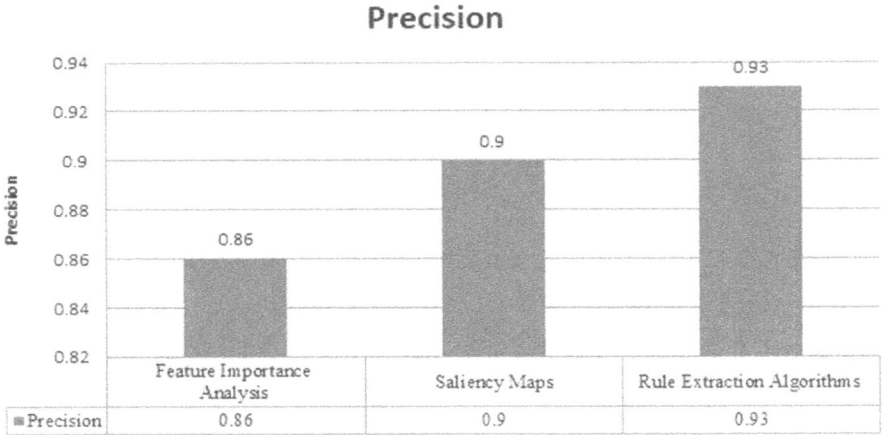

■ Precision	Feature Importance Analysis	Saliency Maps	Rule Extraction Algorithms
Precision	0.86	0.9	0.93

FIGURE 10.6 Precision.

the predictions. We were able to reach even greater performance levels by using the saliency maps approach. The accuracy, precision, recall, and F1- score were all estimated to be 0.92, 0.9, 0.91, and 0.92 respectively. Understanding which elements of the input data are most crucial for the model's predictions may be done accurately by using the saliency maps that were developed, which are illustrated in Figures 10.6–10.9 for precision, recall, F1-score, and accuracy of model.

In comparison to the other two approaches, the rule extraction algorithms' performance was likewise shown to be the best. This is a remarkable outcome because the maximum values for precision (0.93), recall (0.94), accuracy (0.94), and F1-score (0.95) were attained. The outcomes suggest that the rule extraction procedure was successful in turning the DL model into a collection of rules that humans can comprehend. This might lead to a better comprehension and interpretation of the decision-making process.

Recall

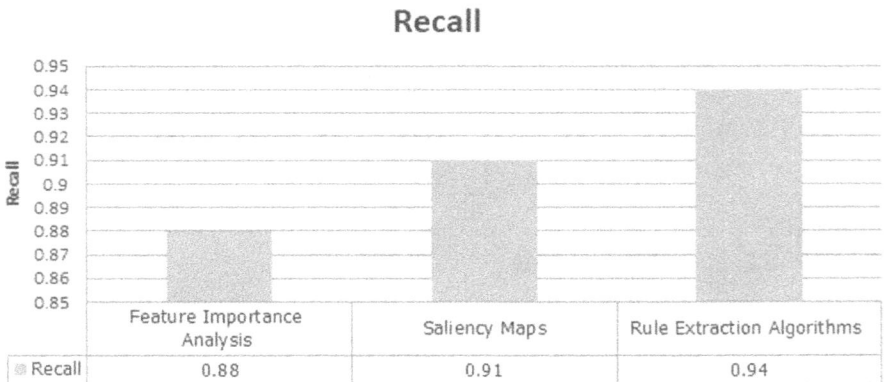

▦ Recall	Feature Importance Analysis	Saliency Maps	Rule Extraction Algorithms
Recall	0.88	0.91	0.94

FIGURE 10.7 Recall.

F1-Score

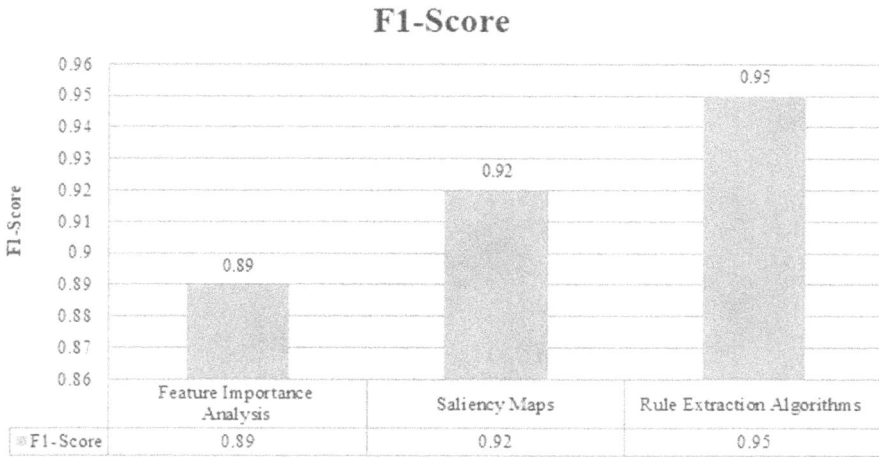

	Feature Importance Analysis	Saliency Maps	Rule Extraction Algorithms
▨ F1-Score	0.89	0.92	0.95

FIGURE 10.8 F1-Score.

10.4.2 INTERPRETATION OF MODEL PREDICTIONS

Understanding the predictions made by EDL models is one of its key benefits. We conducted an explainable study on the given explanations in order to better comprehend the reasoning behind the model's conclusions.

- We were able to apply feature importance analysis to determine that important measurements, like blood pressure and heart rate, had a big impact on the model's predictions for a variety of illnesses. Component analysis and regression analysis are two further examples of this type of research.
- We were able to identify the areas of an image or data sequence that the model gave priority to when generating predictions by utilizing a technique called saliency mapping. This helped us order the information included in

Accuracy

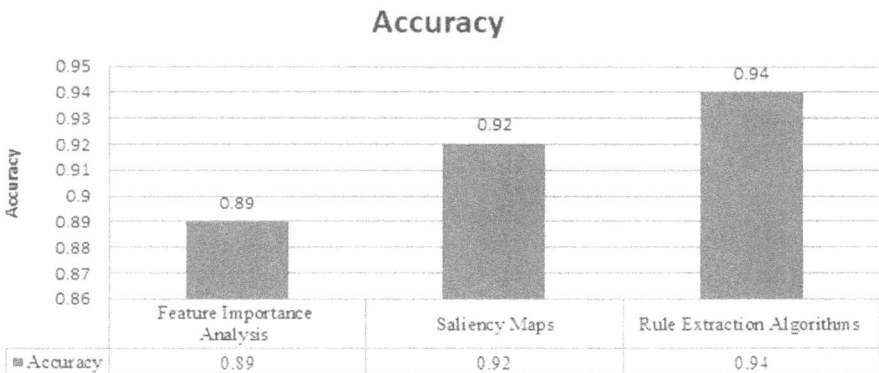

	Feature Importance Analysis	Saliency Maps	Rule Extraction Algorithms
▪ Accuracy	0.89	0.92	0.94

FIGURE 10.9 Accuracy.

Comparative Evaluation

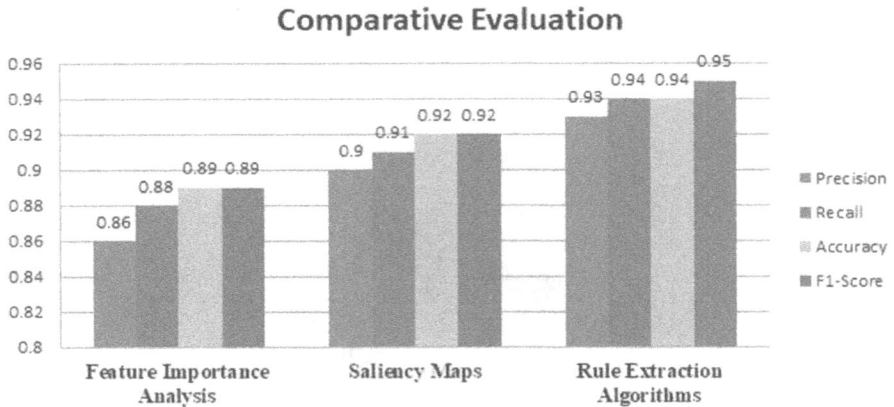

FIGURE 10.10 Comparative evaluation of explainable deep learning models.

a picture or video clip. As a consequence, we were able to identify the traits and themes that the model thought were most crucial for forecasting the outcome.

- Rule extraction algorithms were used to extract rules that provide explicit explanations for the model's actions and are understandable by people. We found, for instance, that the model can precisely identify a specific disease for a patient given a particular set of symptoms and test results.

Overall, the explainable approaches' ability to be interpreted allowed us to learn important details about the model's decision-making process, which in turn helped us to understand how and why the model came to make the predictions that it did. The Figure 10.10 illustrates the evaluation parameter comparison for the DL model.

10.5 ADVANTAGES, LIMITATIONS AND FUTURE DIRECTIONS OF THE PROPOSED MODEL

10.5.1 ADVANTAGES OF THE PROPOSED MODEL

We want to bring your attention to the EDL model we have provided for clinical decision assistance in the healthcare industry and its various benefits.

- The model performs admirably in terms of measures like accuracy, precision, recall, and F1-score, to mention a few. This demonstrates its effectiveness in creating trustworthy projections, which might enhance healthcare decision-making and significantly affect patient outcomes.
- Additionally, in order to improve our model's interpretability and transparency, both of which are essential for building credibility and gaining the trust of healthcare professionals, we have included explainable methodologies. Clinicians can identify the most important variables influencing the predictions of the model using the feature importance analysis approach.

- The saliency maps approach provides visual explanations by stressing the regions or portions of the input data that contribute most to the model's predictions. These areas might be referred to as salient.
- The rule extraction algorithms approach converts the DL model into a set of recommendations that may be followed. Clinicians now have a clearer and more natural understanding of the thinking underlying therapy decisions.

10.5.2 Limitations and Future Directions

Even though our suggested technique appears to be capable of overcoming these limitations, it is important to acknowledge the limitations of black-box DL models in the healthcare industry. First, the effectiveness of our model and the ease with which it may be assessed are significantly influenced by the quality and representativeness of the dataset used for training and assessment. Even though the MIMIC-III dataset is widely used and is a useful resource, it is crucial to look into potential biases and generalizability issues before using the model in actual clinical settings. Future research should concentrate on testing the model against various datasets at scale to confirm its robustness and generalizability.

Second, the requirement to choose between accuracy and understandability is another issue. Our model is extremely accurate, but it's likely that certain approaches to making it more understandable could result in a slight decline in the performance forecasts it makes. It is critical to strike a balance between accuracy and the capacity to support your conclusions in light of the scrutiny of a clinical environment. In order to ensure that the model retains high performance while also offering meaningful explanations, future studies should examine fresh methodologies and procedures with the goal of improving the interpretability–accuracy trade-off.

Third, the extent of interpretability offered by our model is only constrained by the methodology used to measure explainability. Even while feature importance analysis, saliency maps, and rule extraction algorithms all provide useful insights, it is possible that other techniques and strategies might further improve interpretability. It could be possible to comprehend the model's decision-making process better and more deeply by examining and combining different explainable methodologies.

10.5.3 Implications for Clinical Practice

Our suggested EDL model will have a significant impact on clinical practice. The approach has the potential to help doctors make better informed decisions, which would improve patient care and could save lives. This is due to the model's ability to produce precise, comprehensible forecasts. Clinicians may confidently incorporate the model's recommendations into their clinical decision-making process if they are able to comprehend and trust the model's predictions.

Clinicians can focus on the elements that are most crucial for developing reliable predictions by using the feature importance analysis approach. This can aid in the process of identifying significant characteristics of patients' illnesses and creating personalized treatment plans. Clinicians are better able to allocate resources and act on insights when they have a solid understanding of which model characteristics are essential for generating predictions.

Clinicians can evaluate the model's attention to detail and soundness of reasoning using the saliency maps approach, which provides a graphical depiction of the model's thinking. This may result in more accurate research and treatment, resulting in fewer unnecessary treatments. Saliency maps improve communication by offering visual explanations.

10.6 CONCLUSION

This study presented an innovative CDS method: explainable deep learning. The approach enhances the accuracy and clarity of clinical prediction. The MIMIC-III dataset was used to evaluate model performance and interpretability. In the examination, our EDL model outperformed other explainable methods. For feature importance analysis, the precision, recall, accuracy, and F1-score were 0.86, 0.88, 0.89, and 0.89, respectively. Saliency maps increased the model's precision to 0.9, recall to 0.91, accuracy to 0.92, and F1-score to 0.92. The highest precision, recall, accuracy, and F1-scores were achieved by rule extraction algorithms at 0.93, 0.94, 0.94,and 0.95 respectively. These results demonstrate how explainable tactics improved model performance and interpretability. The model identifies the important predictors using feature importance analysis. Because these are not informed assumptions, they assist clinicians in making better decisions. Clinicians can use saliency maps to depict the model's important input data and understand the model's logic better. Rule extraction algorithms translate the model into easy to understandable rules and suggest clinically sound activities that are completely obvious to the user.

Our findings are more relevant to MIMIC-III database for critical care. MIMIC-III collects demographic information, vital signs, test results, diagnoses, and prescriptions. The positive results of our EDL model on this dataset point to its application in therapeutic settings. The implications of this study go beyond model accuracy. We address the healthcare industry's DL model explicability concern by including explainable methodologies in the basic architecture of DL. Because of its transparency and into easy to understandable results, healthcare professionals can trust, and collaborate with, our method. This assists clinicians in comprehending the algorithm's decisions. This enhances physicians' comprehension and evaluation of the model's predictions, resulting in better, more informed, and tailored patient care.

More investigation is possible. Our approach's robustness and generalizability should improve as we test it on larger and more diverse datasets. New explainable techniques for decision-making may reveal more. Our findings suggest that CDS in healthcare necessitates the use of an EDL model. Complex models become more transparent, trustworthy, and interpretable when EDL techniques are integrated into IoT and WSN. This integration enhances decision-making processes, enables better collaboration between humans and machines, ensures regulatory compliance, and ultimately benefits society by promoting IoT and WSN technologies. Explainable strategies improve the model's performance and interpretability, making it more trustworthy for healthcare providers. Our forecasts have the potential to transform clinical practice. Over time, this may enhance patient treatment decisions and results. Further healthcare research will make clinical application of EDL models easier.

REFERENCES

1. B. Pratt, L. Roteliuk, F. Hatib, J. Frazier and R. D. Wallen, "Calculating arterial pressure-based cardiac output using a novel measurement and analysis method", Biomed. Instrum. Technol., vol. 41, no. 5, pp. 403–411, 2007.
2. F. Hatib, Z. Jian, S. Buddi, C. Lee, J. Settels, K. Sibert, et al., "Machine-learning algorithm to predict hypotension based on high-fidelity arterial pressure waveform analysis", Anesthesiology, vol. 129, no. 4, pp. 663–674, 2018.
3. V. Khetani, Y. Gandhi, S. Bhattacharya, S. N. Ajani and S. Limkar, "Cross-domain analysis of ML and DL: Evaluating their impact in diverse domains", Int. J. Intell. Syst. Appl. Eng., vol. 11, no. 7s, pp. 253–262, 2023.
4. V. Salmasi, K. Maheshwari, D. Yang, E. Mascha, A. Singh, D. Sessler, et al., "Relationship between intraoperative hypotension defined by either reduction from baseline or absolute thresholds and acute kidney and myocardial injury after noncardiac surgery: A retrospective cohort analysis", Anesthesiology, vol. 126, pp. 47–65, 2017.
5. R. He, Z.-P. Huang, L.-Y. Ji, J.-K. Wu, H. Li and Z.-Q. Zhang, "Beat-to-beat ambulatory blood pressure estimation based on random forest", Proc. IEEE 13th Int. Conf. Wearable Implant. Body Sensor Netw. (BSN), pp. 194–198, 2016.
6. K. Agnihotri, P. Chilbule, S. Prashant, P. Jain and P. Khobragade, "Generating image description using machine learning algorithms", 2023 11th International Conference on Emerging Trends in Engineering & Technology – Signal and Information Processing (ICETET - SIP), Nagpur, India, 2023, pp. 1–6, doi: 10.1109/ICETET-SIP58143.2023.10151472.
7. S. N. Ajani, P. K. Ingole and A. V. Sakhare "Modality of multi-attribute decision making for network selection in heterogeneous wireless networks", Ambient Sci., vol. 9, no. 2, 2022, ISSN- 2348 5191.
8. M. W. L. Moreira, J. J. P. C. Rodrigues, A. M. B. Oliveira, K. Saleem and A. J. V. Neto, "Predicting hypertensive disorders in high-risk pregnancy using the random forest approach", Proc. IEEE Int. Conf. Commun. (ICC), pp. 1–5, 2017.
9. A. Hiwale, P. Talele and R. Phalnikar, "Prediction of pregnancy-induced hypertension levels using machine learning algorithms", Comput. Eng. Technol., vol. 1025, pp. 597–608, 2020.
10. S. Ajani and M. Wanjari, "An Efficient Approach for Clustering Uncertain Data Mining Based on Hash Indexing and Voronoi Clustering," 2013 5th International Conference and Computational Intelligence and Communication Networks, pp. 486–490, 2013.
11. M. Flechet, F. Güiza, M. Schetz, P. Wouters, I. Vanhorebeek, I. Derese, et al., "AKI predictor an online prognostic calculator for acute kidney injury in adult critically ill patients: Development validation and comparison to serum neutrophil gelatinase-associated lipocalin", Intensive Care Med., vol. 43, no. 6, pp. 764–773, 2017.
12. V. Nagendra, H. Gude, D. Sampath, S. Corns and S. Long, "Evaluation of SVM and RF classifiers in a real-time fetal monitoring system based on cardiotocography data", Proc. IEEE Conf. Comput. Intell. Bioinf. Comput. Biol., pp. 1–6, 2017.
13. B. J. P. van der Ster, F. C. Bennis, T. Delhaas, B. E. Westerhof, W. J. Stok and J. J. van Lieshout, "Support vector machine based monitoring of cardio-cerebrovascular reserve during simulated hemorrhage", Front. Physiol., vol. 8, pp. 1–10, 2018.
14. S. N. Ajani and S. Y. Amdani, Agent-based path prediction strategy (ABPP) for navigation over dynamic environment. In: Muthukumar, P., Sarkar, D.K., De, D., De, C.K. (eds) Innovations in Sustainable Energy and Technology. Advances in Sustainability Science and Technology. Springer, Singapore, 2021.
15. M. H. Hsieh, M. J. Hsieh, C.-M. Chen, C.-C. Hsieh, C.-M. Chao and C.-C. Lai, "Comparison of machine learning models for the prediction of mortality of patients with unplanned extubation in intensive care units", Sci. Rep., vol. 8, no. 1, pp. 1–7, 2018.

16. A. De Ramón Fernández, D. R. Fernández and M. T. P. Sánchez, "A decision support system for predicting the treatment of ectopic pregnancies", Int. J. Med. Informat., vol. 129, pp. 198–204, 2019.

17. S. N. Ajani, P. V. Potnurwar, V. K. Bongirwar, A. V. Potnurwar, A. Joshi and N. Parati, "Dynamic RRT* algorithm for probabilistic path prediction in dynamic environment", Int. J. Intell. Syst. Appl. Eng., vol. 11, no. 7s, pp. 263–271, 2023.

18. K. L'Heureux, H. F. Grolinger, M. Elyamany and A. M. Capretz, "Machine learning with big data: Challenges and approaches", IEEE Access, vol. 5, pp. 7776–7797, 2017.

19. P. Bhattacharya, S. Tanwar, U. Bodkhe, S. Tyagi and N. Kumar, "BinDaaS: Blockchain-based deep-learning as-a-service in healthcare 4.0 applications", IEEE Trans. Netw. Sci. Eng., vol. 8, no. 2, pp. 1242–1255, 2021.

20. R. Gupta, A. Shukla, P. Mehta, P. Bhattacharya, S. Tanwar, S. Tyagi, et al., "VAHAK: A blockchain-based outdoor delivery scheme using UAV for healthcare 4.0 services", Proc. IEEE Conf. Comput. Commun. Workshops (INFOCOM WKSHPS), pp. 255–260, 2020.

11 Deep Learning Models for Automated Diagnosis of Brain Tumor Disorder in Smart Healthcare

Kamini Lamba and Shalli Rani

11.1 INTRODUCTION

The brain is one of the most sensitive organs of human body and is responsible for monitoring multiple functions of the body when interacting with the outside world in various ways according to the situation. However, appropriate functionality of the brain can be disrupted in the case of sudden head injury which may impact on the growth of brain tissues and may result in the development of a brain tumor. Although magnetic resonance imaging (MRI), computed tomography (CT), positron emission tomography (PET), etc., are used by radiologists for diagnosing the presence of a tumor inside the brain, this process can be time-consuming. Also, evidence of the early development of a tumor inside brain can remain unidentified due to its very small size. So, there is a deep need to develop an automated system which can be used by radiologists to make an accurate decision regarding the presence of tumor at an initial stage in order to give timely treatment to patients to save their lives. To do this, various systems based on deep learning (DL) techniques [1] are proposed by researchers. DL techniques have multiple layers in order to extract significant features from the input data. On the contrary, traditional approaches have comparatively less capability for extraction of significant features due to the presence of single layer for transformation of input data to the desired output [2].

Thus, various factors such as multiple hidden layers, their interconnectivity, and capacity for learning features from a given input are a significant difference between conventional and DL-based networks. Other than this, traditional artificial neural networks (ANNs) generally comprise three layers which are given training to achieve representations based on supervised learning for performing particular tasks [3]. On the contrary, deep neural networks (DNNs) have the ability to represent patterns of their respective outcomes based on the given inputs via an unsupervised learning approach [4]. Figure 11.1 illustrates the pattern of a traditional ANN comprising three layers responsible for input, processing, and output (extraction of features). Here, artificial neurons reside at the input layer to identify pixel values of given data, and these are sent to a hidden layer for activation according to their weights for identifying edges (connections) based on which the output layer extracts features for generating output at the final stage. Figure 11.2 represents DNNs with multiple layers

DOI: 10.1201/9781003437079-11

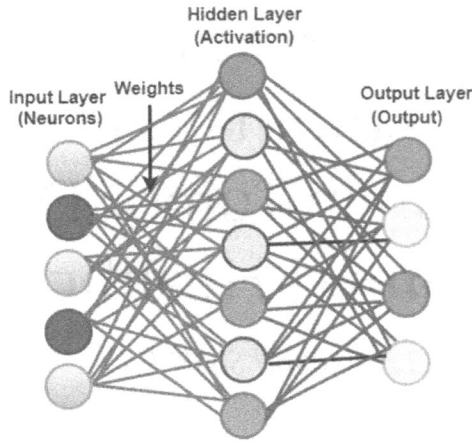

FIGURE 11.1 Basic architecture of a traditional artificial neural network.

for input, processing, and extraction of features, which are further fed to fine-tune the network using back-propagation algorithms to perform end-to-end tasks.

It is also clear that models based on DL help in processing complex patterns, as well as objects, to achieve accurate and efficient outcomes while predicting and classifying the given data. DL networks have the ability to extract significant information based on distinct patterns, which could be a challenge for existing approaches to analyze accurately. In this way, these DL models have revolutionized the way the healthcare sector uses big data, and as a result, there is a high

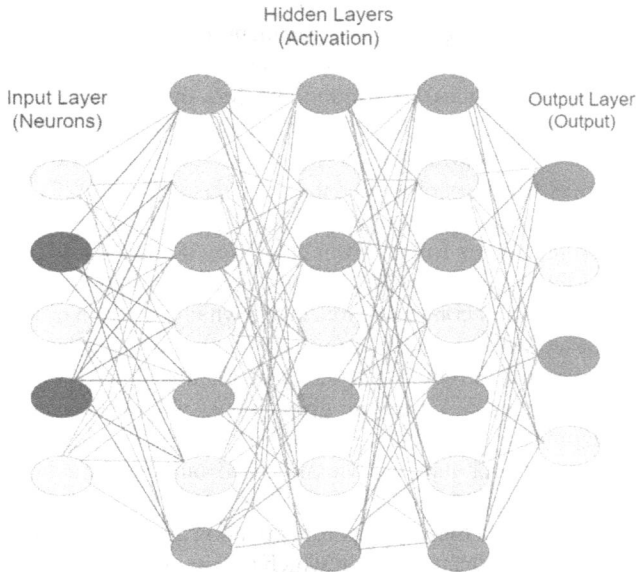

FIGURE 11.2 Basic architecture of a deep neural network.

demand for the emerging technology of DNNs. The dynamic, as well as complicated, nature of associated data such as medical imaging, electronic health records
(EHRs), data extracted from Internet of Things (IoT)-based sensors, wearables,
etc., can be processed by DNNs to drastically improve patient outcomes, including
the early diagnosis of brain tumors.

Additionally, DNNs also support analysis of laboratory procedures, which may
include analyzing chemicals, measuring glucose levels and blood pressure, providing hemoglobin reports, EHRs, clinical notes, specific tests for identifying the presence of disease, along with the presence of drugs in an individual, which could help
medical professionals provide recommendations to their patients for their speedy
recovery. Thus, IoT-based sensors, wearables etc., can be used for efficient collection
of data from multiple sources. Preprocessing on the acquired data keeps the values
of input data between 0 and 1 to perform normalization, and resizing can also be
done at this stage for removal of unwanted noise from the input data and to maintain
uniformity. Moreover, augmentation of data can also be done if considered classes of
data are not balanced to avoid overfitting and underfitting issues in order to achieve
efficient and accurate outcomes in the healthcare sector.

Preprocessed data is generally forwarded to DNNs for extraction of hidden features from the input data. A key parameter of utilizing DL models is that these networks are capable of analyzing input data. Most of the research utilized pretrained
transfer learning models at this point that take comparatively less time to train the
network and to diagnose the presence of a tumor. However, classifiers such as support vector machines, random forest, etc., can be further deployed at the output for
differentiating infected images from healthy ones. After performing such an operation, validation of the model can also be done based on the test data to identify performance metrics such as accuracy, sensitivity, specificity, and F1-score. Based on
such parameters, DL models are generally used by radiologists as an alternate source
of information when making any decisions with respect to tumor presence.

LeCun et al. [1] have summarized a systematic overview comprising architectures
and approaches of DNNs to represent the potential of its applications in healthcare.
Rajkomar et al. [5] discussed the significance of DL in the healthcare sector for analyzing images and providing decision-making processes for clinical support. These
authors also presented in-depth analysis of techniques based on machine learning
to highlight the use of DL in healthcare. Table 11.1 shows a comparison of existing

TABLE 11.1
Comparison of DNN Approaches in Healthcare

Techniques	Features	References
Convolutional Neural Network	Excellent performance in analysis	[1]
Recurrent Neural Network	Ability to model temporal dependency	[2]
Generative Adversarial Network	Generates synthetic data	[4]
Transformer Based Model	Efficient natural language processing tasks	[5]
Long Short-Term Memory	Vanishes gradient issues	[6]
Deep Reinforcement Learning	Handles complex decision-making tasks	[7]

TABLE 11.2

Parameter-Based DL and IoT Approaches

Parameters	IoT Approaches	Deep Learning Approaches
Data Collection	Sensors, wearables	Depends upon preexisting data or requires labeled data
Data Volume	Ability to handle real-time data at large scale	Can handle vast amounts of data
Data Variety	Distinct collection such as environmental factors, vital signs	Structured data
Real-Time Processing	Monitors real-time applications and performs analysis	Needs significant computational time
Complexity	Simple and inexpensive	Complex and requires much infrastructure
Flexibility	Ability to adapt in distinct healthcare settings	Ability to perform distinct healthcare tasks
Interpretability	Depends upon additional analytics	Depends upon features extraction
Training Data	Limited or biased	Needs vast and diverse datasets
Model Updates	May ask for updates in firmware	Possible without hardware updates
Privacy and Security	Concerned	Needs safeguards
Scalability	Limited to device capacity	Requires high-performance computing
Applications	Real-time monitoring, remote patient monitoring, smart healthcare	Diagnosing disorders, medical imaging, drug discovery

approaches of DNNs in healthcare to highlight each technique's main features. DL has shown great potential to offer various advantages as well as the possibility of transforming existing approaches in the healthcare sector.

To give an example, convolutional neural network (CNN) models showed amazing results during analysis of medical images. These automated systems can help in diagnosing numerous diseases, such as tumors, lung cancer, skin cancer [6], Alzheimer's, Parkinson's, skin lesions, etc., at an early stage enabling clinicians to make timely decisions regarding patient health. Even algorithms based on DNNs [7] can process and analyze vast amount of clinical data, which can include clinical notes, EHRs, etc. This capability also allows these systems to extract features from the given data in order to contribute to decision-making processes, which may ultimately result in efficient, as well as accurate, outcomes.

Table 11.2 represents the difference between IoT and DL approaches in healthcare based on key parameters.

11.2 CLASSIFICATION OF DEEP LEARNING MODELS FOR DIAGNOSING BRAIN TUMORS

Numerous DL techniques have been characterized into various algorithms which are required in smart healthcare to provide efficient and accurate outcomes while diagnosing brain tumors. Integration of such IoT and DL technology demonstrates great promise in healthcare via the use of wearables, medical sensors, health monitoring devices, predictive approaches, clinical records management, etc. These classification approaches allow this technology to be used in specific research applications in healthcare, while also highlighting its benefits and challenges.

11.2.1 HYBRID MODELS

Hybrid models analyze images as well as sequence-based medical data, and this collaboration of two or more DL models produces efficient and accurate outcomes for diagnosing brain tumors, along with providing recommendations for treatment. For instance, hybrid models have the ability to analyze clinical notes and EHRs comprised of a structured form of lab results, patient health records, etc., or an unstructured form representing patient symptoms over a period of time.

11.2.2 TRANSFER LEARNING

Transfer learning utilizes the concept of pretrained DL networks for extracting features in terms of pretrained weights based on a large dataset to overcome issues when data is less available. It can also reduce time-consuming processes in diagnosing diseases as the pretraining facility of these networks enables them to learn from past experiences, resulting in improved performance of networks in smart healthcare. Thus, the patterns or features of one model during the training process can be utilized by another model without any training in order to provide quick and efficient outcomes in diagnosis. For instance, a model which is pretrained on MRI images can be provided with a fine-tuning feature to ensure efficient outcomes while diagnosing a brain tumor or performing segmentation of a brain stroke. Thus, these features can be transferred from one place to another for performing the desired operation in healthcare. Figure 11.3 illustrates the effective hybrid model for diagnosing brain tumors in collaboration with style-transfer learning [8] where data has been gathered with the help of IoT sensors, wearables, etc., and has also undergone preprocessing for removal of unwanted distortions or balancing its respective classes with the help of data augmentation, resizing, etc. Also, images are converted into respective pixels with the help of an encoding operation. Training has been provided to the model with the help of the hybrid Alex network which further utilizes transfer learning for extraction of significant features from input data while diagnosing a brain tumor. Obtained output is evaluated based on test data to achieve efficient results in terms of its accuracy, specificity, time, complexity, and sensitivity.

11.2.3 FEDERATED LEARNING

To maintain data privacy, federated learning technology is used in various healthcare industries, as shown in Figure 11.4. Here, training has been provided simultaneously to the multiple interconnected devices on a decentralized server [9] without sharing sensitive information due to the central hub aggregator which leads to privacy of data in the smart healthcare. Moreover, this technology also

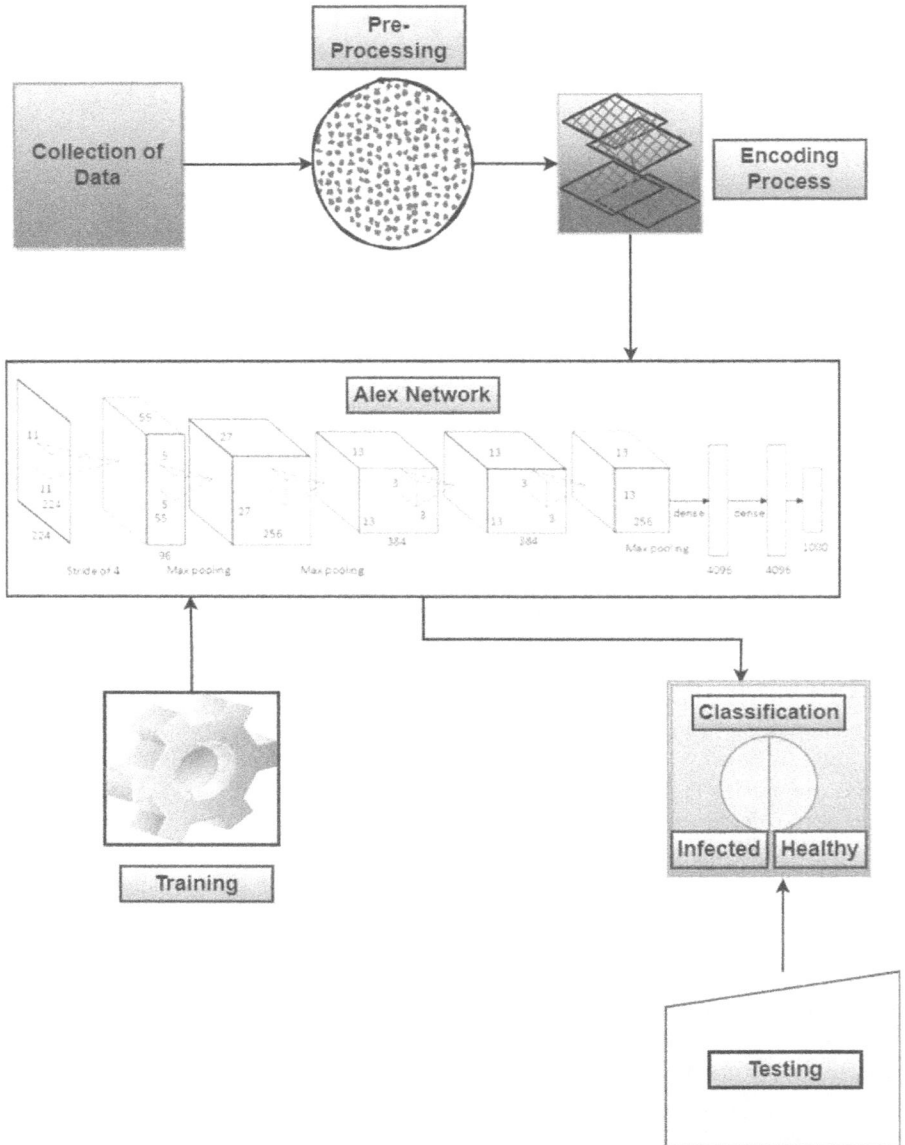

FIGURE 11.3 Basic architecture of an effective hybrid model with style-transfer learning.

minimizes the cost of communicating between multiple devices as training is performed at a local level on edges of devices, which ultimately require less bandwidth consumption too. Thus, the initial global model helps to achieve the privacy of data in smart healthcare and provides efficient outcomes while diagnosing brain tumors due to its property of supporting updates in the local model whenever required and verification of model is performed to ensure effective outcomes in smart healthcare.

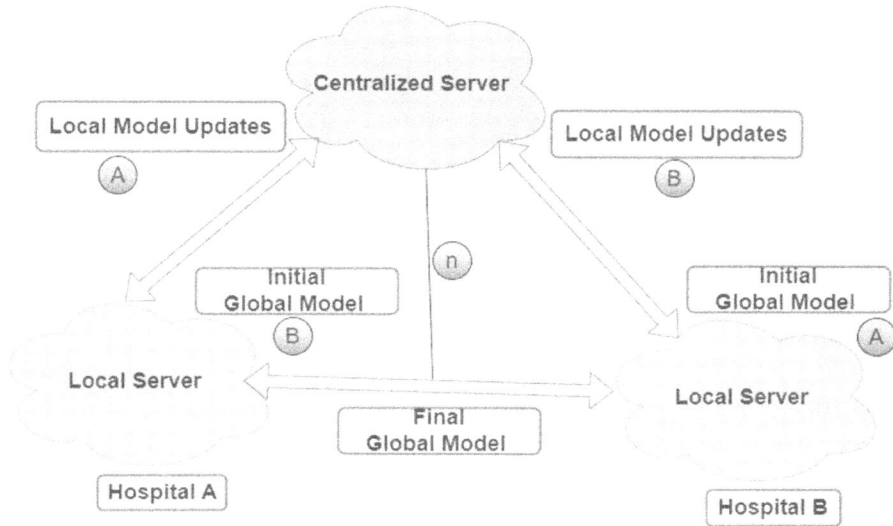

FIGURE 11.4   Basic architecture of federated learning.

11.2.4   REINFORCEMENT LEARNING

Reinforcement learning technology helps to train models using current experiences gained from sending the agent out into the open environment. For each and every action, agents get rewarded for positive action, which enables them to reach their destination, and they receive a penalty for each negative action that moves them away from reaching their goal. This helps them to make intelligent decisions based on the feedback received, which then improves prediction analysis and enhances the performance of networks in smart healthcare.

Based on these classification approaches, Bolhasani et al. [10] discussed benefits, applications, and challenges of multiple DL techniques to be used in smart healthcare for diagnosing disorders at an early stage, health monitoring, risk management, decision-making, etc., in order to have efficient and accurate results.

11.3   DEEP LEARNING ARCHITECTURES FOR DIAGNOSING BRAIN TUMORS

There are various frameworks based on DL models which have been designed to be used in special applications in smart healthcare. These frameworks enable the tasks of analyzing, interpreting, and extracting meaningful information from the given input to enhance and improve patient outcomes in healthcare. The following sections describe some of the DL architectures utilized for diagnosing brain tumors in smart healthcare.

11.3.1   CONVOLUTIONAL NEURAL NETWORKS

CNN architecture, as shown in Figure 11.5, comprises of convolutional, pooling, and fully connected layers. These analyze given data in smart healthcare for the

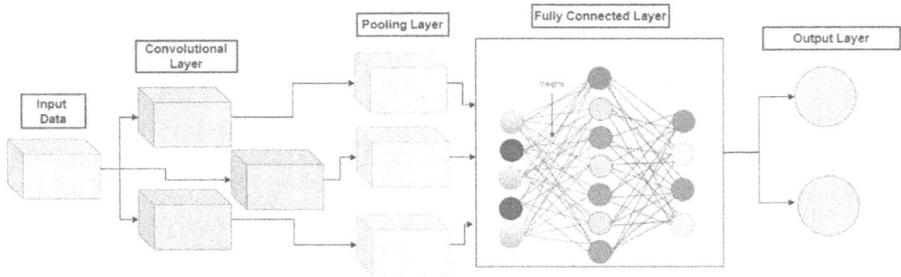

FIGURE 11.5   Basic architecture of convolutional neural networks.

extraction of meaningful information and to provide a clear understanding of the brain tumor at an early stage in order to initiate timely treatment and a risk management plan.

To achieve such tasks, convolutional layers utilize filters to remove noise from images and to detect patterns or features accurately. The pooling layer is responsible for the reduction of high-dimensionality features to minimize complexity of networks to have significant features. Fully connected layers are used for making predictions based on learned features, thus providing efficient outcomes in smart healthcare. These attributes mean that the CNN network can be deployed in multiple applications in healthcare sector, such as identification of Alzheimer's, Parkinson's, or a tumor from MRI images, the classification of skin cancer, glaucoma, diabetes, etc., and diagnose any abnormality contained in scans.

Litjens et al. [11] also discussed various aspects of CNNs in DL models including the analysis of medical images, and conveying the benefits and challenges, and highlighting the efficient and accurate outcomes in smart healthcare.

11.3.2   RECURRENT NEURAL NETWORKS

Recurrent neural networks (RNNs), as shown in Figure 11.6, are used in the analysis of sequence-based data for real-time applications in smart healthcare systems. This is largely due to their ability to capture dependencies as well as variations in lengthy variables collected from IoT devices. These networks have features of feedback connectivity [12] which ultimately allows data to pass from one place to other based on distinct time zones. A hidden layer within these networks is responsible for maintaining the experiences of past data by updating itself, and provides help for future tasks to be performed. Thus, recurrent layers compute output at each step and send signals back to the network for accurate analysis based on a time-series. These capabilities allow RNNs to analyze clinical notes, and also EHRs, for prediction of the progress of the brain tumor and its risk management, and remote patient monitoring in the smart healthcare sectors. Choi et al. [13] also provided studies in which RNNs have been utilized for prediction of medical conditions using EHRs to achieve efficient results in making intelligent decisions while working in smart healthcare.

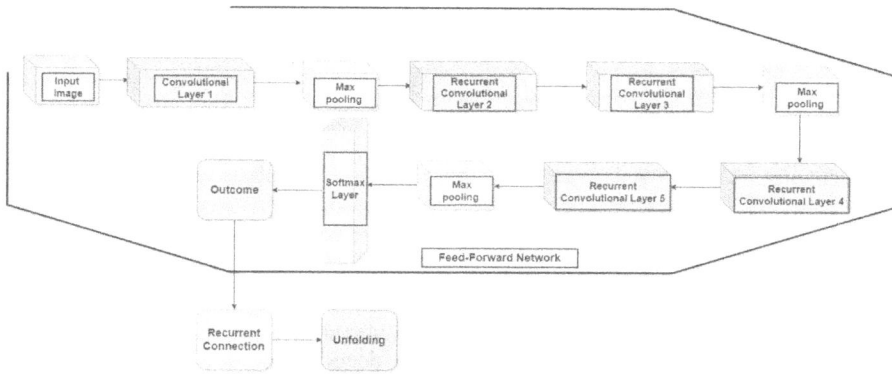

FIGURE 11.6 Basic architecture of recurrent neural networks.

11.3.3 LONG SHORT-TERM MEMORY MODELS

Long short-term memory (LSTM) models, as shown in Figure 11.7, have the ability to tackle dependencies of long-term while analyzing input data based on time-series for clinical notes, EHRs, etc., and these networks are able to resolve issues raised due to vanishing gradient. The memory cells comprises three elements, namely: Input gate, responsible for regulating flow of data toward a memory cell; Forget gate, which helps in monitoring information that needs to be removed from a memory cell; and Output gate used for determination of output value on the base of given input and the state of memory cell. This means it has vast storage capacity for the information to be shared from one network to another, providing portable real-time facilities for patients for diagnosing and treating disorders, which ultimately minimizes the chances of losing medical data, and providing efficient outcomes in smart healthcare. Lipton et al. [14] discussed the various techniques of these models used in collaboration with RNNs in diagnosing disease, including brain tumors, based on sequential medical data where long-term dependencies exist in real-time, and where

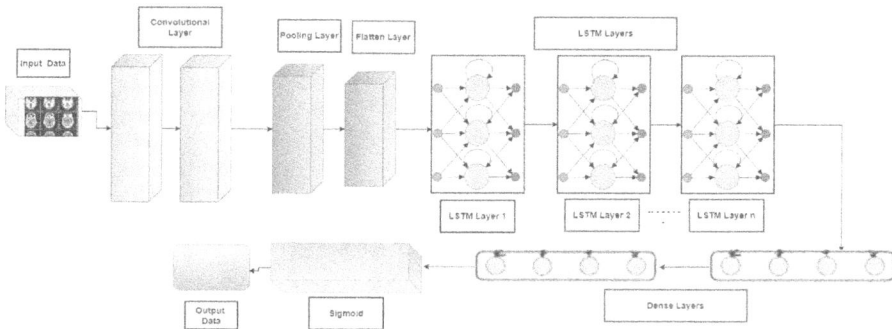

FIGURE 11.7 Basic architecture of long short-term memory model.

the detection of abnormalities from input MRI images is required at the priority level to avoid the risk of losing patients' lives and improving outcomes of patients' health.

11.3.4 AUTOENCODERS

Autoencoder (AE) networks, as shown in Figure 11.8, follow unsupervised learning techniques. AE networks help in the reconstruction of given data and identifying abnormalities lying within signals or images provided as an input in healthcare systems which have been generally collected via IoT sensors or wearables. These networks have the ability to identify any deviation from the actual pattern as they are comprise of an encoding-decoding parameter [15] where encoding is used for mapping input data to capture significant features, and where decoding is used for reconstructing the actual input data based on the representations stored in its network for achieving accurate results. Moreover, AEs feature data imputation that aids in filling in the missing or noisy medical data collected by various IoT sensors or variables on the basis of learned pattern representations. Generation of synthetic data is also possible to attain the help of AEs, and this can be performed to augment data in case of less availability of data, or to balance the classes while diagnosing diseases in smart healthcare. Kim et al. [16] have also summarized the significance of using AEs in healthcare systems for capturing various complex patterns, and identifying any abnormalities, in order to analyze and reconstruct medical data for improvement in patient outcomes, including for diagnosing brain tumors, in smart healthcare applications.

11.3.5 GENERATIVE ADVERSARIAL NETWORK

Generative adversarial networks (GANs), as shown in Figure 11.9, are utilized to generate synthetic medical data for improving pixel resolution of images. This allows these networks to perform further operations based on data augmentation, filling missing or noisy values in input data while training the model to achieve accurate and efficient results. These GANs are generally comprised of two elements, namely: a

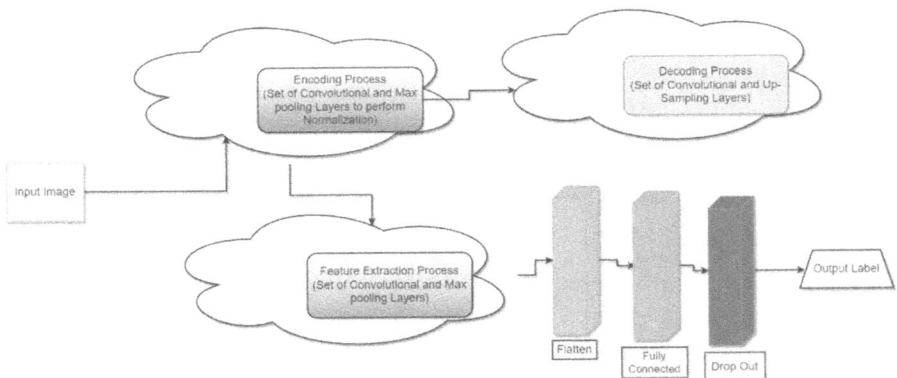

FIGURE 11.8 Basic architecture of a convolutional-based autoencoder (AE).

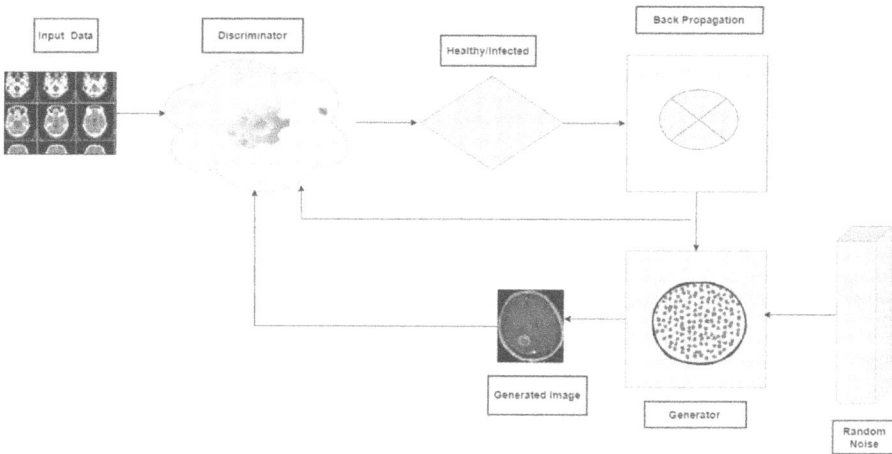

FIGURE 11.9   Basic architecture of generative adversarial network.

generator network, responsible for obtaining synthetic data via mapping of vectors to achieve the required distribution of data; and a discriminator network, responsible for providing feedback to the generator to help it improve its capability for generating more accurate data. Thus, both networks give significant contributions to analyzing and predicting diseases where accessing data is restricted due to regulatory frameworks in healthcare, while preserving data to be transferred from one place to other. Yi et al. [17] have also discussed various studies in which these networks have been used in specific applications of smart healthcare to obtain synthetic data, image transformation, as well as diagnosis of abnormalities, if any. These networks also demonstrate great promise in generating accurate and efficient outcomes while diagnosing brain tumors.

In addition, Tables 11.3–11.5 represent techniques proposed by various researchers for diagnosing brain tumor disorders in smart healthcare along with their respective objectives.

11.4   INNOVATIVE NEW TECHNIQUES FOR DIAGNOSING BRAIN TUMORS IN SMART HEALTHCARE

Much of the research discussed traumatic brain injury [60] along with its origin and causes, and some authors looked into pharmacological preconditioning of the brain [61]. Based on such research, the following are the innovative new techniques responsible for incorporating existing approaches with advanced technology to predict disorder at an early stage, enabling remote patient monitoring and improving patient outcomes:

1. *Internet of Things and wearables*: To monitor patient health remotely, the collaboration of IoT [62] and wearables is required. This allows for data collecting based on real-time applications to make predictive analysis in order to prevent fatal diseases in patients and provide timely treatment.

TABLE 11.3

State-of-the-Art Techniques for Diagnosing Brain Tumors

Years	References	Techniques	Objectives
2020	[18]	DNN-transfer learning	Automated brain tumor classification
2019	[19]	Convolutional neural network, discrete wavelet transform, long short-term memory network	Distinguishing liver and brain tumor
2022	[20]	Deep convolutional neural network	Diagnosing brain tumor
2019	[21]	Deep convolutional neural network with transfer learning	Classification of brain tumor
2022	[22]	Deep learning and VGG16	Analysis of brain tumor
2019	[23]	Convolutional neural network	Classification of brain waves
2019	[24]	Integrated noise-to-image and image-to-image	Data augmentation for diagnosing brain tumor
2018	[25]	Deep neural network	Classification of brain tumor
2017	[26]	Deep neural network	Diagnosing brain cancer
2016	[27]	Contourlet transform and Zernike Moments based data fusion	Diagnosing brain tumor disease
2020	[28]	Integrated optimal wavelet statistical texture and recurrent neural network	Diagnosing and classifying brain tumor disease
2019	[29]	Convolutional neural network	Identification and classification of abnormal brain images
2020	[30]	Deep neural network	Detection and classification of brain tumor
2022	[31]	Generative adversarial network-style transfer	Generate of brain tumor image

2. *Artificial intelligence (AI) and machine learning (ML)*: To make intelligent decisions while diagnosing illnesses and predicting the progression [63] of illness, AI and ML techniques play major role in analyzing a vast amount of data for effective and accurate outcomes.

3. *Blockchain*: To ensure integrity of data, interoperability, management of medical records, blockchain technology [64] enhances data privacy as well as security, while sharing data from one place to another.

4. *Telemedicine and virtual reality (VR)*: To consult experts remotely, this technology [65] provides a way for patients to minimize the need to visit in-person and promotes timely treatment and recommendations where virtual reality helps medical experts in visualizing complex structures in the brain while considering 3-dimensional space to provide clear and better understanding of tumor location, its size and extent. However, telemedicine refers to the process which considers communication technology for providing medical care from distance to provide remote consultations to the patients in underserved areas.

5. *Predictive analytics and big data*: For identification of patterns in order to make accurate and efficient predictions, techniques of predictive analytics and big data [66] come into play.

TABLE 11.4

State-of-the-Art Techniques for Diagnosing Brain Tumors

Years	References	Techniques	Objectives
2020	[32]	Enhanced softmax loss function – deep neural network	Diagnosing brain tumor disease
2021	[33]	DNN – machine learning classifier	Classifying brain tumor disorder
2021	[34]	Deep CNN – fully optimized framework	To perform multi-classification of brain tumor
2020	[35]	Blockwise fine-tuning – transfer learning	To classify brain tumor
2019	[36]	DNN – machine learning algorithm	Brain tumor diagnosis
2019	[37]	CNN – genetic algorithms	Classification of brain tumor grades
2019	[38]	Deep neural network	Brain tumor diagnosis
2019	[39]	Capsule-network	To identify impact of preprocessing on brain tumor classification
2019	[40]	Convolutional neural network	To classify brain images
2021	[41]	Convolutional dictionary learning – local constraint	To classify brain tumor
2021	[42]	3D-CNN and feature selection	To diagnose microscopic brain tumor
2019	[43]	Deep CNN – extensive data augmentation	To classify multi-grade brain tumor

TABLE 11.5

State-of-the-Art Techniques for Diagnosing Brain Tumors

Years	References	Techniques	Objectives
2021	[44]	CNN-SVM	To identify brain tumor disease
2021	[45]	Residual network –global average pooling	To classify multi-class brain tumor
2020	[46]	DNN –robust feature selection	To diagnose and classify multimodal brain tumor
2020	[47]	Integrated VGG16-CNN	To classify brain tumor disease
2018	[48]	Hybrid learning	To diagnose and classify brain tumor
2018	[49]	3D-CNN	To classify pancreatic tumor
2017	[50]	Deep neural network	To segment brain tumor
2019	[51]	Deep neural network	To diagnose and segment brain tumor
2017	[52]	Deep neural network	Theoretical framework to segment brain tumor
2016	[53]	CNN – transfer learning	Theoretical framework to diagnose brain tumor
2019	[54]	Deep neural network	To assess glioma burden for automatic volumetric and bidirectional measurement
2016	[55]	Convolutional neural network	To segment MRI brain images automatically
2019	[56]	Transfer learning – fine tuning	To diagnose and classify brain tumor
2019	[57]	R-CNN	To diagnose and classify brain tumor
2020	[58]	Convolutional neural network	To diagnose brain tumor using MRI scans
2015	[59]	Watershed segmentation	To diagnose brain tumor

11.5 NEW OPPORTUNITIES IN THE 21 ST CENTURY

The following are the new opportunities which have been made available due to the integration of IoT and DL techniques for diagnosing brain tumors in smart healthcare, as shown in Figure 11.10:

1. *Disease diagnosis*: There are various algorithms based on DL that have demonstrated great potential in diagnosing diseases by analyzing medical images given as input data. CNNs have been used in this way and are major contributors to providing efficient and accurate results in healthcare diagnosis and classification of diseases [6].
2. *Health monitoring*: Smart healthcare systems enable collection of physiological data for monitoring patient health using IoT devices [5] which help to diagnose abnormalities, predict health related issues and suggest precautions to maintain good health.
3. *Predictive analysis*: RNNs can analyze data based on times-series from IoT-based devices to identify symptoms of any disorder experienced by an individual at an early stage, predict progress of an illness, and increase the life expectancy of patients [67].
4. *Remote patient monitoring*: DL techniques can monitor patients remotely with the help of IoT wearables or sensors so any health abnormalities can be detected and diagnosed by alerting the patient to consult with a specialist without delay [68].
5. *Data security*: Integration of DL and IoT results in the need to ensure data is secure while being shared from one point to another. The Tensorflow

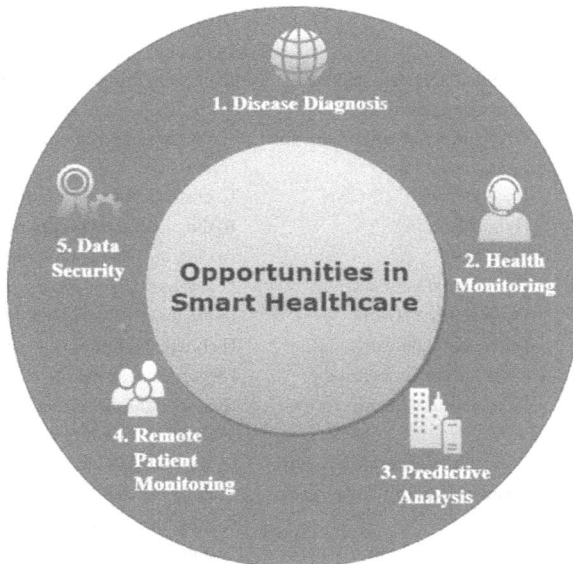

FIGURE 11.10   Opportunities in smart healthcare.

Federated framework has been suggested as a way of providing confidentiality during analysis of any medically sensitive data [69].

11.6 CURRENT CHALLENGES IN IoT-BASED SMART HEALTHCARE

Although the integration of DL techniques with IoT devices has demonstrated great potential in smart healthcare systems, there is an urgent need to address some of the challenges experienced when using this technology, as shown in Figure 11.11. These challenges are as follows:

1. *Data privacy*: Collection of patient data by IoT wearables or sensors has raised concern for privacy of data which may be vulnerable to unauthorized access while being transmitting, leading to security issues for sensitive medical data [70].
2. *Interoperability and standardization*: Most of the systems that comprise IoT sensors may not support interoperability as well as standardization.

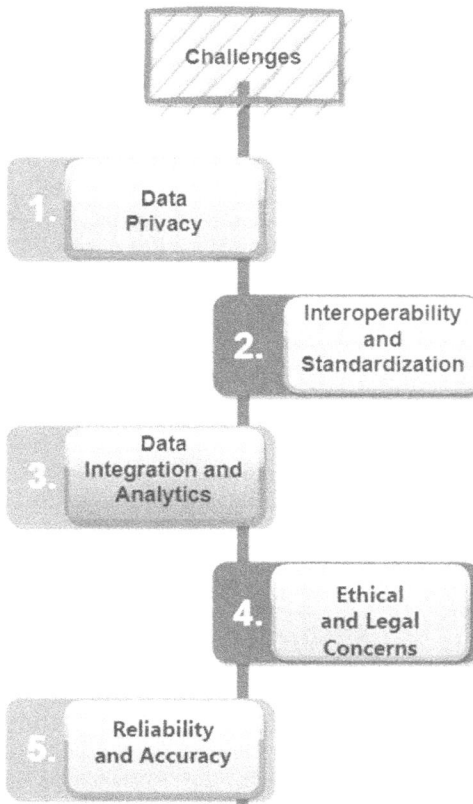

FIGURE 11.11   Current challenges in IoT-based smart healthcare.

This is a key component when sharing information [71] from one place to another, and without this systems may experience challenges in ensuring efficient outcomes, especially in healthcare sector [72].

3. *Data integration and analytics*: Although IoT-based systems have the ability to analyze medical data, as the data exists in various forms such as images, signals, records, etc., this can create challenges for such systems during preprocessing and extracting features and make achieving efficient outcomes more difficult [73].

4. *Ethical and legal concerns*: IoT-based systems for smart healthcare may experience issues regarding ethical and legal concerns from patients whose data is supposed to be shared from one place to another [74]. The system must be transparent from beginning to end, which is quite challenging as it may also lead to bias, [75] and this is against research ethics.

5. *Reliability and accuracy*: Analyzing patient data collected via IoT sensors and wearables may result in low-quality or noisy signals or images which can deteriorate the quality of actual data and lead to unreliable and less accurate data during decision-making [76]. Keeping input data noise-free and reliable, and deploying data validation approaches, are necessary to provide efficient results in the healthcare sector [77].

11.7 CONCLUSION

Integrating IoT with DL demonstrates great potential in smart healthcare when diagnosing brain tumors at an early stage, performing predictive analysis, health monitoring, etc., and ensures efficient and accurate outcomes. To achieve this, there are various IoT-based sensors and wearables responsible for the collection of big data which further rely on DL techniques such as CNNs, RNNs, transfer learning, federated learning, etc., for analyzing medical data and to make the decision-making process more accurate. Moreover, these systems have the ability to monitor patients remotely to avoid any delay in their treatment, which ultimately increases the life expectancy of patients. While working with such systems, there are also a few challenges that need to be addressed, such as data privacy, interoperability and standardization issues, data integration and analytics, ethical and legal concerns, reliability and accuracy, etc. Notwithstanding these challenges, integrated DL-IoT models provide efficient and accurate outcomes to aid radiologists in decision-making processes while diagnosing brain tumors, in order for them to provide timely treatment to patients in the smart healthcare sector.

REFERENCES

1. Y. LeCun, Y. Bengio, and G. Hinton, "Deep learning," *Nature*, vol. 521, no. 7553, pp. 436–444, 2015.
2. Y. Bengio, A. Courville, and P. Vincent, "Representation learning: A review and new perspectives," *IEEE transactions on pattern analysis and machine intelligence*, vol. 35, no. 8, pp. 1798–1828, 2013.
3. D. E. Rumelhart, G. E. Hinton, and R. J. Williams, "Learning representations by back-propagating errors," *Nature*, vol. 323, no. 6088, pp. 533–536, 1986.

4. Y. Bengio *et al.*, "Learning deep architectures for AI," *Foundations and Trends® in machine learning*, vol. 2, no. 1, pp. 1–127, 2009.
5. A. Rajkomar, J. Dean, and I. Kohane, "Machine learning in medicine," *New England journal of medicine*, vol. 380, no. 14, pp. 1347–1358, 2019.
6. A. Esteva, B. Kuprel, R. A. Novoa, J. Ko, S. M. Swetter, H. M. Blau, and S. Thrun, "Dermatologist-level classification of skin cancer with deep neural networks," *Nature*, vol. 542, no. 7639, pp. 115–118, 2017.
7. R. Miotto, F. Wang, S. Wang, X. Jiang, and J. T. Dudley, "Deep learning for healthcare: Review, opportunities and challenges," *Briefings in bioinformatics*, vol. 19, no. 6, pp. 1236–1246, 2018.
8. N. A. Samee, N. F. Mahmoud, G. Atteia, H. A. Abdallah, M. Alabdulhafith, M. S. Al-Gaashani, S. Ahmad, and M. S. A. Muthanna, "Classification framework for medical diagnosis of brain tumor with an effective hybrid transfer learning model," *Diagnostics*, vol. 12, no. 10, p. 2541, 2022.
9. S. Nazir, and M. Kaleem, "Federated learning for medical image analysis with deep neural networks," *Diagnostics*, vol. 13, no. 9, p. 1532, 2023.
10. H. Bolhasani, M. Mohseni, and A. M. Rahmani, "Deep learning applications for IoT in health care: A systematic review," *Informatics in medicine unlocked*, vol. 23, p. 100550, 2021.
11. G. Litjens, T. Kooi, B. E. Bejnordi, A. A. A. Setio, F. Ciompi, M. Ghafoorian, J. A. Van Der Laak, B. Van Ginneken, and C. I. Sánchez, "A survey on deep learning in medical image analysis," *Medical image analysis*, vol. 42, pp. 60–88, 2017.
12. S. Srinivasan, P. S. M. Bai, S. K. Mathivanan, V. Muthukumaran, J. C. Babu, and L. Vilcekova, "Grade classification of tumors from brain magnetic resonance images using a deep learning technique," *Diagnostics*, vol. 13, no. 6, p. 1153, 2023.
13. E. Choi, M. T. Bahadori, A. Schuetz, W. F. Stewart, and J. Sun, "Doctor AI: Predicting clinical events via recurrent neural networks," in *Machine learning for healthcare conference*. PMLR, 2016, pp. 301–318.
14. Z. C. Lipton, D. C. Kale, C. Elkan, and R. Wetzel, "Learning to diagnose with LSTM recurrent neural networks," *arXiv preprint arXiv:1511.03677*, 2015.
15. F. Bashir-Gonbadi, and H. Khotanlou, "Brain tumor classification using deep convolutional autoencoder-based neural network: Multi-task approach," *Multimedia tools and applications*, vol. 80, pp. 19909–19929, 2021.
16. J.-C. Kim, and K. Chung, "Multi-modal stacked denoising autoencoder for handling missing data in healthcare big data," *IEEE access*, vol. 8, pp. 104933–104943, 2020.
17. X. Yi, E. Walia, and P. Babyn, "Generative adversarial network in medical imaging: A review," *Medical image analysis*, vol. 58, p. 101552, 2019.
18. A. Rehman, S. Naz, M. I. Razzak, F. Akram, and M. Imran, "A deep learning-based framework for automatic brain tumors classification using transfer learning," *Circuits, systems, and signal processing*, vol. 39, pp. 757–775, 2020.
19. H. Kutlu, and E. Avcı, "A novel method for classifying liver and brain tumors using convolutional neural networks, discrete wavelet transform and long short-term memory networks," *Sensors*, vol. 19, no. 9, p. 1992, 2019.
20. A. Chattopadhyay, and M. Maitra, "MRI-based brain tumour image detection using CNN based deep learning method," *Neuroscience informatics*, vol. 2, no. 4, p. 100060, 2022.
21. S. Deepak, and P. Ameer, "Brain tumor classification using deep CNN features via transfer learning," *Computers in biology and medicine*, vol. 111, p. 103345, 2019.
22. A. Younis, L. Qiang, C. O. Nyatega, M. J. Adamu, and H. B. Kawuwa, "Brain tumor analysis using deep learning and VGG-16 ensembling learning approaches," *Applied sciences*, vol. 12, no. 14, p. 7282, 2022.
23. S. R. Joshi, D. B. Headley, K. Ho, D. Paré, and S. S. Nair, "Classification of brain-waves using convolutional neural network," in *2019 27th European Signal Processing Conference (EUSIPCO)*. IEEE, 2019, pp. 1–5.

24. C. Han, L. Rundo, R. Araki, Y. Nagano, Y. Furukawa, G. Mauri, H. Nakayama, and H. Hayashi, "Combining noise-to-image and image-to-image GANs: Brain MR image augmentation for tumor detection," *IEEE access*, vol. 7, pp. 156966–156977, 2019.

25. H. Mohsen, E.-S. A. El-Dahshan, E.-S. M. El-Horbaty, and A.-B. M. Salem, "Classification using deep learning neural networks for brain tumors," *Future computing and informatics journal*, vol. 3, no. 1, pp. 68–71, 2018.

26. M. Kachwalla, M. Shinde, R. Katare, A. Agrawal, V. Wadhai, and M. Jadhav, "Classification of brain MRI images for cancer detection using deep learning," *International Journal of Advanced Research in Computer and Communication Engineering*, vol. 3, pp. 635–637, 2017.

27. A. A. Pandian, and R. Balasubramanian, "Fusion of contourlet transform and zernike moments using content based image retrieval for MRI brain tumor images," *Indian Journal of Science and Technology*, vol. 9, no. 29, pp. 1–8, 2016.

28. S. S. Begum, and D. R. Lakshmi, "Combining optimal wavelet statistical texture and recurrent neural network for tumour detection and classification over MRI," *Multimedia tools and applications*, vol. 79, pp. 14009–14030, 2020.

29. P. M. Krishnammal, and S. S. Raja, "Convolutional neural network based image classification and detection of abnormalities in MRI brain images," in *2019 International Conference on Communication and Signal Processing (ICCSP)*. IEEE, 2019, pp. 0548–0553.

30. A. Raj, A. Anil, P. Deepa, H. A. Sarma, and R. N. Chandran, "Brainnet: A deep learning network for brain tumor detection and classification," in *Advances in communication systems and networks: Select proceedings of ComNet 2019*. Springer, 2020, pp. 577–589.

31. D. Mukherjee, P. Saha, D. Kaplun, A. Sinitca, and R. Sarkar, "Brain tumor image generation using an aggregation of GAN models with style transfer," *Scientific reports*, vol. 12, no. 1, p. 9141, 2022.

32. S. Maharjan, A. Alsadoon, P. Prasad, T. Al-Dalain, and O. H. Alsadoon, "A novel enhanced softmax loss function for brain tumour detection using deep learning," *Journal of neuroscience methods*, vol. 330, p. 108520, 2020.

33. J. Kang, Z. Ullah, and J. Gwak, "MRI-based brain tumor classification using ensemble of deep features and machine learning classifiers," *Sensors*, vol. 21, no. 6, p. 2222, 2021.

34. E. Irmak, "Multi-classification of brain tumor MRI images using deep convolutional neural network with fully optimized framework," *Iranian Journal of science and technology, transactions of electrical engineering*, vol. 45, no. 3, pp. 1015–1036, 2021.

35. B. Anilkumar, and P. R. Kumar, "Tumor classification using block wise fine tuning and transfer learning of deep neural network and KNN classifier on MR brain images," *International Journal of Emerging Trends in Engineering Research*, vol. 8, no. 2, pp. 574–583, 2020.

36. M. Siar, and M. Teshnehlab, "Brain tumor detection using deep neural network and machine learning algorithm," in *2019 9th international conference on computer and knowledge engineering (ICCKE)*. IEEE, 2019, pp. 363–368.

37. A. K. Anaraki, M. Ayati, and F. Kazemi, "Magnetic resonance imaging-based brain tumor grades classification and grading via convolutional neural networks and genetic algorithms," *Biocybernetics and biomedical engineering*, vol. 39, no. 1, pp. 63–74, 2019.

38. P. T. Selvy, V. Dharani, and A. Indhuja, "Brain tumour detection using deep learning techniques," *International Journal of Scientific Research in Computer Science, Engineering and Information Technology*, vol. 169, 175, 2019.

39. R. V. Kurup, V. Sowmya, and K. Soman, "Effect of data pre-processing on brain tumor classification using CapsuleNet," in *ICICCT 2019–System reliability, quality control, safety, maintenance and management: Applications to electrical, electronics and computer science and engineering*. Springer, 2020, pp. 110–119.

40. A. E. Boustani, M. Aatila, E. E. Bachari, and A. E. Oirrak, "MRI brain images classification using convolutional neural networks," in *Advanced Intelligent Systems for Sustainable Development (AI2SD'2019) Volume 4 – Advanced intelligent systems for applied computing sciences*. Springer, 2020, pp. 308–320.

41. X. Gu, Z. Shen, J. Xue, Y. Fan, and T. Ni, "Brain tumor MR image classification using convolutional dictionary learning with local constraint," *Frontiers in neuroscience*, vol. 15, p. 679847, 2021.

42. A. Rehman, M. A. Khan, T. Saba, Z. Mehmood, U. Tariq, and N. Ayesha, "Microscopic brain tumor detection and classification using 3D CNN and feature selection architecture," *Microscopy research and technique*, vol. 84, no. 1, pp. 133–149, 2021.

43. M. Sajjad, S. Khan, K. Muhammad, W. Wu, A. Ullah, and S. W. Baik, "Multi-grade brain tumor classification using deep CNN with extensive data augmentation," *Journal of computational science*, vol. 30, pp. 174–182, 2019.

44. S. Deepak, and P. Ameer, "Automated categorization of brain tumor from MRI using CNN features and SVM," *Journal of ambient intelligence and humanized computing*, vol. 12, pp. 8357–8369, 2021.

45. R. L. Kumar, J. Kakarla, B. V. Isunuri, and M. Singh, "Multi-class brain tumor classification using residual network and global average pooling," *Multimedia tools and applications*, vol. 80, pp. 13429–13438, 2021.

46. M. A. Khan, I. Ashraf, M. Alhaisoni, R. Damaševičius, R. Scherer, A. Rehman, and S. A. C. Bukhari, "Multimodal brain tumor classification using deep learning and robust feature selection: A machine learning application for radiologists," *Diagnostics*, vol. 10, no. 8, p. 565, 2020.

47. O. N. Belaid, and M. Loudini, "Classification of brain tumor by combination of pretrained VGG16 CNN," *Journal of information technology management*, vol. 12, no. 2, pp. 13–25, 2020.

48. D. Amrapur, "Computer based diagnosis system for tumor detection & classification: A hybrid approach," *International journal of pure and applied mathematics*, vol. 118, no. 7, pp. 33–43, 2018.

49. X. Chen, Y. Chen, C. Ma, X. Liu, and X. Tang, "Classification of pancreatic tumors based on MRI images using 3D convolutional neural networks," in *Proceedings of the 2nd International Symposium on Image Computing and Digital Medicine*, 2018, pp. 92–96.

50. M. Havaei, A. Davy, D. Warde-Farley, A. Biard, A. Courville, Y. Bengio, C. Pal, P. M. Jodoin, and H. Larochelle, "Brain tumor segmentation with deep neural networks," *Medical image analysis*, vol. 35, pp. 18–31, 2017.

51. S. Sajid, S. Hussain, and A. Sarwar, "Brain tumor detection and segmentation in MR images using deep learning," *Arabian journal for science and engineering*, vol. 44, pp. 9249–9261, 2019.

52. Z. Akkus, A. Galimzianova, A. Hoogi, D. L. Rubin, and B. J. Erickson, "Deep learning for brain MRI segmentation: State of the art and future directions," *Journal of digital imaging*, vol. 30, pp. 449–459, 2017.

53. H.-C. Shin, H. R. Roth, M. Gao, L. Lu, Z. Xu, I. Nogues, J. Yao, D. Mollura, and R. M. Summers, "Deep convolutional neural networks for computer-aided detection: CNN architectures, dataset characteristics and transfer learning," *IEEE transactions on medical imaging*, vol. 35, no. 5, pp. 1285–1298, 2016.

54. K. Chang, A. L. Beers, H. X. Bai, J. M. Brown, K. I. Ly, X. Li, J. T. Senders, V. K. Kavouridis, A. Boaro, and C. Su *et al.*, "Automatic assessment of glioma burden: A deep learning algorithm for fully automated volumetric and bidimensional measurement," *Neurooncology*, vol. 21, no. 11, pp. 1412–1422, 2019.

55. P. Moeskops, M. A. Viergever, A. M. Mendrik, L. S. De Vries, M. J. Benders, and I. Išgum, "Automatic segmentation of MR brain images with a convolutional neural network," *IEEE transactions on medical imaging*, vol. 35, no. 5, pp. 1252–1261, 2016.

56. Z. N. K. Swati, Q. Zhao, M. Kabir, F. Ali, Z. Ali, S. Ahmed, and J. Lu, "Brain tumor classification for MR images using transfer learning and fine-tuning," *Computerized medical imaging and graphics*, vol. 75, pp. 34–46, 2019.
57. E. Avşar, and K. Salçin, "Detection and classification of brain tumours from MRI images using faster R-CNN," *Tehnički glasnik*, vol. 13, no. 4, pp. 337–342, 2019.
58. S. Sarkar, A. Kumar, S. Chakraborty, S. Aich, J.-S. Sim, and H.-C. Kim, "A CNN based approach for the detection of brain tumor using MRI scans," *Test engineering and management*, vol. 83, pp. 16580–16586, 2020.
59. P. Shanthakumar, and P. Ganesh Kumar, "Computer aided brain tumor detection system using watershed segmentation techniques," *International journal of imaging systems and technology*, vol. 25, no. 4, pp. 297–301, 2015.
60. K. Thapa, H. Khan, T. G. Singh, and A. Kaur, "Traumatic brain injury: Mechanistic insight on pathophysiology and potential therapeutic targets," *Journal of molecular neuroscience*, vol. 71, no. 9, pp. 1725–1742, 2021.
61. A. K. Rehni, T. G. Singh, A. S. Jaggi, and N. Singh, "Pharmacological preconditioning of the brain: A possible interplay between opioid and calcitonin Gene related peptide transduction systems," *Pharmacological reports*, vol. 60, no. 6, p. 904, 2008.
62. Y. Yuehong, Y. Zeng, X. Chen, and Y. Fan, "The Internet of Things in healthcare: An overview," *Journal of industrial information integration*, vol. 1, pp. 3–13, 2016.
63. E. J. Topol, "High-performance medicine: The convergence of human and artificial intelligence," *Nature medicine*, vol. 25, no. 1, pp. 44–56, 2019.
64. P. Zhang, D. C. Schmidt, J. White, and G. Lenz, "Blockchain technology use cases in healthcare," in *Advances in computers*. Elsevier, 2018, vol. 111, pp. 1–41.
65. D. M. Hilty, K. Randhawa, M. M. Maheu, A. J. McKean, R. Pantera, M. C. Mishkind, and A. Rizzo, "A review of telepresence, virtual reality, and augmented reality applied to clinical care," *Journal of technology in behavioral science*, vol. 5, pp. 178–205, 2020.
66. X.-W. Chen, and X. Lin, "Big data deep learning: Challenges and perspectives," *IEEE access*, vol. 2, pp. 514–525, 2014.
67. D. S. Rajeswari, A. L. Simha, M. N. Subhani, B. Shivaleelavathi, and V. Yatnalli *et al.*, "A review on remote health monitoring sensors and their filtering techniques," *Global transitions proceedings*, vol. 2, no. 2, pp. 392–401, 2021.
68. C. Thapa, and S. Camtepe, "Precision health data: Requirements, challenges and existing techniques for data security and privacy," *Computers in biology and medicine*, vol. 129, p. 104130, 2021.
69. M. K. Hasan, T. M. Ghazal, R. A. Saeed, B. Pandey, H. Gohel, A. Eshmawi, S. Abdel-Khalek, and H. M. Alkhassawneh, "A review on security threats, vulnerabilities, and counter measures of 5G enabled internet-of-medical-things," *IET communications*, vol. 16, no. 5, pp. 421–432, 2022.
70. J. Qi, P. Yang, G. Min, O. Amft, F. Dong, and L. Xu, "Advanced Internet of Things for personalised healthcare systems: A survey," *Pervasive and mobile computing*, vol. 41, pp. 132–149, 2017.
71. H. Bhatia, S. N. Panda, and D. Nagpal, "Internet of Things and its applications in healthcare – a survey," in *2020 8th International Conference on Reliability, Infocom Technologies and Optimization (Trends and Future Directions)(ICRITO)*. IEEE, 2020, pp. 305–310.
72. A. Ebenezer, and S. Durga, "Big data analytics in healthcare: A survey," *Journal of engineering & applied sciences*, vol. 10, no. 8, pp. 3645–3650, 2015.
73. N. Kumar, R. K. Kaushal, S. N. Panda, and S. Bhardwaj, "Impact of the Internet of Things and clinical decision support system in healthcare," in *IoT and WSN based smart cities: A machine learning perspective*. Springer, 2022, pp. 15–26.
74. N. Y. Philip, J. J. Rodrigues, H. Wang, S. J. Fong, and J. Chen, "Internet of Things for in-home health monitoring systems: Current advances, challenges and future directions," *IEEE journal on selected areas in communications*, vol. 39, no. 2, pp. 300–310, 2021.

75. H. Naz, R. Sharma, N. Sharma, and S. Ahuja, "IoT-inspired smart healthcare service for diagnosing remote patients with diabetes," in *Machine learning for edge computing.* CRC Press, 2022, pp. 97–114.
76. M. S. Patel, D. A. Asch, and K. G. Volpp, "Wearable devices as facilitators, not drivers, of health behavior change," *JAMA*, vol. 313, no. 5, pp. 459–460, 2015.
77. P. Verma, R. Tiwari, W.-C. Hong, S. Upadhyay, and Y.-H. Yeh, "Fetch: A deep learning-based fog computing and IoT integrated environment for healthcare monitoring and diagnosis," *IEEE access*, vol. 10, pp. 12548–12563, 2022.

12 6G and Distributed Computing for IoT

A Survey

*Abhishek Hazra, Mainak Adhikari,
and Lalit Kumar Awasthi*

12.1 INTRODUCTION

The Internet of Things (IoT) is a rapidly evolving field that integrates various physical devices and systems, such as sensors, cameras, and actuators, with the Internet. It enables real-time data collection, transmission, and analysis. IoT has potential applications in smart cities, power grids, smart agriculture, transportation systems, and healthcare [1]. For instance, in the healthcare industry, IoT implementation can enhance patient care, increase efficiency, and reduce costs by tracking healthcare staff and patients, monitoring vital signs, and managing chronic diseases [2]. IoT also brings benefits to industries like optimizing traffic flow, improving crop yields, and managing power grids more efficiently. To fully leverage the advantages of IoT, a robust network infrastructure is crucial, encompassing communication and computation capabilities that can handle the large volume of data generated by IoT devices. This includes advanced technologies like beyond 5G/6G communication, cloud computing, and fog computing. Additionally, robust security and privacy measures are necessary to safeguard sensitive data [1].

The evolution of the internet, from peer-to-peer to the world wide web, mobile to social, and now to the IoT, has been driven by rapid networking advancements. Fifth generation (5G) wireless networks, with their high bandwidth (10 Gbps), low latency (1–100 ms), and lower computational cost, offer the ideal infrastructure for executing IoT applications that require various quality-of-service (QoS) parameters. The Tactile Internet, another significant advancement in Internet technology, enables ultra-low latency machine-to-machine and human-to-machine communication, ensuring network availability, security, and reliability for data transmission [3]. With these advancements, IoT devices can now transmit data with minimal latency and high reliability to computational servers. Smaller IoT applications can be executed locally on connected sensors or fog devices, while larger applications necessitate robust network infrastructure, storage, and computation resources to meet QoS parameters [4]. For demanding IoT applications, data transmission to high computational devices such as cloud data centers, which possess ample resource capacity and storage, is required.

One of the main limitations of cloud computing is the use of virtual machine (VM) technology, which combine programs onto a single server [5]. Each VM program has its own operating system (OS) image, leading to memory and storage overhead.

DOI: 10.1201/9781003437079-12

Containers, on the other hand, are self-contained units that package software and its dependencies. Similar to VM programs, containers enable multiple programs to share resources on a single server [6]. However, containers virtualize resources on the OS level, providing a lightweight solution with quick application deployment and fine-grained resource sharing. Organizations increasingly rely on container technologies like Docker and Linux Containers to deploy IoT applications [7]. These containers can be deployed in private data centers or virtualized public cloud environments.

In recent years, IoT applications have also been deployed in traditional cloud environments. However, a serverless infrastructure model has emerged, offering the ability to run IoT, web, or event-driven applications [8]. This serverless environment delivers cloud-based services directly to users, eliminating the need for providers to maintain and manage computing resources. This model, known as function-as-a-service (FaaS), is an alternative to the infrastructure-as-a-service (IaaS) in cloud environments.

Cloud computing has appeared as the leading technology for addressing ever increasing demands due to its massive storage and processing capabilities [9]. Transferring IoT applications from users' devices to a centralized cloud environment is less advantageous and less suitable for latency-intensive IoT applications [10], such as smart grid and connected vehicles [11], fire detector [12], and content delivery [13]. In a multicloud environment, the CDCs are often distributed around the globe and comprise multiple heterogeneous resources [14]. However, executing dependent IoT applications on multiple CDCs may increase transmission and response time due to the overhead of intercloud communication [15, 16].

The fog computing paradigm has been designed as a solution to these disadvantages of the traditional cloud environment. With fog computing, latency-intensive applications can run with minimal delay on local or on nearby fog devices (e.g. network edge devices, sensors, sink nodes, laptops, desktops, and tablets), increasing the possibilities for IoT resources. A multilevel computing architecture, combined with cloud-level execution, IoT-level execution, and fog-level execution, is illustrated in Figure 12.1. Additionally, current literature shows that cloud data centers are not viable for executing IoT applications, whereas fog devices can be used as an alternative [17–19].

Multi-access edge computing (MEC) is a cutting-edge network architecture that addresses some of the challenges associated with cloud computing. In addition to enhancing customer experiences, MEC enables cloud computing capabilities to be functional at the edge of the network, resulting in lower latency, more efficient network operations, and more efficient service delivery. MEC can address various cloud computing challenges by performing specific tasks in real time or near real-time at the network edge [20]. It is known that latency is a major problem associated with cloud computing because it takes longer to send and receive data, particularly for real-time applications. MEC overcomes this by bringing cloud computing resources closer to end-users, reducing data travel distance, lowering network congestion, and enhancing application responsiveness and efficiency. Local data processing and storage capabilities also help to decrease the cost of data transportation to and from the cloud [21]. Moreover, existing literature supports the idea that MEC is a breakthrough network architecture that can help cloud computing overcome latency, data transportation costs, and network congestion challenges, resulting in better performance and customer service.

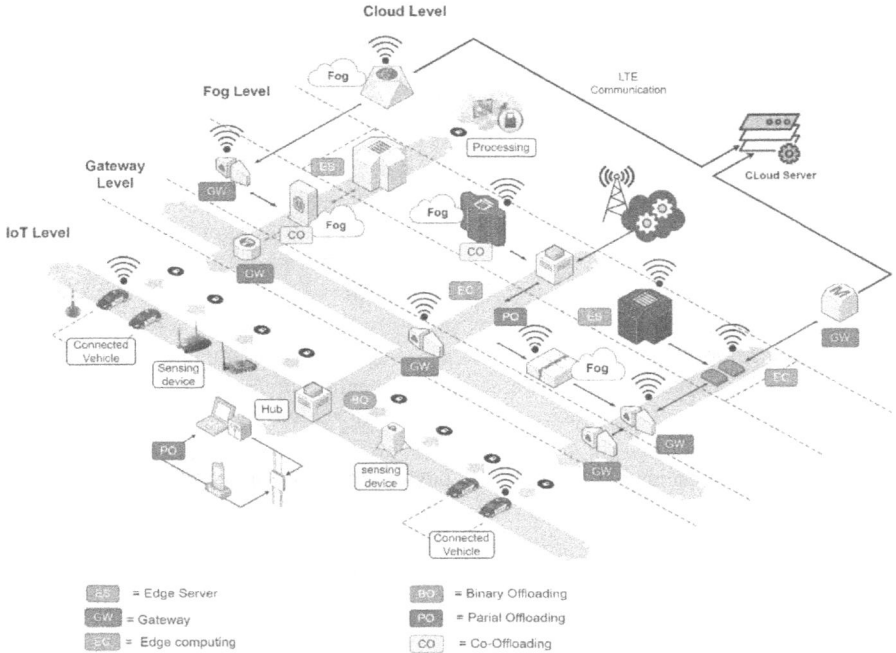

FIGURE 12.1 IoT and fog–cloud architecture.

In order to support these emerging technologies, we need fast and ultrareliable networking technology that can carry enormous bandwidth and transfer data with minimal delay [22]. Furthermore, current cellular technology cannot transmit data to rural or remote areas, even with 5G technology [23]. Hence, the sixth generation of cellular technology, also popularly known as 6G, is expected to handle massive datasets for several applications such as industrial IoT, smart transportation, satellite communication, rural area coverage, extended reality, etc. [24, 25].

12.1.1 Requirements of 6G in IoT

Smart IoT applications require faster data transfer speeds, reduced latency, and more reliable connections, which necessitate the development of 6G wireless networks. Existing 4G and 5G technologies are increasingly constrained in handling rising traffic because of the rising number of connected devices and the amount of data they generate. However, state-of-the-art 6G is expected to provide the necessary enhancements in data rates, latency, reliability, capacity, and security to support the growing number of IoT devices and applications and to meet the increasing demands of the IoT market [26]. A comparison among the computing devices is shown in Table 12.1. A summary of the advantages of using 6G is a follows:

Increased data rates: 6G is expected to provide enhanced data rates of 1 Tbps, significantly higher than current 5G networks [27]. This will enable faster data transfer and more responsive IoT applications.

TABLE 12.1

Comparison between 6G and the Existing Communication Technologies

Parameter	4G	5G	6G
Standardization	3GPP	3GPP	3GPP
BS density	8–10 BS/km^2	40–50 BS/km^2	100 BS/km^2
User experience	20 Mb/s	100 Mb/s	>1 Gb/s
Data/packet rate	100 Mbps – 1 Gbps	20 Gb/s	>1 Tb/s
End-to-end delay	60 ms	10 ms	<0.001–0.01 ms
Traffic density	1 Mb/s/m^2	10 Tb/s/km^2	>100 Tb/s/km^2
Connection density	100 thousand/km^2	1 million/km^2	>10 million/ km^2
Spectral efficiency	1.5–3 × relative to 3G	3–5 × relative to 4G	5 × relative to 5G
Processing delay	80 ns	50 ns	10 ns
Frequency band	2–8 GHz	30–300 GHz	0.3–3 THz
Uplink data rate	2–6 Mb/s	10 Gb/s	>1 Tb/s
Downlink data rate	1–2 Mb/s	20 Gb/s	>5 Tb/s
Error rate	10^{-3}	10^{-5}	10^{-9}
Period	2010–2020	2020–2030	2030–2040
AI support	Partial	Full	Next generation AI
Device type	Smartphone, smartwatch	Sensor and actuator, unmanned aerial vehicle (UAV), extended reality (XR) equipped	UHD screen, UAV, autonomous vehicle, connected robotics, multisensory XR
Application	Telemedicine, video conference	Augmented reality/ Virtual reality (AR/VR), telemedicine, Internet of vehicles (IoV) communication	Holographic society, tactile internet, Industry 5.0, machine type communication, satellite communication, ocean technology
Promising technology	LTE, LTE-Advanced broadband, Wi-Fi	mmWave, massive MIMO, network slicing, full duplex, duel connectivity	THz communication, blockchain-based spectrum sharing, quantum ML, ultra-massive MIMO, edge AI, energy transfer and harvesting
Standard	LTE-A/IMT-A	IMT-2020	IMT-2030

Lower latency: 6G is expected to achieve ultra-low latency of less than 1 ms end-to-end data transmission, significantly lower than current 5G cellular networks [28]. This will enable real-time communication and control in IoT systems.

Improved reliability: 6G is expected to provide more reliable connections, with higher availability and robustness in interference and other wireless challenges.

Increased capacity: 6G is expected to provide increased capacity to support the growing number of IoT devices and applications. 6G is expected to use advanced technologies such as mmWave (millimeter-wave) bands, massive MIMO, and beamforming to increase capacity and improve coverage. Moreover, 6G will utilize new spectrum bands, such as the terahertz band, for data transmission. Using machine learning (ML) and artificial intelligence (AI) techniques may also optimize the use of network resources and improve the efficiency of the network.

Enhanced security: 6G is expected to provide enhanced security features to protect IoT devices and networks from cyber threats and attacks. 6G will use advanced technologies like quantum key distribution (QKD), homomorphic encryption, and blockchain [29] to increase security. Aside from using 5G security enhancements and AI-based security solutions, 6G may incorporate advanced security protocols and technologies.

12.1.2 SCOPE OF THIS SURVEY

Recently, IoT has seen significant growth in the number of interconnected devices and the variety of applications they enable. However, the standard cloud model to sustain these applications is no longer sufficient to meet the IoT demands. The emergence of 6G networks and distributed computing technologies presents an opportunity to address the boundaries of conventional cloud computing and support the growing needs of IoT. Research in this area can focus on several key areas, including resource allocation optimization, security and privacy, edge computing, real-time decision-making, network slicing, energy efficiency, interoperability, and reliability [30]. These research areas have the possibility to significantly enhance the performance and scalability of IoT applications and enable new use cases that were not previously possible. As a result of these advancements, we present a systematic literature review of IoT technology and its expansion in various application domains. In addition, we consider two essential technologies that help to boost the vision of IoT: 6G communication and distributed computing technology. We have also highlighted the current need and future demand for emerging IoT applications.

12.1.3 OUTLINE OF KEY CONTRIBUTIONS

The key objective of this chapter is to provide a research direction by combining the advancements of 6G wireless communication technology and distributed computing paradigms. The chapter significantly contributes to recent IoT, distributed computing, and 6G communication literature. The major contributions are as follows:

- This survey provides an outline of the key components and objectives of IoT as presented in recent literature. Specifically, we mention the requirements of IoT and several IoT objectives that are crucial for the version of digital society.
- Next, we examine the 6G wireless communication and the state-of-the-art emerging distributed technologies, such as cloud computing, containerization, fog computing, and serverless computing, for IoT applications.

Furthermore, a list of IoT applications that utilizes these technologies to achieve QoS is also provided.

- This survey delves into the significant challenges associated with 6G networks and distributed computing technologies, particularly when transmitting real time IoT data to remote computing devices for processing and analysis.
- Finally, we illustrate the future research directions of distributed computing and 6G-enabled IoT in computation offloading, scalability, security, trust, and privacy management.

The rest of the sections are organized as follows: Section 2 summarizes IoT along with its objectives and resource requirements. Section 3 illustrates the realization of 6G and distributed computing for IoT applications. Section 4 focuses on emerging 6G-aware computing technologies for massive IoT applications. Section 5 describes the future research directions for IoT applications on 6G-aware networks. Finally, we conclude the chapter with Section 6.

12.2   REQUIREMENTS AND OBJECTIVES OF IoT

IoT has become an umbrella term for many applications with a few common characteristics. The European Commission defines IoT as a technology that enables the connection of people and things anytime, anywhere, with anything and anyone, ideally using any path and service. This highlights that IoT encompasses a collection of devices with diverse characteristics connected over the internet, offering promising applications in smart cities, smart transportation, remote surgery, smart agriculture, smart water and waste management, smart energy, and grid management. It is evident that most devices worldwide are equipped with various sensor nodes capable of processing data locally or transmitting it to computational servers such as fog devices or cloud data centers (CDC) for further processing. IoT applications generate a significant volume of latency-critical data and require local or low-latency processing on computational servers. It is also challenging for IoT devices to analyze, process, and offload data based on the application's size and resource requirements. Moreover, IoT leverages underlying technologies such as sensor networks, communication technologies, pervasive computing, and embedded devices to transform traditional devices into smart objects. IoT is a rapidly evolving field with the potential to have a significant impact on various domains, including cities, homes, and healthcare [31]. IoT technology has the potential to boost the global economy and enhance our quality of life.

One of the prominent applications of IoT technology is in the field of smart homes, where home automation and consumer electronics are employed. These devices enable residents to efficiently utilize household gadgets and enhance safety measures within their living environment. Smart home technology encompasses a wide range of devices, from simple smoke sensors to advanced systems like temperature and humidity detectors, fire detectors, light sensors, vibration sensors, and water level sensors. These devices can be controlled remotely through mobile applications, such as the HS-SR501 PIR Infrared and velocity sensor module, which allows for scheduling of lights and air conditioning. Additionally, programmable thermostats aid in energy management by regulating temperatures based on a predefined schedule. Security devices like fire detection and buzzer alarm sensor modules can safeguard homes from intruders. Other applications of IoT technology in smart homes include door locks, access control, carbon monoxide detectors, and smoke detectors. All these applications indicate that the smart home market is highly competitive, with dozens of vendors worldwide offering a wide range of home automation devices tailored to the customer's lifestyle. This allows for an easy selection of devices that best suit the needs and preferences of homeowners. However, it is essential to note that while IoT technology has the potential to bring about significant benefits, it also poses privacy and security concerns that must be addressed.

IoT technology has progressed from being limited to smart homes to being implemented in smart cities. Smart cities can produce a wide range of IoT applications best executed locally, on edge devices, or fog devices that provide low latency and context awareness. In healthcare, IoT technology is essential to the real world and requires collaboration with various entities, including insurance, government, logistics, hospitals, and pharmacies. Telemedicine and mobile health are considered essential aspects of 6G technology, and various IoT medical sensors monitor health-related information and track records for future purposes. Edge and fog computing is widely used in IoT technology to achieve performance goals. Vehicle-to-everything (V2X) also benefits from 6G technology, which supports vehicle-to-grid, vehicle-to-vehicle, vehicle-to-IoT devices, and vehicle-to-computational servers [32]. The V2X concept requires a reliable communication network to send IoT data to other devices or servers, including vehicle maintenance, autonomous and semi-autonomous driving, and vehicle infotainment. This provides continuous vehicle sensing, real-time traffic monitoring, improved network security, and supporting infotainment applications [33]. Additionally, humans want to interact with the devices of the outside world based on data aggregated by IoT objects using mixed reality combined with VR and AR technologies. In these technologies, data is offloaded from computational servers with minimal latency by combining computer-generated information with the real world.

12.2.1 6G Realization for IoT

This 6G-enabled IoT technology needs to address several challenges to enable effective and efficient operations. These include effectively processing and analyzing real-time data generated by IoT devices and meeting QoS objectives such as minimum latency, computational time, energy consumption, and maximum resource utilization. Researchers and developers continuously work on these new paradigms, such as cloud computing, fog computing, serverless computing, and counter-based technology, to meet these objectives [34]. Still, each has its own merits and demerits. Some providers have developed new platforms based on multiple paradigms for more effective processing of IoT applications. For example, Amazon has created the Amazon Lambda platform, which combines cloud and serverless computing, and the Fargate platform, which uses container-based technology and serverless computing for event-driven and resource-based applications. Here we look at the QoS objectives in more detail:

Latency is a critical metric for assessing the implementation of 6G-enabled IoT systems, particularly for delay-sensitive applications prevalent in smart environments such as smart homes, smart cities, smart transportation, smart waste management, and smart water management. These delay-sensitive applications are best executed locally, near the IoT devices such as edge and fog nodes, to minimize communication overhead and latency [35]. Furthermore, latency is a key indicator of IoT devices' overall round-trip time and computation time. Minimizing latency, therefore, is imperative for optimizing the performance of these devices. The utilization of edge and fog computing technologies in a distributed environment can reduce latency in real-time IoT applications by processing data locally or distributing it among devices. However, applications that require significant computational resources should be assigned to CDCs, even though this may result in increased latency due to the centralized nature of these CDCs, as they can meet the resource requirements of these applications.

Computation time is the computation time of 6G-enabled IoT applications is a crucial factor that determines their performance. The transmission time, also known as delay, plays a significant role in the computation time. This time is affected by the distance between computational servers and IoT devices. Each IoT device has a limited processing capacity and may offload specific applications to computational servers. In cases where the applications are executed within the IoT device, the transmission time is zero [34]. However, when offloading data to a computational server, the transmission time increases with the distance. To optimize this, researchers are exploring fog and edge computing technology. This technology involves placing computational devices closer to the IoT devices, thereby minimizing the transmission time. Additionally, researchers are developing intelligent offloading strategies that prioritize local or edge computation to reduce overall computation and transmission time. This enhances the enactment of the 6G-enabled IoT applications and reduces the dependence on the CDC.

Energy consumption is the implementation of 6G-enabled IoT applications results in energy consumption during processing within the resources of computational servers, referred to as computational energy, as well as the power required for transmitting data through a reliable network, referred to as transmission energy. The transmission energy of IoT applications is contingent upon the distance between the IoT device and the computational server. One strategy for minimizing energy consumption is to execute applications locally within the IoT device, thereby eliminating the need for transmission energy [36]. However, offloading specific data to computational servers close to close may also be advantageous in minimizing transmission energy. Incorporating fog and edge computing technology can potentially optimize transmission energy in 6G-enabled applications by placing fog or edge devices close to IoT devices. Ultimately, the overall energy consumption of IoT applications is determined by both the computational energy required by the assigned computational server and the transmission energy needed to transmit data to that server. Currently, researchers are exploring the development of intelligent offloading strategies that would allow for the execution of most applications locally or on edge or fog nodes, thereby reducing overall energy consumption.

Resource utilization is critical for computation servers, particularly for processing 6G-enabled IoT applications. These servers assign central processing unit (CPU), memory, and disk space resources to IoT applications as VM instances or containers. While some IoT devices can process small applications locally, utilizing resources efficiently, larger, more complex tasks are typically executed on local computation servers, such as edge nodes or fog nodes. These servers use container-based virtualization techniques for delay-intensive or event-driven applications due to their faster deployment and increased parallelism while maintaining proper resource utilization. However, it is worth noting that some service providers still rely on VM instances for virtualization techniques, as it provides more security and privacy for the application, albeit at the cost of decreased resource utilization compared to container-based virtualization. For resource-intensive applications, CDCs are typically utilized, which may use a combination of container-based and VM-based virtualization techniques, assigning resources based on the specific requirements or properties of the application. In summary, proper resource utilization of computing resources increases parallelism among applications and ensures efficient use of resources, making it a key consideration for computation servers.

Load balancing is a crucial aspect of computation servers for efficiently processing 6G-nabled IoT applications with minimal waiting and computation times. It involves distributing the load among active computational servers, allowing for the processing of 6G-enabled applications with minimal delay. With the advent of the latest technologies, such as serverless computing cloud computing, and fog computing, scheduling and processing 6G-enabled IoT applications with minimal computation delay have become possible. The system manager plays a key role in deploying IoT applications to the most suitable computing environment, which can efficiently handle the applications and improve system performance while satisfying mixed QoS constraints. For instance, serverless computing is well-suited for event-driven applications, fog/edge nodes are perfect for time-critical applications, and resource-intensive and request-based applications are typically executed in a cloud environment [37]. Additionally, fog/edge nodes can distribute applications among neighboring edge and fog devices and process them independently unless offloading them to the CDC. This strategy reduces overall transmission delay and computation time. Proper load balancing of computational servers also improves the efficient utilization of computing resources [38].

Security and privacy is the sustainability of data communications between IoT devices and computation servers heavily depends upon the guarantees of security and privacy [39]. It is imperative that accurate information and data transactions are not disrupted, which can only be achieved through the provision of robust security and privacy measures. Without such guarantees, information and data compromise are risks through two distinct avenues. First, intruders may disrupt the sensing policy's deployment strategy and usage pattern, thereby compromising the privacy and security of the network. Second, sensing applications and data from the environment may be changed, leading to a degradation in data transfer performance to users or IoT objects for specific tasks. The security and privacy of smart IoT objects is a critical concern, particularly regarding providing security and privacy for computation servers, including trustworthy fog and edge devices, protecting against potential cyberattacks on smart things, ensuring authentication and trust, and securing data and networks. In the context of a smart city, cyberattacks targeting IoT devices such as edge and fog devices can significantly impact network performance, data analysis, and service delivery from CDCs to IoT devices. This can result in incorrect decisions and responses during emergencies or disasters and even lead to the failure of devices. Network security is also essential to protect IoT devices from sniffer attacks, jamming attacks, and other security threats during data transfer to computation servers [40]. These concerns highlight the need for robust cybersecurity measures to ensure the smooth functioning of smart cities and IoT-enabled technologies.

Context discovery and awareness is one of the key goals of 6G-enabled IoT applications running on various computation servers is to gain context. Local nodes, such as IoT devices, edge, and fog devices, can infer context information, including location, environmental conditions, nearby computer devices, and their current status, among other things. This context data is especially useful for understanding the current state of active computation devices in an area. By utilizing context data, efficient and effective data processing policies can be designed, mainly when several IoT users or fog or edge devices with similar characteristics are nearby. Ideally, the IoT devices and local gateways of the IoT devices should be context-aware and intelligent enough to automatically select the best-fit devices for each application based on their location and capacity. Another important aspect of context awareness is cooperative and opportunistic sensing. IoT devices should be able to balance their workload with neighboring devices in a location and ensure no resources are wasted, resulting in increased parallelism among the applications and efficient resource utilization. Additionally, context data significantly impacts data collection and fusion strategies.

12.3   REALIZATION OF 6G AND DISTRIBUTED COMPUTING

Realizing 6G and distributed computing for IoT applications is a current research demand. On one side, 6G minimized transmission delay to almost zero. On the other side, distributed computing boosts the execution performance of IoT applications. There are several aspects of this realization. Below we will discuss the requirement of distributed computing, 6G wireless communication, the combination of 6G-distributed computing, and the IoT application areas that need these technologies for scalability, reliability, robustness, and sustainability.

12.3.1 DISTRIBUTED COMPUTING FOR IoT APPLICATIONS

Realtime IoT applications are becoming more standard and AI-assisted, and this is where we need support from distributed computing paradigms. Distributed computing helps to enhance scalability, security, efficiency, and decision-making processes. Compared to traditional/centralized computing, distributed computing paradigms support high mobility, underwater communication, and satellite communication and have proven to be an ideal solution for implementing IoT-based applications. The need for distributed computing paradigms for IoT applications is explained below:

Scalability: Distributed computing allows for the dynamic allocation of resources, enabling IoT applications to scale up or down as needed to accommodate device density or data volume changes.

Fault tolerance: IoT systems are typically designed in a layered structure, making a fault tolerance system challenging. Distributed computing architectures allow for the replication of data and services across multiple devices, providing increased fault tolerance and improved reliability of IoT applications.

Improved security: Security in IoT is a critical issue, and handling security is even more challenging due to its complexity, dynamicity, and device heterogeneity. Distributed computing allows for the decentralization of data storage and processing, making it more difficult for hackers to target a single point of weakness and improve IoT application security.

Increased efficiency: Distributed computing allows for the efficient use of resources, enabling IoT applications to use multiple devices to perform complex computations, resulting in faster processing times.

Enhanced data processing: IoT represents data collection, gathering, and processing in real-time. Distributed computing allows for the parallel processing of large amounts of data, enabling faster and more accurate analysis of data generated by IoT devices.

Edge computing: Distributed computing allows for processing data nearer to the origin, lowering the amount of data required to be transferred over the network and enabling faster response times for control systems [41].

Real-time decision-making: Realtime IoT applications are primarily trained AI/ML techniques with large datasets. Distributed computing allows for the processing of data and decision-making in real-time, which is essential for IoT applications such as autonomous vehicles and industrial control systems [1].

Cost-electiveness: Cost optimization is a significant issue for large-scale IoT deployment. Industries, from small enterprises to large production companies, face this issue due to IoT cost overheads while deploying. Distributed computing allows for the efficient use of resources, enabling IoT applications to be deployed and run at a lower cost than traditional centralized computing architectures.

12.3.2 6G COMMUNICATION FOR IOT APPLICATIONS

Sixth-generation (6G) wireless communication is the successor to 5G wireless technology. It is anticipated to deliver substantially more increased capacity and much more low latency than earlier generations. Existing literature shows that one of the critical objectives of 6G is to uphold one-microsecond latency in data transmission, which is 1,000 times faster than existing 5G networks. The capability to operate in untapped radio frequencies and use cognitive technologies such as AI, visible light communication, larger frequency, and quantum computing is also expected to play a substantial role in the capabilities of 6G networks. In addition, 6G helps to obtain energy efficiency and low energy consumption for IoT data transmission. The essential benefits of 6G communication over the IoT network are illustrated in Figure 12.2. As corporations and industries strive to reach net-zero emission targets, 6G is a crucial enabler for sustainable digitalization.

The need for 6G wireless technology for IoT applications is explained as follows:

- *High-speed communication*: 6G networks are anticipated to deliver much higher data transfer speeds than current 4G and 5G networks, enabling faster and better communication between IoT devices.
- *Low-latency communication*: 6G networks are predicted to supply much lower latency than current networks, enabling real-time communication between IoT devices and faster response times for control systems. Specifically, higher scalability helps to obtain a better user satisfaction rate.
- *High-capacity networks*: 6G networks are expected to provide much higher capacity than current networks, enabling the connection of a larger number of IoT devices.

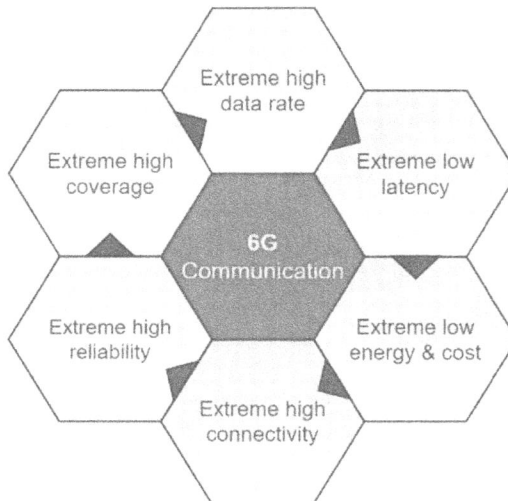

FIGURE 12.2   Requirements of 6G in IoT technology. (Source: NTT Docomo)

- *Improved security*: 6G networks are expected to provide improved security features, such as network slicing and secure communication protocols, which will be essential for the protection of sensitive data generated by IoT applications.
- *Enhanced reliability*: 6G networks are expected to provide enhanced reliability, with features such as network redundancy, which will be essential for the smooth operation of critical IoT applications, such as autonomous vehicles and industrial control systems.
- *Increased energy efficiency*: 6G networks are expected to be more energy efficient, which will be important for battery-powered IoT devices with limited power resources.
- *Improved data transfer*: 6G networks are expected to provide enhanced data transfer and processing capabilities, which will be essential for IoT applications that require large amounts of data to be analyzed and acted upon in real-time.

12.3.3   6G AND DISTRIBUTED COMPUTING FOR IoT APPLICATIONS

IoT applications suffer from both computing and communication prospects. This is due to the simple and lightweight design of IoT architecture. Thus, the need for faster and more reliable data transmission technologies will ultimately maintain the lifetime of IoT devices. Another aspect is the support of distributed technologies so that IoT devices can take the aid of on-demand processing facilities with AI support. The combination of 6G and distributed computing technologies is enabling the scalability, latency reduction, reliability, security, and flexibility required to support future IoT applications, discussed as follows:

- *Scalability*: IoT applications generate large amounts of data and require real-time processing, which can be challenging for traditional centralized computing models. Distributed computing technologies, such as edge computing and fog computing, enable data to be processed closer to the source, reducing the need for data transfer and enabling efficient scalability to handle large amounts of data.
- *Latency reduction*: 6G networks provide ultra-reliable and low latency communication, which is essential for real-time communication and control in IoT systems. Distributed computing technologies enable data to be processed closer to the source, reducing the need for data transfer and further reducing latency.
- *Reliability*: Distributed computing technologies enable data to be processed and stored at multiple locations, providing redundancy and increasing the overall reliability of the system. 6G networks also provide more reliable connections with higher availability and robustness in the presence of interference and other wireless challenges.
- *Security*: Distributed computing enhance system security by enabling the processing and storage of data at various locations, introducing redundancy measures that contribute to the robustness of the overall security

architecture. 6G networks also provide enhanced security features to protect IoT devices and networks from cyber threats and attacks.

- *Flexibility*: Distributed computing technologies enable data processing to be performed at the edge, enabling more flexible deployment of IoT applications, and reducing the dependency on centralized cloud computing.

12.3.4 IoT APPLICATIONS THAT REQUIRE 6G AND DISTRIBUTED COMPUTING

Distributed computing has been widely adopted in large amounts of IoT applications. On the other hand, the need for 6G-supported applications is also increasing rapidly. This subsection will briefly highlight IoT applications that demand combined support from 6G and distributed computing paradigms. These applications include smart transportation, healthcare, agriculture, industrial applications, etc., discussed as follows:

- *Smart city*: 6G and distributed computing technologies are necessary for smart cities to handle the large amounts of data generated by IoT devices such as cameras, sensors, and traffic lights. In such scenarios, edge computing can process the sensor-generated data locally, reducing the need for data transfer and providing real-time insights.
- *Industry 5.0*: Distributed computing technologies can be used to process data from sensors and machines in industrial environments, providing real-time insights and enabling predictive maintenance [42]. In addition, 6G networks are needed for the reliable and low-latency communication required for control systems in industrial environments to make efficient decisions.
- *Smart healthcare*: Distributed computing technologies can be used to process data from wearable devices and medical equipment, enabling real-time monitoring of patients' vital signs and providing early warning of potential health issues [43]. Besides, 6G networks are needed for the secure and reliable communication required for telemedicine and remote surgery.
- *Connected vehicles*: 6G networks are needed for the high-speed and low latency communication required for autonomous vehicles and V2X communication. Further, distributed computing technologies can be used to process sensor data from vehicles, enabling real-time traffic management and improved safety.
- *Augmented reality/virtual reality*: 6G networks are needed for the high-speed and low-latency communication required for real-time interaction with virtual and augmented environments. On the other hand, distributed computing technologies can be used to process data from sensors and cameras, enabling accurate tracking and rendering of virtual objects.
- *Smart grid*: 6G networks are needed for the reliable communication required for remote management and control of power grids. Distributed computing technologies can be used to process sensor data from power plants and substations, enabling real-time monitoring and management of energy consumption.

- *Smart agriculture*: 6G networks are needed for the reliable communication required for remote management and control of agricultural equipment and monitoring of crops, where distributed computing technologies can be used to process sensor data, enabling real-time monitoring of soil moisture, temperature, and other environmental factors.

12.4 6G-ENABLED DISTRIBUTED COMPUTING TECHNOLOGIES

Over time, several computing paradigms have been introduced to support and scale IoT applications. However, the increase in enormous IoT applications and dynamic service demands drastically increases the need for distributed computing paradigms. Besides, incorporating 6G technology into distributed computing paradigms significantly improves user handling and satisfaction rates. This section focuses on popular distributed computing paradigms supporting 6G technology for faster data transmission.

12.4.1 Cloud Computing

IoT technology connects billions of devices, including smart appliances, industrial equipment, and wearable devices, to computer-based systems, enabling them to analyze and act on data in real-time [44]. Soon, 6G wireless technology will provide faster and more reliable communication for these connected devices, enabling the development of new and innovative IoT applications. Cloud computing is a key enabler for 6G-powered IoT, as it allows for many distributed resources linked together by a fast network, as illustrated in Figure 12.3. Three are mainly three different types of cloud services that are relevant for IoT: Infrastructure-as-a-service (IaaS), software -as-a-service (SaaS), and platform-as-a-service (PaaS). SaaS permits users to access and use software applications hosted on remote servers without installing and managing them on local devices. PaaS provides a computing platform for developing and deploying 6G-enabled IoT applications, including operating systems, execution environments, and web servers. IaaS offers access to virtual servers and other computing resources for resource-intensive and demand-based IoT applications [45]. Examples of PaaS cloud services include Amazon Elastic Beanstalk, Cloud Foundry, and Google App Engine, while Amazon EC2 and Microsoft Azure are examples of IaaS cloud services. By using these cloud services, developers can build and deploy 6G-powered IoT applications quickly and easily without the need to manage the underlying hardware and software infrastructure.

12.4.2 Fog Computing

To ease the burden of CDCs and overcome the issues raised by cloud computing, fog computing has emerged as an alternative computing platform in a distributed environment [46]. Fog computing facilitates a near-edge computing architecture where devices are positioned nearer to IoT devices, which reduces latency and optimizes various QoS parameters. A standard three-layer fog computing architecture

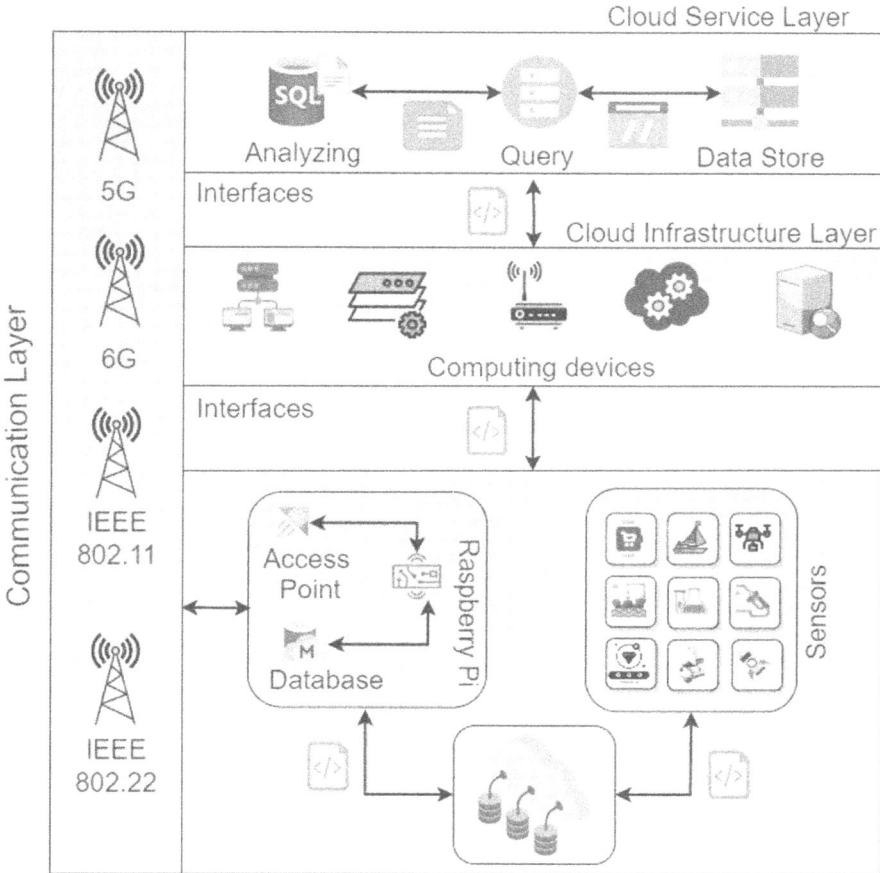

FIGURE 12.3   IoT–cloud layered model.

is illustrated in Figure 12.4. This technology is most appropriate for delay-intensive IoT applications, which demand the interpreted results within the minimum time. Cisco first initiated the fog computing concept, which helps expand cloud computing to the edge of the network [47]. The word "fog" is used because fog devices make cloud resources more accessible to IoT devices or users, i.e. from core to edge computing which helps to provide various reliable services in a short period. Like the traditional CDC, fog devices yield a virtualized platform that furnishes processing, storage, and networking services to IoT devices and users. According to existing articles, a fog environment consists of many ubiquitous, heterogeneous, and decentralized computing devices that can execute and hold data locally using various wireless communication modes. This is without the impact of a third party [48]. The OpenFog Consortium, on the other hand, states that fog computing supports system-level horizontal architecture, which can supply computing resources, including computation and storage capacity, and a reliable communication network to connect with

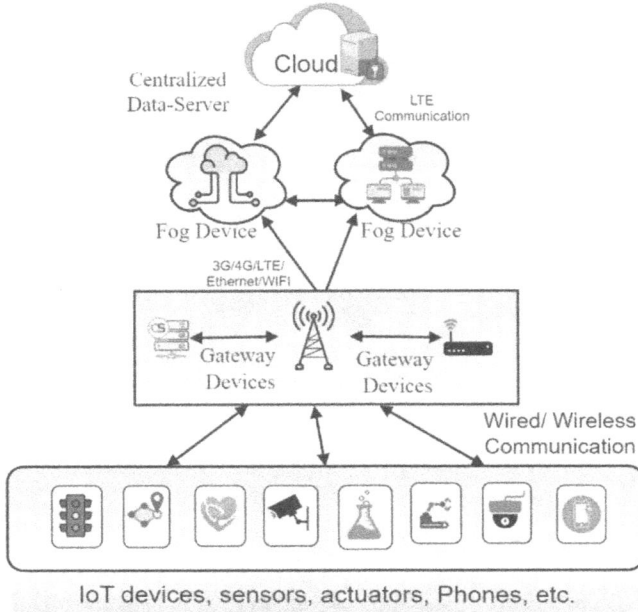

FIGURE 12.4 Standard fog computing architecture.

IoT devices and CDCs for large applications. There are four main characteristics of fog computing: (a) sustaining heterogeneous resources, (b) geographical distribution, (c) supporting real-time and delay-intensive applications, and (d) providing minimal latency [49].

12.4.3 SERVERLESS COMPUTING

In serverless computing, applications can be developed without worrying about the underlying infrastructure management [50]. In the context of 6G-enabled IoT applications, serverless computing can provide a cost-effective and scalable solution for handling event-driven workloads. With serverless computing, developers can focus on writing code and business logic while the cloud provider allocates and manages resources such as servers, storage, and networking [51]. Several platforms and technologies are available for implementing serverless computing in a cloud environment, including FaaS, in which a function (a piece of software) is deployed to meet the specific needs of an IoT application. Each function can respond to a single IoT request and be executed on an ideal cloud server for the application. Serverless cloud environments, such as AWS Lambda, Google Cloud Function (GCF), and Azure Function, can process background cloud applications, such as message queueing, and direct HTTP requests [52]. Other cloud services that can be used in a serverless environments include Amazon SNS, Google Cloud Pub/Sub, and DynamoDB. Figure 12.5 illustrates a brief architecture of serverless computing.

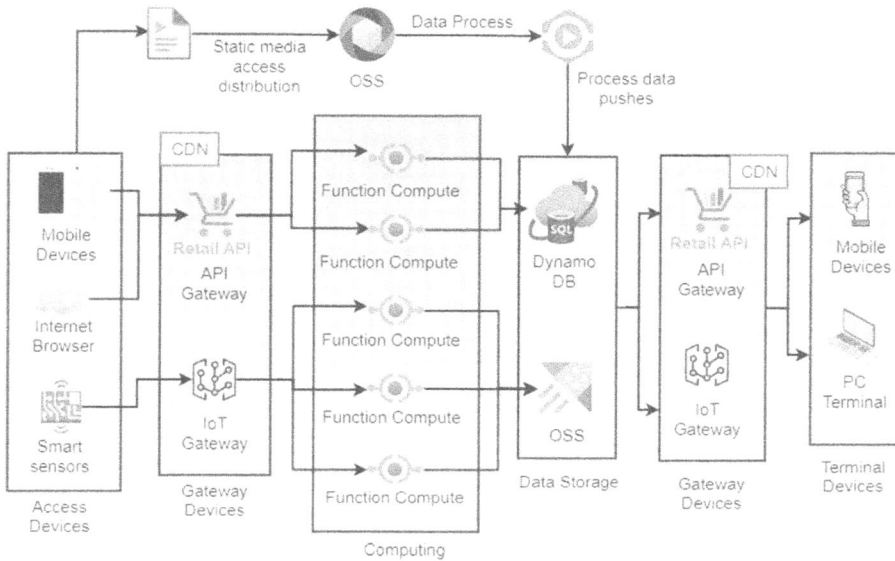

FIGURE 12.5 Components and functionalities of serverless computing.

12.4.4 CONTAINER-BASED COMPUTING

Virtualization is a critical technology that enables the deployment of cloud environments in a distributed context. Virtualization allows service providers to assign multiple VM instances to tasks or applications without the user's knowledge, improving resource efficiency and reducing energy consumption. In the past, cloud providers offered resources to users in the form of VM instances, which virtualized computing resources at the hardware level. Based on bare-metal or hypervisor technology, each VM instance has its own operating system (OS) and shares the processing resources of a host server. A hypervisor is a piece of software or firmware that provides the ability to deploy and operate VM instances. The major benefit of VM instances is that they allow for the virtualization of server resources and increased parallelism of tasks. This improves the utilization of resources, as shown in Figure 12.6. For example, a single physical server could host multiple VM instances, each running a different application or workload, thereby maximizing the utilization of the server's resources.

12.4.5 MULTI-ACCESS EDGE COMPUTING

MEC is a network architecture that enables computing resources and services to be delivered to the edge of a network closer to the end-user. MEC aims to reduce latency, improve network efficiency, and enhance the user experience by providing applications and services closer to the end user. MEC works by deploying small data centers (edge nodes or servers) at the network's edge, typically within the mobile operator's radio access network (RAN) [53]. These edge nodes are equipped with

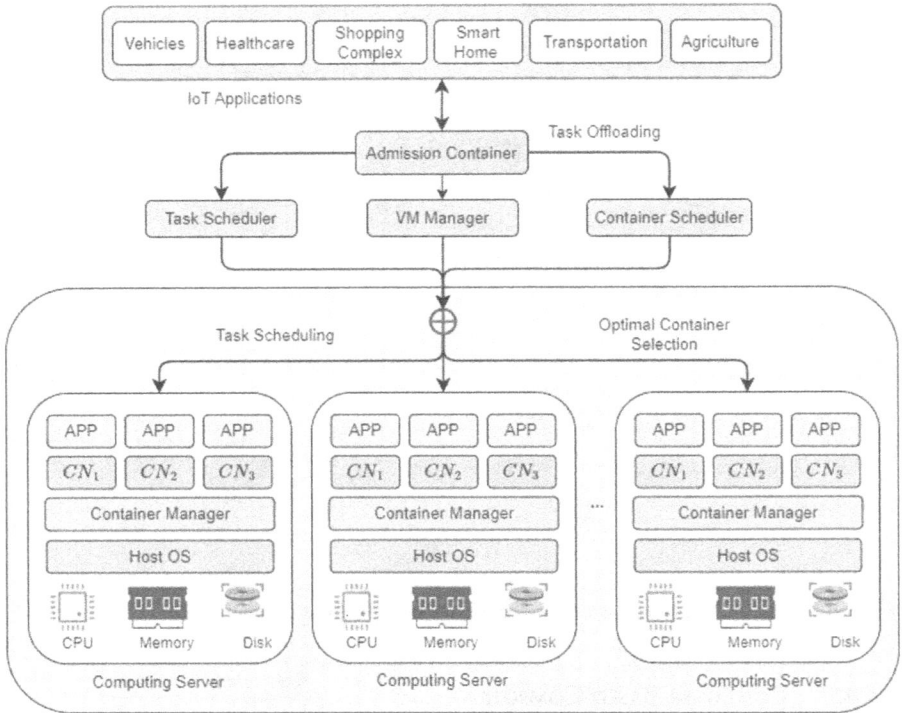

FIGURE 12.6 Components and functionalities of container-based scheduling.

computing, storage, and networking resources that enable them to process data locally. This reduces the need to send data back and forth to a centralized core or cloud infrastructure. By reducing the distance between edge devices and the cloud, MEC allows edge devices to access cloud services and applications more efficiently. With MEC, computing resources are brought closer to the network edge, enabling real-time data processing essential for applications requiring low latency, such as self-driving cars. MEC has numerous applications across different industries [54]. MEC can be utilized in the healthcare industry for real-time patient monitoring, remote consultations, and telemedicine. In the manufacturing industry, MEC can be used for predictive maintenance, real-time monitoring of factory equipment, and supply chain optimization. MEC can be used in the transportation industry for real-time traffic monitoring, autonomous vehicles, and street management. A summary of related contributions is summarized in Table 12.2.

12.5 FUTURE RESEARCH DIRECTIONS

This section gives a promising unexplored pathway for future works to process and analyzes 6G communication in IoT applications with emerging distributed environments, including cloud computing, fog computing, containers, and serverless environments.

TABLE 12.2
Performance Analysis among the Existing Research

Existing Research	Topic	T1	T2	T3	T4	T5	Limitation
[24]	Service deployment strategy	×	✓	✓	×	×	6G communication
[56]	Task offloading strategy	×	✓	×	×	×	Partial offloading
[32]	Service placement for vehicles	×	✓	×	×	✓	Resource optimization
[51]	Efficient data processing	×	×	✓	×	×	Load balancing
[55]	Energy optimized offloading	✓	✓	✓	×	×	IoT controller
[52]	Fog-level execution	✓	×	×	×	×	6G communication
[3]	Scheduling and security on grid	×	✓	✓	×	×	Resource optimization
[30]	Load balancing and scheduling	×	✓	✓	×	×	Blockchain
[43]	Edge intelligence and offloading	×	×	✓	×	✓	Load balancing
[35]	QoE in fog computing	✓	×	×	×	×	Partial offloading
[26]	Reliability and security	✓	✓	✓	✓	✓	Resource optimization
[21]	Offloading in fog networks	✓	✓	✓	✓	✓	Edge intelligence
[7]	Distributed computing environment	✓	✓	✓	✓	✓	IoT controller
[6]	Distributed communication	✓	✓	✓	✓	✓	Blockchain
[4]	Advanced communication protocols	✓	✓	✓	✓	✓	6G communication
[40]	Transmission control in 6TiSCH	✓	✓	✓	✓	✓	Load balancing
[57]	Edge resource federation	✓	✓	✓	✓	✓	Resource optimization
[58]	Smart healthcare system	✓	✓	✓	✓	✓	Partial offloading
[59]	Task execution and trust	✓	✓	✓	✓	✓	Load balancing
[60]	Industrial service provisioning	✓	✓	✓	✓	✓	Resource optimization
[61]	Communication, resource allocation	✓	✓	✓	✓	✓	6G communication
[62]	Renewable system for the cloud	✓	✓	✓	✓	✓	Partial offloading
[63]	AI and serverless computing	✓	✓	✓	✓	✓	Reliability
[64]	Security in Healthcare 4.0	✓	✓	✓	✓	✓	Resource optimization
[65]	Smart healthcare system	✓	✓	✓	✓	✓	Load balancing
[66]	AI and IoT for the smart city	✓	✓	✓	✓	✓	Reliability
[67]	Secure protocols for vehicles	✓	✓	✓	✓	✓	Partial offloading

"✓" incorporated the features into the research.
"×" didn't incorporate the features into the research.
"T1" signifies cloud computing.
"T2" signifies fog computing.
"T3" signifies serverless computing.
"T4" signifies container-based computing.
"T5" signifies multi-access edge computing.

12.5.1 ENERGY OPTIMIZED OFFLOADING

Cloud computing is a mature centralized computing approach that utilizes many computers and consumes significant electricity, resulting in increased costs for service providers and negative environmental impacts. In contrast, fog devices supported by 6G technology are dispersed globally with little resource capacity and consume minimal energy when providing services to IoT applications. The energy consumption of computing devices is directly proportional to CO_2 emission rate and temperature, which also impact the environment. Furthermore, unlike VM instances, containers have minimal resource requirements and consume minimal energy [55]. Thus, developing energy-efficient offloading strategies is crucial in fog and cloud computing for decreasing energy utilization and minimizing the CO_2 emission rate from computing devices. One possible strategy is to integrate fog-enabled ultra-high communication technology into distributed paradigms, while another approach could be designing energy-efficient scheduling techniques to compute IoT applications locally, ultimately optimizing execution delay and transmission costs.

12.5.2 TASK SCHEDULING WITH DWINDLING RESOURCE REQUIREMENTS

Using distributed computing resources efficiently is the primary objective of IoT applications. This involves balancing the resource requirements of various sensors and the time constraints of each application. The critical challenge of distributed IoT task scheduling with diminishing resource requirements is identifying an optimal computing device with sufficient resource capacity to meet task deadlines while maximizing resource efficiency. For instance, patient monitoring data should be offloaded to local fog devices for faster processing while adhering to 6G limitations. Conversely, weather prediction and agriculture data should be transmitted to the cloud server for future storage and analysis [47]. Thus, there is a need to develop balanced decision-making systems based on the importance of IoT applications and their subsequent requirements. The use of ML-based decision-making systems can increase control of the network. On the other hand, DL-based techniques such as deep reinforcement learning can be utilized in such scenarios to predict environmental uncertainties and estimate the next action with high accuracy.

12.5.3 DYNAMIC QoS REQUIREMENTS

Maintaining QoS requirements for 6G-enabled IoT applications, such as delay deadlines and budget constraints, is challenging for heuristic or approximate methods due to the growing demands and diverse resources on distributed computing devices. Cloud servers and fog nodes, which are dynamic and unstable, may not be well-suited to traditional computing paradigms like serverless computing, fog computing, and cloud computing due to their static QoS limits. An alternative would be to describe the implementation of the system by a degree of QoS values and to determine the service provider's ability to deliver more realistic QoS parameters based on those values. For example, instead of specifying strict deadlines and budgets, a range of values for completing the IoT application could be provided to achieve optimal QoS.

Additionally, the probability of achieving computing challenges could be used to portray the value of the QoS parameters. Dynamic QoS management not only meets the many goals of 6G-enabled IoT applications but also optimizes the utilization of computational resources, resulting in better performance and resource efficiency.

12.5.4 WORKLOAD PREDICTION AND RUN-TIME ESTIMATION

The most advanced computing devices have large-scale components that must satisfy QoS requirements. To improve the system's resource consumption and operational conditions, it is essential to comprehend the feature needs and patterns of the workload. To achieve this, it is necessary to analyze the workload and predict the target computing device for deploying tasks, especially in distributed environments. Workload prediction should be dynamic due to the dynamic nature of computational resources, enabling selecting of a more appropriate computing device and estimating the run-time of a 6G-enabled IoT application based on actual parameters. Incorporating previous experience into the performance model can improve the prediction technique. Moreover, algorithms should rely less on run-time predictions to produce high-quality schedules. By using this approach, the most appropriate target computing devices for 6G-enabled IoT applications with the lowest resource consumption can be identified. Using AI models in such situations will allow for a better understanding of the network load and the effective distribution of excess workload from one computing device to nearby devices using optimal prediction techniques.

12.5.5 RESOURCE PREDICTION AND TASK SCHEDULING

In distributed computing, scheduling handles incoming service requests, deciding which requests to run next and how many requests to run in parallel. An efficient scheduling algorithm provides proportional system execution by preserving the lowest global state information, as global state information (CPU load) is proportionate to network overhead. Due to the dynamic nature of the current networks, identifying sufficient resources for 6G-enabled IoT applications is another major issue. So, resource prediction is an essential issue in terms of scheduling 6G-enabled IoT applications. By utilizing an efficient and intelligent resource prediction technique, the accuracy of system performance can be improved. Based on resources and IoT workloads, this technique could assist in selecting the right computer hardware for each IoT application [48]. Intelligent resource prediction techniques can also minimize transmission time and cost and maximize resource use while satisfying QoS requirements.

12.5.6 RELIABLE NETWORK COMMUNICATION

Nowadays, 6G communication technology in the distributed environment uses heterogeneous networking technologies to transmit IoT data, increasing the excessive bandwidth demand of the network [68]. Moreover, efficient utilization of distributed resources and minimizing the 6G bandwidth wastage are still issues for emerging wireless technologies. A 6G network is particularly necessary for IoT devices to deliver large datasets to a serverless environment or cloud servers. Nevertheless, the available bandwidth is insufficient to send huge datasets to specific computers. So, researchers should devise efficient strategies for utilizing the bandwidth between devices and reducing the transmission time as much as possible. This could also make IoT applications use less energy during transmission, eventually reducing CO_2 and temperature emission rates.

12.5.7 RESOURCE SCALABILITY AND FLEXIBILITY

The scalability of resources and operational flexibility are also significant for emerging distributed computing technologies. Over time, millions of IoT devices with beyond 5G or 6G connectivity have generated a lot of data that needs to be stored, processed, and analyzed within the given time bound. But delay-restricted IoT applications with 6G connectivity should follow a specific bandwidth and be processed near the edge networks [57]. On the other hand, computation-intensive tasks can be executed in any given period based on resource availability. So, turning on all the computers when necessary will increase the whole system's performance and boost the performance of many QoS parameters, such as energy use, CO_2 and temperature emissions, and the use of resources. To improve how well the QoS parameters work, service providers should intelligently scale up or down the number of computers. So, researchers and developers must pay attention to different auto-scaling strategies for new 6G-enabled distributed computing paradigms to improve the network's performance.

12.6 CONCLUSION

This chapter has presented a systematic literature review of developing 6G wireless communication and distributed computing technologies for various IoT applications. We discussed the requirement for 6G communication for future IoT applications. Second, we examined several IoT objectives that require advanced computation and communication technologies for scalability and robustness. However, sensing data from IoT objects and transmitting them to a centralized cloud environment is less valuable. To overcome such issues, numerous emerging paradigms, such as counter-based computing, serverless computing, and fog computing, and their intersections have come into play. On the other hand, dynamic user service requests and future IoT applications require high bandwidth, low latency, and high QoS. 6G wireless technology and distributed computing paradigms were created to address the ever-increasing bandwidth needs and alleviate the shortcomings of centralized data processing. Fourth, this survey also investigated all the mature distributed technologies and their applicability to IoT applications. Additionally, we highlighted the need for 6G technology in the distributed environment to solve the challenges of delay-optimized parallel execution. Finally, we explored research difficulties and future perspectives for processing IoT applications with these technologies.

ACKNOWLEDGMENT

The authors would like to thank the Communications & Networks Lab, National University of Singapore, for providing the necessary resources and support for this research.

REFERENCES

1. A. Talpur, and M. Gurusamy, "Machine learning for security in vehicular networks: A comprehensive survey," IEEE Communications Surveys & Tutorials, vol. 24, no. 1, pp. 346–379, 2021.
2. D. K. Sah, A. Hazra, R. Kumar, and T. Amgoth, "Harvested energy prediction technique for solar-powered wireless sensor networks," IEEE Sensors Journal, vol. 23, no. 8, pp. 8932–8940, 2023.
3. S. Singh, S. Batabyal, and S. Tripathi, "Security aware dynamic scheduling algorithm (SADSA) for real-time applications on grid," Cluster Computing, vol. 23, no. 2, pp. 989–1005, 2020.
4. A. Kalita, and M. Khatua, "Channel condition based dynamic beacon interval for faster formation of 6TiSCH network," IEEE Transactions on Mobile Computing, vol. 20, no. 7, pp. 2326–2337, 2020.
5. P. K. Donta, S. N. Srirama, T. Amgoth, and C. S. R. Annavarapu, "Survey on recent advances in IoT application layer protocols and machine learning scope for research directions," Digital Communications and Networks, vol. 8, no. 5, pp. 727–744, 2022.
6. S. Dustdar, V. C. Pujol, and P. K. Donta, "On distributed computing continuum systems," IEEE Transactions on Knowledge and Data Engineering, vol. 35, no. 4, pp. 4092–4105, 2022.
7. P. K. Donta, B. Sedlak, V. Casamayor Pujol, and S. Dustdar, "Governance and sustainability of distributed continuum systems: A big data approach," Journal of Big Data, vol. 10, no. 1, pp. 1–31, 2023.
8. M. Malawski, A. Gajek, A. Zima, B. Balis, and K. Figiela, "Serverless execution of scientific workflows: Experiments with HyperFlow, AWS Lambda and Google Cloud Functions," Future Generation Computer Systems, vol. 110, pp. 502–514, 2017.
9. I. F. Akyildiz, A. Kak, and S. Nie, "6G and beyond: The future of wireless communications systems," IEEE Access, vol. 8, pp. 133995–134030, 2020.
10. L. Jiao, R. Friedman, X. Fu, S. Secci, Z. Smoreda, and H. Tschofenig, "Cloud-based computation offloading for mobile devices: State of the art, challenges and opportunities," in 2013 future network & Mobile summit. IEEE, 2013, pp. 1–11.
11. I. Stojmenovic, "Fog computing: A cloud to the ground support for smart things and machine-to-machine networks," in 2014 Australasian telecommunication networks and applications conference (ATNAC). IEEE, 2014, pp. 117–122.
12. S. Yangui, P. Ravindran, O. Bibani, R. H. Glitho, N. B. Hadj-Alouane, M. J. Morrow, and P. A. Polakos, "A platform as-a-service for hybrid cloud/fog environments," in 2016 IEEE International Symposium on Local and Metropolitan Area Networks (LANMAN). IEEE, 2016, pp. 1–7.
13. X. Zhu, D. S. Chan, H. Hu, M. S. Prabhu, E. Ganesan, and F. Bonomi, "Improving video performance with edge servers in the fog computing architecture." Intel Technology Journal, vol. 19, no. 1, 2015.
14. S. Yangui, and S. Tata, "The SPD approach to deploy service-based applications in the cloud," Concurrency and Computation: Practice and Experience, vol. 27, no. 15, pp. 3943–3960, 2015.
15. D. Pop, G. Iuhasz, C. Craciun, and S. Panica, "Support services for applications execution in multi-clouds environments," in 2016 IEEE International Conference on Autonomic Computing (ICAC). IEEE, 2016, pp. 343–348.
16. B. Di Martino, "Applications portability and services interoperability among multiple clouds," IEEE Cloud Computing, vol. 1, no. 1, pp. 74–77, 2014.
17. F. Bonomi, R. Milito, J. Zhu, and S. Addepalli, "Fog computing and its role in the Internet of Things," in Proceedings of the first edition of the MCC workshop on Mobile cloud computing, ACM, 2012, pp. 13–16.

18. A. Hazra, P. Choudhary, and M. S. Singh, "Recent advances in deep learning techniques and its applications: An overview," Advances in Biomedical Engineering and Technology, pp. 103–122.

19. F. Bonomi, R. Milito, P. Natarajan, and J. Zhu, "Fog computing: A platform for Internet of Things and analytics," in Big data and Internet of Things: A roadmap for smart environments. Springer, 2014, pp. 169–186.

20. S. Singh, "A systematic review on security aware real-time task scheduling," Sustainable Computing: Informatics and Systems, p. 100872, 2023.

21. M. Mukherjee, S. Kumar, C. X. Mavromoustakis, G. Mastorakis, R. Matam, V. Kumar, and Q. Zhang, "Latency-driven parallel task data offloading in fog computing networks for industrial applications," IEEE Transactions on Industrial Informatics, vol. 16, no. 9, pp. 6050––6058, 2019.

22. A. Yazar, and H. Arslan, "A waveform parameter assignment framework for 6G with the role of machine learning," IEEE Open Journal of Vehicular Technology, vol. 1, pp. 156–172, 2020.

23. W. Saad, M. Bennis, and M. Chen, "A vision of 6g wireless systems: Applications, trends, technologies, and open research problems," IEEE Network, vol. 34, no. 3, pp. 134–142, 2020.

24. A. Hazra, M. Adhikari, T. Amgoth, and S. N. Srirama, "Stackelberg game for service deployment of IoT-enabled applications in 6G-aware fog networks," IEEE Internet of Things Journal, vol. 8, no. 7, pp. 5185–5193, 2020.

25. I. Tomkos, D. Klonidis, E. Pikasis, and S. Theodoridis, "Toward the 6G network era: Opportunities and challenges," IT Professional, vol. 22, no. 1, pp. 34–38, 2020.

26. X. Li, M. Huang, Y. Liu, V. G. Menon, A. Paul, and Z. Ding, "I/Q imbalance aware nonlinear wireless-powered relaying of B5G networks: Security and reliability analysis," IEEE Transactions on Network Science and Engineering, vol. 8, no. 4, pp. 2995–3008, 2020.

27. B. Ji, Y. Wang, K. Song, C. Li, H. Wen, V. G. Menon, and S. Mumtaz, "A survey of computational intelligence for 6G: Key technologies, applications and trends," IEEE Transactions on Industrial Informatics, vol. 17, no. 10, pp. 7145–7154, 2021.

28. H. Flores, P. Hui, S. Tarkoma, Y. Li, S. Srirama, and R. Buyya, "Mobile code offloading: From concept to practice and beyond," IEEE Communications Magazine, vol. 53, no. 3, pp. 80–88, 2015.

29. M. A. Ferrag, M. Derdour, M. Mukherjee, A. Derhab, L. Maglaras, and H. Janicke, "Blockchain technologies for the Internet of Things: Research issues and challenges," IEEE Internet of Things Journal, vol. 6, no. 2, pp. 2188–2204, 2018.

30. S. Singh, and S. Tripathi, "Slope: Secure and load optimized packet scheduling model in a grid environment," Journal of Systems Architecture, vol. 91, pp. 41–52, 2018.

31. A. Kumari, S. Tanwar, S. Tyagi, and N. Kumar, "Fog computing for healthcare 4.0 environment: Opportunities and challenges," Computers & Electrical Engineering, vol. 72, pp. 1–13, 2018.

32. A. Talpur, and M. Gurusamy, " DRLD-SP: A deep-reinforcement-learning-based dynamic service placement in edge-enabled internet of vehicles," IEEE Internet of Things Journal, vol. 9, no. 8, pp. 6239–6251, 2022.

33. A. Hazra, A. Alkhayyat, and M. Adhikari, "Blockchain-aided integrated edge framework of cybersecurity for Internet of Things," IEEE Consumer Electronics Magazine, vol. 13, no. 1, pp. 97–102, Jan. 2024.

34. R. Buyya, S. N. Srirama, G. Casale, R. Calheiros, Y. Simmhan, B. Varghese, E. Gelenbe, B. Javadi, L. M. Vaquero, M. A. Netto et al., "A manifesto for future generation cloud computing: Research directions for the next decade," ACM Computing Surveys (CSUR), vol. 51, no. 5, pp. 1–38, 2018.

35. R. Mahmud, S. N. Srirama, K. Ramamohanarao, and R. Buyya, "Quality of experience (QoE)-aware placement of applications in fog computing environments," Journal of Parallel and Distributed Computing, vol. 132, pp. 190–203, 2019.
36. U. Bodkhe, S. Tanwar, K. Parekh, P. Khanpara, S. Tyagi, N. Kumar, and M. Alazab, "Blockchain for industry 4.0: A comprehensive review," IEEE Access, vol. 8, pp. 79764–79800, 2020.
37. M. Mukherjee, L. Shu, and D. Wang, "Survey of fog computing: Fundamental, network applications, and research challenges," IEEE Communications Surveys & Tutorials, vol. 20, no. 3, pp. 1826–1857, 2018.
38. T. B. Brown, B. Mann, N. Ryder, M. Subbiah, J. Kaplan, P. Dhariwal, A. Neelakantan, P. Shyam, G. Sastry, A. Askell et al., "Language models are few-shot learners," in Proceedings of the 34th International Conference on Neural Information Processing Systems, ACM, 2020, pp. 1877–1901.
39. M. Mukherjee, R. Matam, L. Shu, L. Maglaras, M. A. Ferrag, N. Choudhury, and V. Kumar, "Security and privacy in fog computing: Challenges," IEEE Access, vol. 5, pp. 19293–19304, 2017.
40. A. Kalita, and M. Khatua, "Opportunistic transmission of control packets for faster formation of 6TiSCH network," ACM Transactions on Internet of Things, vol. 2, no. 1, pp. 1–29, 2021.
41. T. Das, V. Sridharan, and M. Gurusamy, "A survey on controller placement in SDN," IEEE Communications Surveys & Tutorials, vol. 22, no. 1, pp. 472–503, 2019.
42. B. Chen, J. Wan, L. Shu, P. Li, M. Mukherjee, and B. Yin, "Smart factory of industry 4.0: Key technologies, application case, and challenges," IEEE Access, vol. 6, pp. 6505–6519, 2017.
43. S. Deng, H. Zhao, W. Fang, J. Yin, S. Dustdar, and A. Y. Zomaya, "Edge intelligence: The confluence of edge computing and artificial intelligence," IEEE Internet of Things Journal, vol. 7, no. 8, pp. 7457–7469, 2020.
44. W. Shi, and S. Dustdar, "The promise of edge computing," Computer, vol. 49, no. 5, pp. 78–81, 2016.
45. M. Villari, M. Fazio, S. Dustdar, O. Rana, and R. Ranjan, "Osmotic computing: A new paradigm for edge/cloud integration," IEEE Cloud Computing, vol. 3, no. 6, pp. 76–83, 2016.
46. A. Hazra, M. Adhikari, T. Amgoth, and S. N. Srirama, "Intelligent service deployment policy for next-generation industrial edge networks," IEEE Transactions on Network Science and Engineering, vol. 9, no. 5, pp. 3057–3066, 2022.
47. A. Hazra, M. Adhikari, T. Amgoth, and S. Srirama, "Joint computation offloading and scheduling optimization of IoT applications in fog networks," IEEE Transactions on Network Science and Engineering, vol. 7, no. 4, pp. 3266–3278, 2020.
48. A. Hazra, P. Rana, M. Adhikari, and T. Amgoth, "Fog computing for next generation Internet of Things: Fundamental, state-of-the-art and research challenges," Computer Science Review, vol. 48, p. 100549, 2023.
49. "IEEE approved draft standard for adoption of OpenFog Reference Architecture for fog computing," IEEE P1934/D2.0, April 2018, pp. 1–175, 2018.
50. S. Poojara, C. K. Dehury, P. Jakovits, and S. N. Srirama, "Serverless data pipelines for IoT data analytics: A cloud vendors perspective and solutions," in Predictive analytics in cloud, fog, and edge computing. Springer, 2023, pp. 107–132.
51. S. R. Poojara, C. K. Dehury, P. Jakovits, and S. N. Srirama, "Serverless data pipeline approaches for IoT data in fog and cloud computing," Future Generation Computer Systems, vol. 130, pp. 91–105, 2022.
52. S. Sarkar, R. Wankar, S. N. Srirama, and N. K. Suryadevara, "Serverless management of sensing systems for fog computing framework," IEEE Sensors Journal, vol. 20, no. 3, pp. 1564–1572, 2020.

53. Baktir, A. C., Sonmez, C., Ersoy, C., Ozgovde, A., & Varghese, B. (2019). Addressing the challenges in federating edge resources. Fog and Edge Computing: Principles and Paradigms, 25–49.

54. "IEEE draft standard for adoption of OpenFog reference architecture for fog computing," IEEE P1934/D1.0, February 2018, pp. 1–170, 2018. https://img.antpedia.com/standard/files/pdfs_ora/20230616-ieee/IEEE/Std/IEEE%20Std%201934-2018.pdf

55. A. Hazra, M. Adhikari, T. Amgoth, and S. N. Srirama, "Fog computing for energy-efficient data offloading of IoT applications in industrial sensor networks," IEEE Sensors Journal, vol. 22, no. 9, pp. 8663–8671, 2022.

56. A. Hazra, M. Adhikari, T. Amgoth, and S. N. Srirama, "Collaborative AI-enabled intelligent partial service provisioning in green industrial fog networks," IEEE Internet of Things Journal, vol. 10, no. 4, pp. 2913–2921, 15 Feb. 2023.

57. A. Hazra, M. Adhikari, S. Nandy, K. Doulani, and V. G. Menon, "Federated-learning-aided next-generation edge networks for intelligent services," IEEE Network, vol. 36, no. 3, pp. 56–64, 2022.

58. S. Misra, S. Pal, N. Ahmed, and A. Mukherjee, "SDN-controlled resource-tailored analytics for healthcare IoT system," IEEE Systems Journal, vol. 17, no. 2, pp. 1777–1784, June 2023.

59. S. Pratap, P. Dass, and S. Misra, "CoTEV: Trustworthy and cooperative task execution in internet of vehicles," IEEE Transactions on Mobile Computing, 2023.

60. R. Tapwal, S. Misra, and S. K. Pal, "PerBlocks: A reconfigurable blockchain for service provisioning in industrial environment," IEEE Transactions on Industrial Informatics, vol. 20, no. 1, pp. 911–918, Jan. 2024.

61. W. Binghui, N. V. Abhishek, P. Amogh, and M. Gurusamy, "NPRA: A novel predictive resource allocation mechanism for next generation network slicing," in 2023 IEEE 20th Consumer Communications & Networking Conference (CCNC). IEEE, 2023, pp. 716–721.

62. H. Shen, H. Wang, J. Gao, and R. Buyya, "An instability-resilient renewable energy allocation system for a cloud datacenter," IEEE Transactions on Parallel and Distributed Systems, vol. 34, no. 3, pp. 1020–1034, 2023.

63. A. Mampage, S. Karunasekera, and R. Buyya, "Deep reinforcement learning for application scheduling in resource-constrained, multi-tenant serverless computing environments," Future Generation Computer Systems, vol. 143, pp. 277–292, 2023.

64. M. Kumar, H. Raj, N. Chaurasia, and S. S. Gill, "Blockchain inspired secure and reliable data exchange architecture for cyber-physical healthcare system 4.0," Internet of Things and Cyber-Physical Systems, 2023.

65. M. Golec, S. S. Gill, A. K. Parlikad, and S. Uhlig, "HealthFaaS: AI based smart healthcare system for heart patients using serverless computing," IEEE Internet of Things Journal, vol. 10, no. 21, pp. 18469–18476, 1 Nov. 2023.

66. M. Bansal, I. Chana, and S. Clarke, "UrbanEnQoSPlace: A deep reinforcement learning model for service placement of real-time smart city IoT applications," IEEE Transactions on Services Computing, vol. 16, no. 4, pp. 3043–3060, 1 July–Aug. 2023.

67. S. Shamshad, K. Mahmood, U. Shamshad, I. Hussain, S. Hussain, and A. K. Das, "A provably secure and lightweight access control protocol for EI-based vehicle to grid environment," IEEE Internet of Things Journal, vol. 10, no. 18, pp. 16650–16657, 15 Sept. 2023.

68. A. Gomes, J. Kibi lda, A. Farhang, R. Farrell, and L. A. DaSilva, "Network sharing for reliable networks: A data-driven study," in ICC 2020 - 2020 IEEE International Conference on Communications (ICC), IEEE 2020, pp. 1–6.

13 Designing a Decentralized IoT-WSN Architecture Using Blockchain Technology

Vaibhav V. Gijare, Satpalsing D. Rajput, Srinivas Ambala, Mrunal Swapnil Aware, Wankhede Vishal Ashok, Yatin Gandhi, and Imran Memon

13.1 INTRODUCTION

Two fast expanding areas that make it possible for objects to connect and share data via the Internet are the Internet of Things (IoT) and Wireless Sensor Networks (WSN). However, centralized designs for such systems, as shown in Figure 13.1, can have weak areas in terms of security and redundancy [1].

These problems can be solved by developing a blockchain-based, distributed architecture for IoT-WSNs. When planning such an architecture [2], the following steps need to be kept in mind:

i. **Initial steps** should focus on defining the problem statement and the use cases that the architecture will support. You can learn more about the system's needs and limitations from this.

ii. **Network topology** must be chosen based on the requirements of the use case. A star, mesh, or hybrid topology may be employed, depending on the specifics.

iii. **Blockchain platform selection** consider aspects including scalability, security, and pricing while deciding on a blockchain platform. In addition to IOTA and Hyperledger Fabric, Ethereum is another popular choice.

iv. **Create a smart contract** specifies the parameters under which information can be shared, accessed, and validated. The consensus process for verifying blockchain transactions should also be detailed in the smart contract.

v. **Connect the sensors** connect the sensors to the network and make sure they can talk to each other and the blockchain. LoRaWAN and NB-IoT are two examples of long-range, low-power wireless technologies that could be used for this purpose.

vi. **Deploy blockchain nodes** across the network to make data validation and storage more efficient. These nodes could be hosted in the cloud, on gateways, or on devices in the network's periphery.

DOI: 10.1201/9781003437079-13

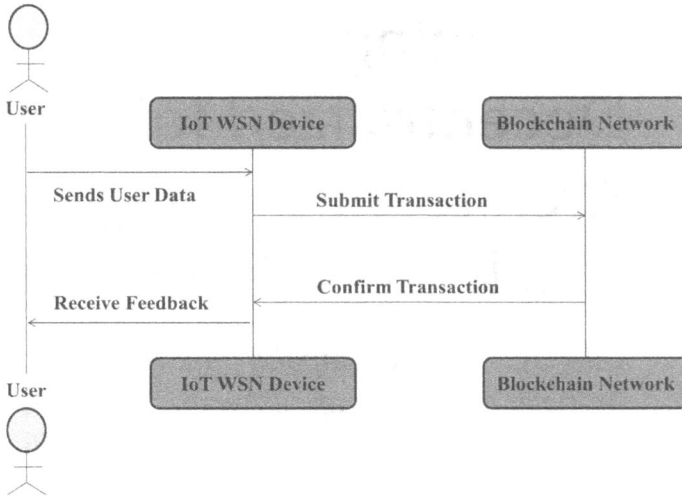

FIGURE 13.1 Overview of IoT-WSN architecture.

vii. **Protect the network** by putting in place stringent security measures that will deter intrusion attempts. Encryption, restricted access, and unusual behavior detection are all possibilities.

viii. **Test the system** to make sure the system works as intended, you should test it extensively. Keep an eye out for slowdowns or scaling problems and fix them as they appear.

Designing a blockchain-based, distributed IoT-WSN architecture is a complex and time-consuming endeavor. Such an architecture can offer an impenetrable and trustworthy foundation for IoT and WSN applications by harnessing the strength of blockchain technology [3–9].

13.1.1 Background

Various gadgets are now able to connect and share data thanks to the rapid development of the IoT and Wireless Sensor Networks (WSN), as illustrated in Figure 13.2. WSNs and IoT devices have found uses in many different areas, including medicine, smart cities, farming, and manufacturing [10].

These systems typically employ centralized architectures, which can introduce security holes and single points of failure. The centralized method calls for a governing body to oversee the flow of information and guarantee the system's safety and dependability [11]. When there are several parties involved with competing priorities, adopting this strategy might be difficult.

13.1.2 Motivation

Using blockchain technology's decentralized approach, problems with traditional centralized IoT and WSN designs can be overcome. The blockchain is a distributed

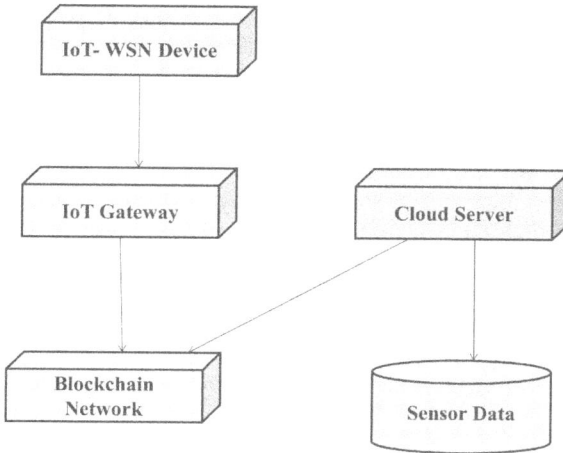

FIGURE 13.2   Working flow diagram of IoT-WSN architecture using blockchain technology.

ledger that facilitates trustless and traceable record-keeping of transactions with no need for intermediaries. There are several ways in which blockchain technology might improve IoT and WSN infrastructures:

i. By removing the need for a central processing unit, or a centralized server, a decentralized design can improve the system's security, reliability, and resilience.

ii. All users may see the data's whole transaction history and history of changes because of the blockchain's transparency.

iii. Data stored on the blockchain cannot be changed after it has been recorded, guaranteeing the data's authenticity and integrity.

iv. Data exchange and transfer can be automated and the need for middlemen reduced with the use of smart contracts, which establish the rules and circumstances for these actions.

Using blockchain technology, a decentralized IoT-WSN architecture may ensure the authenticity and integrity of shared data while allowing for the participation of a wide range of stakeholders in the ecosystem.

13.1.3   RESEARCH OBJECTIVES

When developing a blockchain-based, decentralized IoT-WSN architecture, the following questions should be addressed:

1. The goal of this work is to build [12] a decentralized architecture for IoT-WSN systems utilizing blockchain technology, which will increase the system's security, reliability, and resilience by doing away with the vulnerable central hub.

2. The goal of this work is to provide the parameters of a blockchain-based distributed IoT-WSN architecture, including the desired network topology, blockchain platform, and consensus process.

3. The goal of this work is to use blockchain technology to design the optimal network topology for a decentralized IoT-WSN architecture, taking into account all relevant criteria and restrictions.

4. For a decentralized IoT-WSN architecture, it is important to compare and contrast available blockchain solutions in terms of key criteria including scalability, security, and cost.

5. To create and deploy blockchain-based smart contracts that specify parameters for information exchange between nodes in a distributed IoT and wide area network (WAN) setting.

6. The goal is to use blockchain technology to seamlessly include the sensors into the decentralized IoT-WSN architecture and enable seamless inter-sensor and inter-blockchain communication.

7. To place blockchain nodes strategically across the network for optimal data validation and storage.

8. The goal is to use blockchain technology to create a decentralized IoT-WSN architecture that is impenetrable to assaults and other forms of tampering.

9. The goal of continuously monitoring and fine-tuning the distributed IoT-WSN network built on blockchain technology is to maximize its performance and scalability.

10. The goal is to determine where and how blockchain technology could be used to benefit a decentralized IoT-WSN architecture, such as in the healthcare, smart city, agricultural, and manufacturing sectors.

13.1.4 CONTRIBUTIONS

Multiple contributions are planned from our proposed study of developing a decentralized IoT-WSN architecture with blockchain technology, and these are as follows:

1. The elimination of a system's weak [13] link through the use of a decentralized architecture based on blockchain technology improves the system's security, reliability, and resilience.

2. Determination of what must be included and what must be excluded from a blockchain-based distributed IoT-WSN architecture, including the desired network topology, blockchain platform, and consensus process.

3. The optimal network topology for a blockchain-powered, distributed IoT-WSN architecture is determined by analyzing these factors.

4. Considerations like as scalability, security, and cost will be used to determine which blockchain platform is best suited for a decentralized IoT-WSN architecture.

5. Data sharing and transfer in a distributed IoT-WSN architecture are governed by smart contracts that have been designed and implemented using blockchain technology.

6. Using blockchain technology to include the sensors into the distributed IoT-WSN architecture and guarantee their interoperability.
7. The process of deciding where on the network to place blockchain nodes for the purposes of data validation and storage.
8. Defenses against assaults and unauthorized access to the distributed IoT-WSN architecture employing blockchain technology have been implemented.
9. Through careful monitoring and adaptation, the performance and scalability of the distributed IoT-WSN architecture may be maximized using blockchain technology.
10. Possible applications and advantages of a blockchain-powered, distributed IoT-WSN architecture are identified, such as in the fields of medicine, urban planning, agriculture, and manufacturing.

By providing a safe, transparent, and decentralized way to address the issues associated with traditional centralized designs, the study of developing a decentralized IoT-WSN architecture utilizing blockchain technology can aid in the development of IoT and WSN systems [14]. The suggested architecture has the potential to facilitate the participation of several stakeholders in the ecosystem, safeguarding the data's integrity and authenticity for better, more informed decision-making.

13.1.5 STATE-OF-THE-ART COMPONENTS

This section outlines of some of the most important factors that are often considered when creating a decentralized IoT-WSN using blockchain technology.

Table 13.1 offers a concise comparison of the cutting-edge components and functionalities that are incorporated into the architecture of the IoT-WSN. It is crucial to bear in mind that the selection of technology is in part dictated by the use cases, specifications, and limits that are present. This is something that must be always kept in mind. Recent developments in these fields are being made with the idea of expanding the research that has already been done in order to accomplish the overarching goal of improving the effectiveness, safety, and performance of IoT-WSN systems.

13.2 IoT-WSN ARCHITECTURE

13.2.1 OVERVIEW OF IoT-WSN ARCHITECTURE

The term "Internet of Things" (IoT) refers to a system of interconnected computing devices, sensors, and physical items that can exchange data via the internet. An example of an IoT network [12] is Wireless Sensor Networks (WSNs) collect data from the real world and send it wirelessly to a data center where it can be processed and analyzed is shown in Figure 13.3.

The dynamic and rapidly evolving subject of IoT–Wide Area Network (IoT-WSN) architecture presents both opportunities and challenges. As technological advancements continue, there will be a growing demand for IoT-WSN systems that are increasingly sophisticated and intelligent [19]. In order to manage the massive amounts of data that are produced by the ever-increasing number of connected

TABLE 13.1
Comparative Study of the Components of a Decentralized IoT-WSN System

Technology	Description	Advantages	Limitations
Communication Protocols [15]	Zigbee	Low-power, low-cost, and reliable communication protocol suitable for small-scale deployments.	Limited range and bandwidth, may not be suitable for large-scale deployments or long-range communication.
	Bluetooth Low Energy (BLE)	Low-power, short-range wireless protocol commonly used in IoT devices.	Easy to implement, widely supported by mobile devices. Limited range and bandwidth.
	Wi-Fi	High-speed, long-range wireless protocol with broad industry support.	Wide availability, high data rates. Requires higher power consumption compared to other protocols.
	LoRaWAN	Low-power, long-range protocol suitable for wide-area deployments.	Wide coverage, low-power consumption. Limited data rates and throughput.
	MQTT	Lightweight publish-subscribe messaging protocol for IoT devices.	Efficient, supports low-power devices, and has broad industry adoption.
Data Analytics [16]	Machine Learning	Utilizes algorithms to automatically learn from data and make predictions or decisions.	Enables intelligent decision-making and automation. Requires labeled training data and computational resources.
	Big Data Processing	Techniques for processing and analyzing large volumes of data to extract valuable insights.	Enables real-time analytics and scalable processing. Requires powerful computing infrastructure.
	Edge Computing	Processing and analysis of data at or near the edge of the network, reducing latency and bandwidth requirements.	Enables real-time processing and decision-making. Limited computational resources compared to the cloud.
Security [17]	Blockchain	Distributed ledger technology that ensures immutability, transparency, and tamper-proof data storage.	Enhances data security, trust, and transparency. Scalability and performance limitations.
	Secure Communication	Cryptographic protocols and algorithms for secure data transmission between devices and systems.	Protects data integrity and confidentiality. Adds computational overhead and complexity.
	Access Control	Mechanisms to manage and enforce authorized access to IoT-WSN systems and data.	Prevents unauthorized access and data breaches. Requires robust authentication and authorization mechanisms.

(Continued)

TABLE 13.1 *(Continued)*
Comparative Study of the Components of a Decentralized IoT-WSN System

Technology	Description	Advantages	Limitations
Energy Efficiency [18]	Energy Harvesting	Techniques to capture and utilize ambient energy sources to power IoT devices, reducing the need for battery replacement.	Extends device lifespan and reduces maintenance. Limited by available energy sources and power requirements.
	Low-Power Design	Design strategies that optimize power consumption in IoT devices and networks, including sleep modes and duty cycling.	Prolongs battery life and reduces energy consumption. May impact real-time responsiveness.
	Energy-Efficient Communication	Protocols and algorithms that minimize energy consumption in wireless communication.	Reduces energy usage and extends network lifespan. May sacrifice data rate or latency.

devices, one area of concentration is on improving the scalability of the architecture. Alternative approaches to scalability include the implementation of efficient data aggregation mechanisms, the optimization of network protocols, and the use of processing and storage resources hosted in the cloud.

Energy efficiency is yet another essential aspect of the architecture of IoT-WSN networks. It is essential to achieve optimal energy consumption if one want to extend

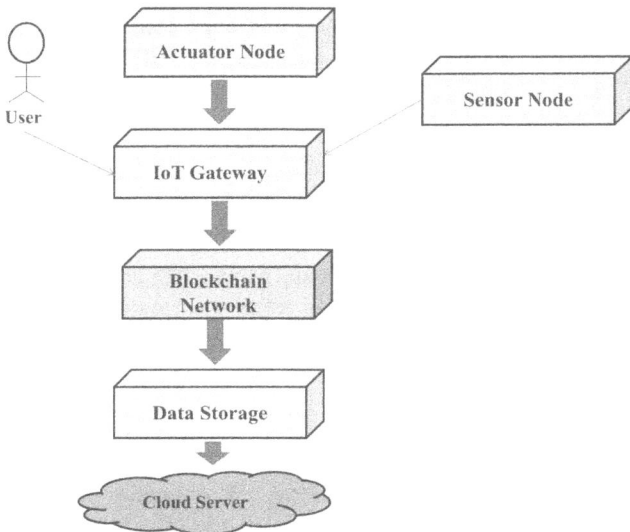

FIGURE 13.3   Basic block diagram of IoT-WSN architecture.

the lifespan of sensor nodes, which are frequently powered by batteries, and reduce the amount of maintenance that must be performed. In order to make IoT-WSN systems use less energy, researchers and engineers are always looking into innovative approaches such as energy harvesting, adaptive power management, and duty cycling. These initiatives attempt to ensure the network's continued survival over the long run by striking a balance between the collection of data and the conservation of energy. IoT-WSN design places a premium on security and privacy protections in large part due to the sensitive nature of the data that is being gathered and shared. It is necessary to implement stringent security measures such as encryption, authentication, and access control in order to protect the data from being accessed in an unauthorized manner and to maintain its integrity. Anonymization of data and differential privacy are two examples of privacy-protecting techniques that can be of assistance in addressing privacy concerns and maintaining compliance with data protection legislation.

Interoperability is an additional challenge that the architecture of the IoT-WSN needs to overcome. As a result of the fact that systems are constructed out of disparate components sourced from a variety of manufacturers, it is necessary to provide seamless integration and communication between these components. It is possible to improve interoperability by developing and implementing standardized protocols, data formats, and application programming interfaces (APIs). This will make it much easier to incorporate new pieces of hardware and services into the ecosystem of IoT-WSN networks. Real-time processing capabilities are becoming increasingly necessary for a growing number of applications, particularly those that require immediate actions or reactions. For instance, in industrial automation, making quick decisions based on real-time sensor data can help improve production operations, hence reducing the need for costly downtime. In order to satisfy real-time requirements, the architecture of IoT-WSN needs algorithms for efficiently processing data, communication protocols with low latency, and edge computing technologies that enable data processing at the edge of the network. The IoT-WSN architecture offers a lot of potential for revolutionizing different industries and enabling new kinds of applications.

13.2.2 System Architecture and Components

IoT-WSN's system design includes sensor nodes, a gateway device, communication protocols, a cloud platform, data analytics and processing, actuator nodes, and a way to keep data safe and private. Sensor nodes gather information, data gathering units put it all together, and communication methods make it easy to send the data to the cloud for research and decision-making.

    i. *Sensor nodes*: These are very small electronic devices that contain sensors and are used to collect data about the environment. These sensors are able to measure a wide variety of factors, such as temperature, humidity, pressure, light, sound, and motion, amongst others. Communication between sensor nodes and other nodes or gateway devices is typically done wirelessly and uses very little power.

ii. *Gateway devices*: These serve as a bridge between sensor nodes and external networks such as the internet or cloud platforms. Gateway devices are also known as bridge devices. After receiving the information from a variety of sensor nodes, they make it possible for the data to be transmitted to other systems in order to undergo further processing and analysis. Gateways are responsible for a variety of tasks, including the collection of data, the filtering of data, and the translation of protocols.

iii. *Communication protocols*: The IoT-WSN architecture includes communication protocols, which define the parameters and requirements for the transmission of data between the various components.

iv. *Cloud platforms*: These are an essential component of the IoT-WSN architecture because they provide the capabilities of storage, computing power, and data analytics. They receive the data from the gateways and preserve it, making it possible to do historical study as well as monitoring in real time. Cloud platforms also provide application programming interfaces (APIs) and tools for developers, allowing them to build apps on top of the IoT-WSN architecture.

v. *Data analytics and processing*: These are the elements that, when combined with the obtained data, operate to provide useful insights and accelerate the decision-making process. Using techniques such as machine learning (ML), statistical analysis, and data mining, patterns can be discovered, anomalies can be identified, and predictions can be formed. These insights make it feasible to make proactive decisions, improve efficiency, and optimize processes, making it possible to optimize processes and boost efficiency.

vi. *Actuator nodes*: In addition to sensors, the architecture of an IoT-WSN could also include actuator nodes. These are responsible for directing physical systems or devices in line with the data they receive. Actuator nodes provide for the activation or deactivation of devices, the modification of settings, and the transmission of commands to other connected systems.

vii. *Mechanisms for data security and privacy*: The architecture of IoT-WSN requires security and privacy components to ensure the integrity, confidentiality, and privacy of the data that is being gathered and exchanged. Encryption, authentication, access control, and secure communication protocols are some of the safeguards that have been implemented as part of these measures. Their purpose is to guarantee that only authorized parties can access and transmit data.

13.2.3 Limitations of IoT-WSN Architecture

Despite the many advantages, the IoT-WSN architecture does have some restrictions that must be considered. Key restrictions of IoT-WSN architecture include the following [20–26]:

i. *Limited battery life*: The short battery life of the sensors is a major drawback of the IoT-WSN design. Battery life is a major concern for many IoT devices and wireless sensors. When sensors need to be placed in inconvenient or inaccessible areas, this can be a major obstacle.

ii. *Limited range*: Another drawback of the IoT-WSN design is that wireless sensor networks have a limited range. Although numerous wireless technologies exist for making this sensor-to-network connection, each has its own range and reliability restrictions. This can be difficult in situations where many sensors need to be spread out over a wide region or if there is a lot of signal interference.

iii. *Security challenges*: Significant security concerns may also arise from the IoT-WSN architecture. Eavesdropping, jamming, and spoofing are just a few of the assaults that can be launched against wireless sensor networks. Data breaches and cyberattacks are just two of the security concerns that might arise from using cloud computing platforms for processing and storing data.

iv. *Complexity*: The complexity of designing, implementing, and managing an IoT-WSN infrastructure is a real issue. Wireless networking, data analytics, the cloud, and cybersecurity are just a few of the areas of knowledge needed. This is a problem for businesses who lack the personnel or means to put in place and maintain a sophisticated IoT-WSN architecture.

v. *Cost*: One final disadvantage of the IoT-WSN design is its potential high price tag. It can be costly to establish a large-scale IoT-WSN infrastructure, especially if sensors need to be placed in inaccessible or remote areas. Additionally, continuing expenditures associated with network maintenance and upgrades can be quite high.

13.2.4 SCALABILITY

The scalability of an IoT-WSN's design is crucial, as it dictates how well the system can accommodate future expansion and modifications. A scalable design should be able to process more information and more traffic without sacrificing speed or dependability. Achieving scalability in an IoT-WSN architecture requires paying attention to the following details [27–32]:

i. *Network infrastructure*: IoT-WSN architecture relies heavily on the network infrastructure, which must be prepared to handle ever-increasing data and traffic loads. To do this, a flexible network topology should be implemented, and scalable wireless technologies like Lora WAN or Zigbee should be used.

ii. *Cloud computing*: Scalability in IoT-WSN architecture can be accomplished with the help of cloud computing platforms. Data storage and analysis on the cloud allows businesses to rapidly grow their infrastructure to deal with rising data and traffic demands. In addition to scalability, cloud platforms can offer variable pricing structures to meet the varying demands of businesses.

iii. *Processing of data*: Achieving scalability requires designing the data processing architecture to accommodate growing data loads. This is possible with the use of distributed computing platforms like Apache Spark, which allow for the concurrent processing of enormous amounts of data.

iv. *Security*: Safeguarding the system from ever-increasing threats necessitates a security architecture that can expand with the system. Scalable security technologies like blockchain can be utilized to create a trustworthy and decentralized system for storing and processing information.

v. *Automation*: Finally, automation is a potent tool for making IoT-WSN design scalable. Data collection and analysis are two examples of mundane jobs that can be automated, relieving pressure on human operators and allowing systems to grow without requiring extra manpower.

A holistic strategy that considers the network infrastructure, data processing, and security architecture is necessary for attaining scalability in IoT-WSN design. Organizations can guarantee that their IoT-WSN architecture will be able to develop and scale with their needs if they design it with scalability in mind from the start.

13.3   BLOCKCHAIN TECHNOLOGY

The blockchain is a decentralized, digital ledger system that is used to store and track transactions in a way that is both safe and transparent. Blockchain technology is also known as distributed ledger technology (DLT). Its initial goal was to provide support for the cryptocurrency known as Bitcoin; however, since that time, it has been used by a variety of other industries and sectors for a wide range of reasons. A network of computers, known as nodes, is responsible for verifying and adding each transaction to a "block" of data that makes up a blockchain [33]. A permanent and unchangeable record of all the transactions that have taken place on the network is created when a block is finished and then added to a chain of blocks that came before it. Because of this, it is extremely difficult, if not impossible, to tamper with or otherwise alter the data that is kept in a blockchain. Because blockchain technology is decentralized, there is no requirement for a centralized authority to check or authenticate transactions. This is one of the most important advantages of using this technology. Because of this, it is more secure than traditional centralized systems and has a higher level of resistance to hackers and fraud. In addition to its application in the realm of cryptocurrencies, the blockchain technology is also finding applications in other domains, including as the management of supply chains, the authentication of identities, and the administration of digital assets [34].

13.3.1   OVERVIEW OF BLOCKCHAIN TECHNOLOGY

Blockchain technology is a distributed ledger that keeps all your transactions and data safe and visible. Each block in the chain contains data and is connected to the one before it through a link, as shown in Figure 13.4.

Data contained in a block in a blockchain is protected against tampering by cryptographic hash functions. Because of the distributed nature of the blockchain, no single entity has influence over the system. Instead, it is kept running by a distributed network of nodes that work together to verify and append new blocks [35]. Because of its decentralized nature, this technology is well-suited for implementation in IoT-WSN systems due to its security, transparency, and immutability [36].

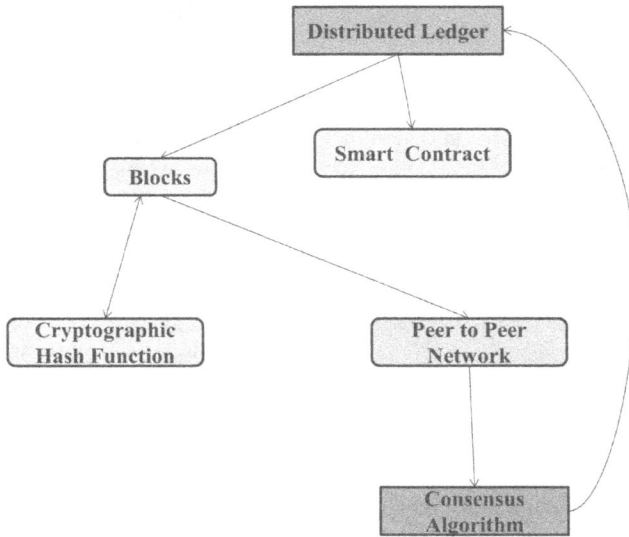

FIGURE 13.4 Overview of blockchain technology.

13.4 PROPOSED DECENTRALIZED IoT-WSN ARCHITECTURE

13.4.1 OVERVIEW OF PROPOSED ARCHITECTURE

The proposed design is a distributed IoT-WSN design that makes use of block-chain technology to improve the system's security, scalability, and dependability. Perception, blockchain, and applications are the three tiers of the architecture, as illustrated in Figure 13.5.

IoT-WSN architecture relies heavily on decentralization [37]. It describes a setup where all nodes, or computers, in a network have the same amount of control over the whole system. Without a central hub or weak link, a decentralized IoT-WSN network can better take care of its own maintenance and security [38]. Each node in a decentralized IoT-WSN network can act as a data transmitter and receiver. The nodes in a mesh network establish connections with one another using wireless means of data transmission. Even if any of the nodes fail or become unavailable, this network structure guarantees that the data can be transferred effectively and securely. The network verifies and approves any changes to the ledger before they are applied to all copies of the ledger stored on all nodes.

A distributed IoT-WSN design has various advantages [39]. First, it improves the system's adaptability and scalability. The design can process massive amounts of data and respond to shifting conditions since there is no one weak link. Second, it raises the system's fault-tolerance and reliability since the nodes may work together to always keep the data safe and accessible. Third, the network's decentralized design makes it less vulnerable to intrusion and better protects users' personal data.

Blockchain's safe and transparent data storage and sharing capabilities make it a great foundation on which to build decentralized IoT-WSN architectures [40]. Nodes

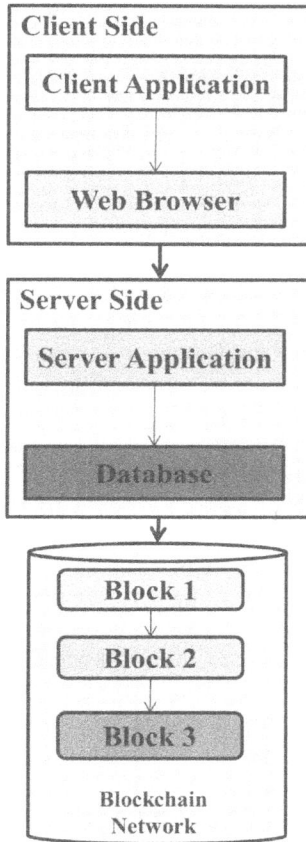

FIGURE 13.5 Block overview of proposed IoT-WSN architecture.

in a blockchain-based, decentralized IoT-WSN architecture can employ smart contracts to automate routine tasks and financial transactions. Smart contracts are digitally signed agreements that can be kept on the blockchain and triggered to carry out predefined activities based on predetermined conditions. As a result, network nodes can work together toward shared goals decentralized from any one controlling body. Data from IoT devices and wireless sensors can be collected, stored, and analyzed on a decentralized architecture that is secure, dependable, and scalable [41]. Decentralized IoT-WSN designs, which make use of distributed ledger technologies like blockchain, can provide a transparent and secure foundation on which to create applications for a variety of sectors.

13.4.2 BLOCKCHAIN-BASED IoT-WSN ARCHITECTURE

Collecting, storing, and analyzing data from IoT devices and wireless sensors is made easier with the blockchain-based IoT-WSN architecture [42].

In order to gather, store, and analyze data from IoT devices and wireless sensors in a safe and scalable manner, a decentralized IoT-WSN system based on blockchain technology is presented. Several crucial parts make up the architecture, including:

i. *IoT devices and wireless sensors*: The system gathers information about its physical surroundings using a network of IoT devices and wireless sensors. Bluetooth, Zigbee, and LoRaWAN are just some of the wireless protocols used to link these gadgets to the Internet.

ii. *Gateway devices*: In order to provide data from a wide variety of sensors and IoT devices to the blockchain network, gateway devices are employed. Data is encoded on these devices in a blockchain-friendly format and, before being sent to the network, may undergo extra preprocessing or analysis.

iii. *Blockchain network*: In the proposed architecture, the blockchain network plays a vital role. It processes and verifies transactions and stores all data gathered by Internet of Things devices and sensors. Each node in the network holds its own copy of the blockchain, making it a distributed ledger.

iv. *Consensus algorithm*: This is used to verify trades and stop double spending, making it a crucial part of keeping the blockchain secure. The suggested architecture employs a proof-of-stake consensus method, which provides incentive for token holders to stake their assets and validate transactions inside the network.

v. *Smart contracts*: Data validation and processing are just two examples of the tasks that can be automated with the help of "smart contracts." Blockchain-based smart contracts are pre-written code that can be set to run automatically when specific circumstances are satisfied.

vi. *Data analytics*: Blockchain data is analyzed using data analytics tools to draw conclusions. In real time, these programs may analyze sensor data, search for patterns, and send out alerts when criteria are met.

The overall goal of the suggested blockchain-based architecture for a decentralized IoT-WSN system is to furnish a trustworthy, extensible, and distributed framework for the gathering, storing, and analyzing of data generated by IoT devices and WSNs, as shown in Figure 13.6. The system can provide a safe and trustworthy platform for several uses in sectors including manufacturing, transportation, and agriculture [19, 43, 44].

13.4.3 Decentralization of IoT-WSN Architecture

When it comes to the architecture of an IoT-WSN, decentralization is a cornerstone. Each component in the network has the same amount of say and control over the whole system; this is what is meant by decentralization. Self-organizing, resilient, and secure with no central point of failure are hallmarks of a decentralized IoT-WSN architecture. Each network node in a distributed IoT-WSN architecture can act as a data source and sink. A mesh network is formed when the nodes are linked via wireless communication. With this configuration, data can still be sent quickly and safely even if some of the nodes in the network go down in Figure 13.7.

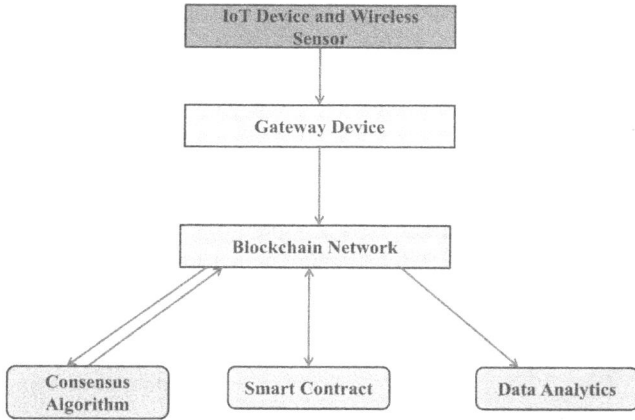

FIGURE 13.6   Basic diagram of blockchain-based IoT-WSN architecture.

Distributed IoT-WSN architecture has several benefits. First, it makes the system more flexible and scalable. The architecture can handle enormous data amounts and changing situations since there is no weak link. Second, because nodes collaborate to ensure data authenticity and accessibility, it improves system fault-tolerance and dependability. Third, the decentralized network makes it less susceptible to infiltration and helps users protect their data. Blockchain and distributed ledgers are often utilized for decentralization. Blockchain is a distributed database administered by nodes. Each node stores a copy of the ledger, and network agreement is needed to modify it 45]. Blockchain is ideal for decentralized IoT-WSN designs because to its

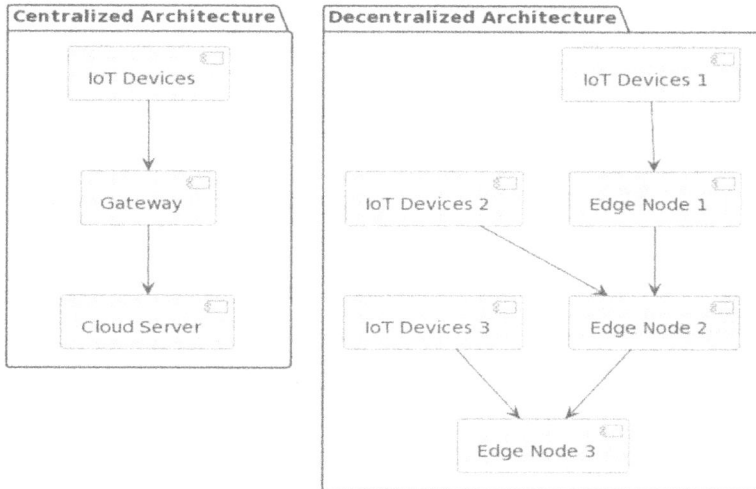

FIGURE 13.7   Block diagram of decentralization in IoT-WSN architecture compared to centralized architecture.

security and transparency. In blockchain-based decentralized IoT-WSN architecture, smart contracts automate operations and transactions. When specific conditions are satisfied, smart contracts (blockchain contracts that may be designed to do particular tasks automatically) will automatically execute the contract. Network nodes may collaborate without a leader. Decentralization is vital to IoT-WSN infrastructure. Decentralized architecture is secure, reliable, and scalable for collecting, storing, and analyzing IoT and wireless sensor data. Through distributed ledger technologies like blockchain, decentralized IoT-WSN architectures may offer a transparent and secure foundation for creating applications in numerous domains [46].

13.4.4 SECURITY OF DECENTRALIZED IoT-WSN ARCHITECTURE

The security of IoT-WSN infrastructure is paramount. Data transferred and stored on the network must be safeguarded from unauthorized access, breaches, and assaults, especially considering the growing number of IoT devices and wireless sensors. Some of the most crucial safety features of the IoT-WSN design are covered here [47–49]:

   i. *Authentication and authorization*: Two of the most important parts of IoT-WSN security are authentication and authorization. These safeguards are in place to ensure that only authorized gadgets and users have access to the network and its data. Authorization specifies the privileges of a user or device, whereas authentication verifies their identity. Digital signatures, public-key cryptography, and secure communication protocols like Transport Layer Security (TLS) and Secure Sockets Layer (SSL) are used in IoT-WSN architectures to accomplish this goal.
   ii. *Data encryption*: Another important part of IoT-WSN security is the use of encryption on sent data. Encryption is the process of changing data into an unreadable format to keep it private. For the purpose of keeping information private during transit and storage, IoT-WSN designs employ encryption methods such the Advanced Encryption Standard (AES), Rivest-Shamir-Adleman (RSA), and elliptic curve cryptography (ECC).
   iii. *Access control*: Assuring that only authorized devices and users have access to the network and the data transmitted over it, access control is a key security feature. Firewalls, intrusion prevention systems, and network segmentation are only some of the access control measures commonly used in IoT-WSN designs to limit network access and prevent security breaches.
   iv. *Device management*: IoT-WSN security also includes the administration of connected devices. Provisioning, monitoring, and maintaining network-connected devices fall under this umbrella term. The security and integrity of IoT-WSN networks relies on device management tools to keep all connected devices patched and secure.
   v. *Trust and reputation management*: Security in IoT-WSN infrastructures is not complete without trust and reputation management. These safeguards guarantee the integrity of data transmitted across a network and aid in establishing and maintaining trust between devices and users. To determine

which devices and users can be trusted and to spot and stop criminal activity, trust and reputation management platforms are implemented in IoT-WSN designs.

Authentication, authorization, data encryption, access control, device management, trust and reputation management, and so on are only some of the security mechanisms that are integral to the IoT-WSN architecture. IoT-WSN architectures that take advantage of these safeguards can ensure the privacy and integrity of all data sent over and stored in the network.

13.4.5 PRIVACY OF DECENTRALIZED IoT-WSN ARCHITECTURE

The architecture of an IoT-WSN places a premium on privacy for its connected devices. The need to protect the privacy of data communicated and stored in a network grows in tandem with the proliferation of IoT devices and wireless sensors. Here we discuss the main privacy issues with the IoT-WSN architecture, and how to fix them, as follows:

   i. *Data anonymization*: Safeguarding personal information by obliterating any trace of a person's identity from the data. Data anonymization in IoT-WSN architectures can be accomplished using methods like k-anonymity, l-diversity, and t-closeness.
  ii. *Data minimization*: Limiting the collection and storage of private information through networked gadgets and sensors, therefore safeguarding an individual's privacy. Data reduction in IoT-WSN architectures is accomplished by collecting only the data needed to complete a task and blocking the collection and transmission of any extraneous data.
 iii. *Data encryption*: As with security, this is used to protect the privacy of data during transmission and storage on a network. Elliptic curve cryptography (ECC), Rivest-Shamir-Adleman (RSA), and the Advanced Encryption Standard (AES) are all viable options for implementing encryption in IoT-WSN infrastructures.
  iv. *Privacy policies*: Organizations are required by law to have privacy policies that detail their procedures for collecting, storing, and using customer information. Users can be made aware of the collection and use of their data as well as the safeguards in place to keep their privacy intact using privacy rules in IoT-WSN designs. In order to gain users' trust and guarantee their privacy, these policies are essential.

The suggested architecture provides a decentralized and secure data management system, which can improve the security, scalability, and reliability of IoT-WSN systems. The potential uses and future avenues of research for the proposed architecture will be discussed in the next section. Data anonymization, data minimization, encryption, and privacy policies are all ways that the IoT-WSN architecture strives to protect users' personal information. By taking these precautions, IoT-WSN architecture can ensure that users' personal information is secure while it is being transmitted over, and stored on, the network.

13.5   CHALLENGES AND LIMITATIONS

While the suggested architecture has many benefits, it also has several drawbacks that must be worked around in IoT-WSN architecture. Among these difficulties are:

i. *Limited resources*: Devices in an IoT-WSN typically have limited resources, such as processing speed, memory, and battery life. Since blockchain technology typically demands substantial computational resources and energy consumption, its implementation on these devices can be difficult.

ii. *Scalability issues*: The blockchain layer's high computational resource requirements to validate and add new blocks to the chain also raise concerns about the scalability of the proposed design. Additionally, the size of the blockchain grows as the number of devices and transactions increases, making it more difficult to administer and store.

iii. *Interoperability*: Because devices and applications may utilize multiple protocols and standards, the proposed architecture may have trouble integrating and communicating with other systems due to interoperability concerns.

iv. *Security risks*: Blockchain technology improves the security of IoT-WSN systems, yet it is still vulnerable to cyberattacks. Security flaws in smart contracts, 51% attacks, and other forms of cybercrime continue to be a major danger to the system's foundation.

v. *Governance problems*: Due to the decentralized nature of the proposed architecture, governance may prove difficult to implement. It can be difficult, especially in large-scale systems, to ensure the cooperation and coordination of many nodes and stakeholders.

To combat these issues, researchers can work to perfect lightweight blockchain protocols and consensus algorithms for IoT-WSN devices. In addition, the scalability and interoperability of the architecture can be improved by the development of standards and protocols for interoperability. Finally, making ensuring the architecture can be maintained and expanded requires building governance models and processes that promote effective coordination and collaboration.

13.6   FUTURE RESEARCH DIRECTIONS

The suggested architecture is an innovative solution to the problems that have plagued previous attempts to build an IoT-WSN system. Nonetheless, more study and development are required. Here are a few potential areas for further study:

i. *Protocol optimization in the blockchain*: As mentioned earlier, IoT-WSN devices with limited resources may struggle to use blockchain technology. Optimized blockchain protocols and consensus algorithms well-suited to IoT-WSN devices can be the focus of future study.

ii. *Standards for interoperability*: These standards must be developed, as this is still a major obstacle for IoT-WSN networks. The suggested architecture's scalability and compatibility can be improved by the creation of interoperability standards and protocols.

iii. *Improved security*: Blockchain technology improves the security of IoT-WSN systems, yet it is still vulnerable to cyberattacks. To counteract these security risks, future studies can concentrate on improving and creating new security measures.

iv. *Integration of AI and ML*: When AI and ML are combined, the resulting architecture gains efficiency and intelligence. Integration of these technologies may improve the architecture's performance, and future studies can investigate this possibility.

v. *Real-world implementation and testing*: This can shed light on the feasibility, scalability, and performance of the suggested architecture. In order to assess the architecture's efficacy and locate areas for development, future studies can concentrate on its implementation and testing in real-world circumstances.

vi. *Energy efficiency*: Implementing blockchain technology can increase the problem of excessive energy usage by IoT-WSN devices. In the future, scientists can work on blockchain protocols optimized for low-power, low-data-rate IoT-WSN devices.

vii. *Improved privacy*: Blockchain technology's distributed ledger and tamper-proof record-keeping can help address the growing concern over personal information in IoT-WSN systems. To further improve the privacy of the design, future studies can investigate the possibility of including privacy-enhancing technologies such homomorphic encryption and differential privacy.

viii. *Edge computing integration*: This type of integration can improve the suggested architecture's speed and efficiency. Integrating edge computing to facilitate real-time data processing and decision-making is a promising area for future study.

ix. *Decentralized architecture*: This allows sustainability and scalability to be improved by creating incentive systems to encourage participation and contribution from different nodes and stakeholders using the blockchain. Incentives for use in IoT-WSN systems that can be implemented using blockchain technology can be a topic of future study.

The proposed architecture can be further optimized and improved by investigating these lines of inquiry to meet the challenges and overcome the limits of current IoT-WSN designs.

13.7 CONCLUSION

Traditional designs for the IoT-WSN system have some drawbacks that could be improved through the implementation of a decentralized IoT-WSN architecture that makes use of blockchain technology. In order to enhance the scalability, privacy protection, and overall security of IoT-WSN systems, the proposed architecture takes advantage of blockchain technology's decentralized and tamper-resistant characteristics.

Following a review of IoT-WSN architecture and its constraints, we then presented an introduction to blockchain technology and discussed its possible uses in IoT-WSN systems. After that, we described the proposed decentralized IoT-WSN architecture, which included an explanation of its primary characteristics, as well as its advantages and disadvantages. Although the suggested design offers a novel method to address the issues of conventional IoT-WSN architectures, we suggested that those challenges and constraints still need to be addressed before the architecture can be considered fully viable. Optimizing blockchain protocols, developing interoperability standards, enhancing security and privacy mechanisms, integrating AI and ML techniques, implementing and testing the architecture in real-world scenarios, and investigating other potential enhancements such as energy efficiency and edge computing, are all areas that could be the focus of future research.

In conclusion, the decentralized IoT-WSN architecture that we have proposed offers a viable alternative to overcome the issues given by standard IoT-WSN architectures. It is possible, with additional study and development, to further optimize and improve the architecture in order to make it possible to build IoT-WSN systems that are safe, scalable, and efficient for a wide variety of applications.

REFERENCES

1. A. Perrig, J. Stankovic, and D. Wagner, "Security in wireless sensor networks," Communications of the ACM, vol. 47, no. 6, pp. 53–57, 2004.
2. T. Zia and A. Zomaya, "Security issues in wireless sensor networks," in 2006 International Conference on Systems and Networks Communications (ICSNC'06), pp. 40–40, IEEE, 2006.
3. J. P. Walters, Z. Liang, W. Shi, and V. Chaudhary, "Wireless sensor network security: A survey," Security in Distributed, Grid, Mobile, and Pervasive Computing, vol. 1, no. 367, p. 6, 2007.
4. S. Singh and H. K. Verma, "Security for wireless sensor network," International Journal on Computer Science and Engineering, vol. 3, no. 6, pp. 2393–2399, 2011.
5. K. Agnihotri, P. Chilbule, S. Prashant, P. Jain and P. Khobragade, "Generating image description using machine learning algorithms," in 2023 11th International Conference on Emerging Trends in Engineering & Technology - Signal and Information Processing (ICETET - SIP), Nagpur, India, 2023, pp. 1–6, doi: 10.1109/ICETET-SIP58143.2023.10151472.
6. V. Khetani, Y. Gandhi, S. Bhattacharya, S. N. Ajani, and S. Limkar, "Cross-domain analysis of ML and DL: Evaluating their impact in diverse domains," International Journal of Intelligent Systems and Applications in Engineering, vol. 11, no. 7s, pp. 253–262, 2023.
7. A. Stanciu, "Blockchain based distributed control system for edge computing," in 2017 21st International Conference on Control Systems and Computer Science (CSCS), pp. 667–671, IEEE, 2017.
8. A. Banafa, "The Industrial Internet of Things (IIoT): Challenges, requirements and benefits," in Secure and Smart Internet of Things (IoT): Using Blockchain and AI, River Publishers, 2018, pp.7–12.
9. E. Karafiloski and A. Mishev, "Blockchain solutions for big data challenges: A literature review," in IEEE EUROCON 2017-17th International Conference on Smart Technologies, pp. 763–768, IEEE, 2017.
10. S.-Y. Wang, Y.-J. Hsu, and S.-J. Hsiao, "Integrating blockchain technology for data collection and analysis in wireless sensor networks with an innovative implementation," in 2018 International Symposium on Computer, Consumer and Control (IS3C), Taichung, Taiwan, 2018, pp. 149–152, doi: 10.1109/IS3C.2018.00045.
11. T. Yang, Q. Guo, X. Tai, H. Sun, B. Zhang, W. Zhao, and C. Lin, "Applying blockchain technology to decentralized operation in future energy internet," in 2017 IEEE Conference on Energy Internet and Energy System Integration (EI2), pp. 1–5, 2017.

12. S. Hu, C. Cai, Q. Wang, C. Wang, X. Luo, and K. Ren, "Searching an encrypted cloud meets blockchain: A decentralized, reliable and fair realization," in IEEE INFOCOM, pp. 792–800, 2018.
13. P. K. Sharma, N. Kumar, and J. H. Park, "Blockchain-based distributed framework for automotive industry in a smart city," IEEE Transactions on Industrial Informatics, vol. 15, no. 7, pp. 4197–4205, 2018.
14. S. N. Ajani, P. V. Potnurwar, V. K. Bongirwar, A. V. Potnurwar, A. Joshi, and N. Parati, "Dynamic RRT* algorithm for probabilistic path prediction in dynamic environment," International Journal of Intelligent Systems and Applications in Engineering, vol. 11, no. 7s, pp. 263–271, 2023.
15. J. Zhang, N. Xue, and X. Huang, "A secure system for pervasive social network-based healthcare," IEEE Access, vol. 4, pp. 9239–9250, 2016.
16. D.-e-S. Agha, F. H. Khan, R. Shams, H. H. Rizvi, and F. Qazi, "A secure crypto base authentication and communication suite in wireless body area network (WBAN) for IoT applications," Wireless Personal Communications, vol. 103, no. 4, pp. 2877–2890, 2018.
17. Z. Shahbazi and Y.-C. Byun, "Towards a secure thermal-energy aware routing protocol in wireless body area network based on blockchain technology," Sensors, vol. 20, no. 12, p. 3604, 2020.
18. S. Kushch and F. Prieto-Castrillo, "Blockchain for dynamic nodes in a smart city," in 2019 IEEE 5th World Forum on Internet of Things (WF-IoT), pp. 29–34, IEEE, 2019.
19. S. D. Erokhin and S. D. Makhrov, "Neural mechanisms in wireless sensor networks," in Proceedings of the 2013 International Conference on Cyber-Enabled Distributed Computing and Knowledge Discovery, pp. 340–343.
20. N. H. Kim, S. M. Kang, and C. S. Hong, "Mobile charger billing system using lightweight blockchain," in APNOMS, pp. 374–377, 2017.
21. E. Munsing, J. Mather, and S. Moura, "Blockchains for decentralized ¨ optimization of energy resources in microgrid networks," in 2017 IEEE conference on control technology and applications (CCTA), pp. 2164–2171, 2017.
22. N. Z. Aitzhan and D. Svetinovic, "Security and privacy in decentralized energy trading through multi-signatures, blockchain and anonymous messaging streams," IEEE Transactions on Dependable and Secure Computing, vol. 15, no. 5, pp. 840–852, 2016.
23. S. N. Ajani, P. Khobragade, M. Dhone, B. Ganguly, N. Shelke, and N. Parati, "Advancements in computing: emerging trends in computational science with next-generation computing," International Journal of Intelligent Systems and Applications in Engineering, vol. 12, no. 7s, pp. 546–559, 2023.
24. S. Y. Wang, Y. J. Hsu, and S. J. Hsiao, "Integrating blockchain technology for data collection and analysis in wireless sensor networks with an innovative implementation," in International Symposium on Computer, Consumer and Control (IS3C), Taichung, Taiwan, 2018, pp. 149–152, doi: 10.1109/IS3C.2018.00045. 2018.
25. S. Ajani and M. Wanjari, "An Efficient Approach for Clustering Uncertain Data Mining Based on Hash Indexing and Voronoi Clustering," in 2013 5th International Conference and Computational Intelligence and Communication Networks, Mathura, India, 2013, pp. 486–490, doi: 10.1109/CICN.2013.106.
26. Puthal, N. Malik, S. P. Mohanty, E. Kougianos, and C. Yang, "The blockchain as a decentralized security framework [future directions]," IEEE Consumer Electronics Magazine, vol. 7, no. 2, pp. 18–21, 2018.
27. K. Fan, Y. Ren, Y. Wang, H. Li, and Y. Yang, "Blockchain-based efficient privacy preserving and data sharing scheme of content–centric network in 5G," IET Communications, vol. 12, no. 12, pp. 527–532, 2018.
28. W. Yin, Q. Wen, W. Li, H. Zhang, and Z. Jin, "An anti-quantum transaction authentication approach in blockchain," IEEE Access, vol. 6, pp. 5393–5401, 2018.

29. J. Gu, B. Sun, X. Du, J. Wang, Y. Zhuang et al., "Consortium blockchain-based malware detection in mobile devices," IEEE Access, vol. 6, pp. 12118–12128, 2018.

30. P. Fairley, "Blockchain world-feeding the blockchain beast if bitcoin ever does go mainstream, the electricity needed to sustain it will be enormous," IEEE Spectrum, vol. 54, no. 10, pp. 36–59, 2017.

31. B. T. Baker, R. F. Silva, V. D. Calhoun, A. D. Sarwate and S. M. Plis, "Large scale collaboration with autonomy: Decentralized data ICA," in 2015 IEEE 25th International Workshop on Machine Learning for Signal Processing (MLSP), Boston, MA, USA, 2015, pp. 1–6, doi: 10.1109/MLSP.2015.7324344.

32. M. E. Peck, "Blockchain world-do you need a blockchain? This chart will tell you if the technology can solve your problem," IEEE Spectrum, vol. 54, no. 10, pp. 38–60,

33. K. Xie, W. Luo, X. Wang, D. Xie, J. Cao et al., "Decentralized context sharing in vehicular delay tolerant networks with compressive sensing," 2016 IEEE 36th International Conference on Distributed Computing Systems (ICDCS), Nara, Japan, pp. 169–178, 2016.

34. K. Kotobi and S. G. Bilen, "Secure blockchains for dynamic spectrum access: A decentralized database in moving cognitive radio networks enhances security and user access," IEEE Vehicular Technology Magazine, vol. 13, no. 1, pp. 32–39, 2018.

35. C. Vaidya, P. Khobragade, and A. Golghate, "Data leakage detection and security in cloud computing," GRD Journals Global Research Development Journal for Engineering, vol. 1, no. 12, pp. 137–140, 2016.

36. Z. Wang, Y. Tian, and J. Zhu, "Data sharing and tracing scheme based on blockchain," in 2018 8th International Conference on Logistics, Informatics and Service Sciences (LISS), Toronto, ON, Canada, 2018, pp. 1–6, doi: 10.1109/LISS.2018.8593225.

37. A. Dorri, M. Steger, S. S. Kanhere and R. Jurdak, "Blockchain: A distributed solution to automotive security and privacy," IEEE Communications Magazine, vol. 55, no. 12, pp. 119–125, 2017.

38. K. Kotobi and S. G. Bilén, "Blockchain-enabled spectrum access in cognitive radio networks," in 2017 Wireless Telecommunications Symposium (WTS), Chicago, IL, USA, 2017, pp. 1–6, doi: 10.1109/WTS.2017.7943523.

39. P. Khobragade, L. G. Malik, "A review on data generation for digital forensic investigation using datamining", IJCAT International Journal of Computing and Technology, vol. 1, no. 3, pp. 372–375, 2014.

40. D. R. Dandekar and P. R. Deshmukh, "Energy balancing multiple sink optimal deployment in multi-hop wireless sensor networks," in 2013 3rd IEEE International Advance Computing Conference (IACC), pp. 408–412, IEEE, 2013.

41. T. K. Jain, D. S. Saini, and S. V. Bhooshan, "Increasing lifetime of a wireless sensor network using multiple sinks," in 2014 11th International Conference on Information Technology: New Generations, pp. 616–619, IEEE, 2014.

42. Z. Cui, X. U. E. Fei, S. Zhang, X. Cai, Y. Cao, W. Zhang, and J. Chen, "A hybrid Blockchain-based identity authentication scheme for multi-WSN," IEEE Transactions on Services Computing, vol. 13, no. 2, pp. 241–251, 2020.

43. Z. Noshad, A. Javaid, M. Zahid, I. Ali, and N. Javaid, "Node recovery in wireless sensor networks via blockchain," in International Conference on P2P, Parallel, Grid, Cloud and Internet Computing, pp. 94–105, Springer, Cham, 2019.

44. A. Mateen, N. Javaid, and S. Iqbal, "Towards energy efficient routing in blockchain based underwater WSNs via recovering the void holes," 2019 (Doctoral dissertation, MS thesis, COMSATS University Islamabad (CUI), Islamabad 44000, Pakistan).

45. N. Islam, Towards a Secure and Energy Efficient Wireless Sensor Network using Blockchain and a Novel Clustering Approach, 2018.

46. D. P. Abreu, K. Velasquez, M. Curado, and E. Monteiro, "A resilient internet of things architecture for smart cities," Annals of Telecommunications, vol. 72, no. 1–2, pp. 19–30, 2017.

47. I. D. Buldin, M. G. Gorodnichev, S. S. Makhrov, and E. N. Denisova, "Next generation industrial blockchain-based wireless sensor networks," in 2018 Wave Electronics and Its Application in Information and Telecommunication Systems (WECONF), pp. 1–5. IEEE, 2018.

48. M. A. Ferrag, M. Derdour, M. Mukherjee, A. Derhab, L. Maglaras and H. Janicke, "Blockchain Technologies for the Internet of Things: Research Issues and Challenges," IEEE Internet of Things Journal, vol. 6, no. 2, pp. 2188–2204, 2019, doi: 10.1109/JIOT.2018.2882794.

49. M. A. Ferrag, L. Maglaras, and A. Ahmim, "Privacy-preserving schemes for ad hoc social networks: A survey," IEEE Communications Surveys & Tutorials, vol. 19, no. 4, pp. 3015–3045, 2017.

14 Privacy-Preserving Machine Learning on Non-Co-Located Datasets Using Federated Learning
Challenges and Opportunities

Jyoti L. Bangare, Nilesh P. Sable,
Parikshit N. Mahalle, and Gitanjali Shinde

14.1 INTRODUCTION

In the age of big data and distributed computing, it is becoming more common to find datasets that are not physically located in the same location. Since these datasets are spread across several locations (i.e. they are non-co-located) and are hosted by diverse organizations, combining and analyzing them presents a variety of difficulties. Recent developments in ML methods, particularly federated learning, provide interesting new options for overcoming these difficulties while preserving the confidentiality and privacy of the data [1, 2].

14.1.1 OVERVIEW OF NON-CO-LOCATED DATASETS

Datasets that are hosted by multiple companies or situated across the globe are considered non-co-located datasets. Information obtained from various sources, such as sensors, mobile devices, and web sources, may be included in this. The quantity of information included in the non-co-located datasets may be useful for applications in the fields of healthcare, finance, transportation, and many others. However, these datasets are harder to combine and interpret because of their scattered nature [3].

14.1.2 CHALLENGES OF INTEGRATING AND ANALYZING NON-CO-LOCATED DATASETS

Integrating and analyzing datasets that are not physically located together presents a number of issues. A key difficulty is data heterogeneity, or variances in datasets in terms of file format, structure, semantics, and quality. Effective data combining and analysis is hampered by these differences. Additionally, data ownership and access rights can be complex, needing cooperation and agreement from a number of

DOI: 10.1201/9781003437079-14

stakeholders [4]. Data governance adds to the difficulty by potentially including legal issues and data sharing regulations.

14.1.3 THE MACHINE LEARNING APPROACHES FOR PRIVACY PRESERVATION IN IoT

Due to the sensitive and private data that IoT devices acquire, privacy preservation is a crucial issue in the Internet of Things (IoT) environment. The privacy issues in IoT devices are largely addressed by machine learning (ML) techniques, as shown in Figure 14.1. Here are some typical ML techniques for protecting privacy in IoT:

> *Federated learning*: IoT devices can train models cooperatively via federated learning without transferring raw data. Instead, individual devices train local models locally, and only model changes or gradients are sent to a centralized server or aggregator. By ensuring that the sensitive data stays on the device, this method preserves privacy while allowing for model advancements.
> *Differential privacy*: This is a technology that ensures privacy by introducing noise to the data. ML algorithms can use differential privacy measures in the context of IoT to introduce controlled noise into the data prior to training the models. Due to this noise, it is difficult to distinguish between individual data points and protecting privacy while still enabling efficient model training.
> *Homomorphic encryption*: Calculations can be made on encrypted data without having to first decrypt it. This method enables the training of ML models using encrypted data in IoT. The model is trained on the encrypted data,

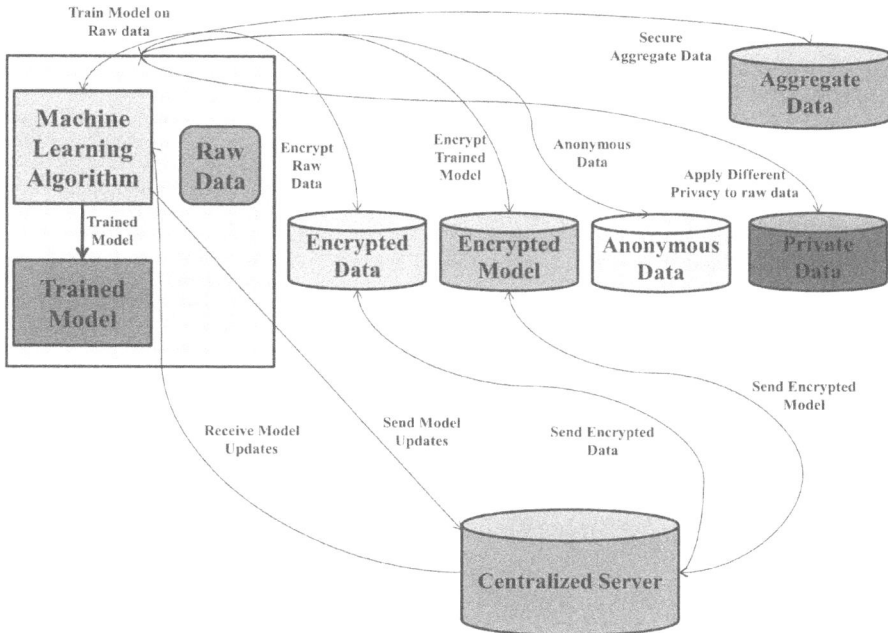

FIGURE 14.1 Machine learning approaches for privacy preservation in IoT.

which is then transmitted back to the IoT devices after being educated on a central server. By doing this, privacy is protected and the data is guaranteed to remain encrypted during the training process.

Secure multi-party computation (SMPC): SMPC is a cryptographic method that permits several parties to work together to compute a result without disclosing their individual inputs. ML operations on encrypted data can be carried out using SMPC in the context of IoT. Without disclosing the underlying data, each IoT device encrypts its own data and works together to compute the model changes. This strategy promotes collaborative learning while ensuring privacy.

Secure aggregation: When combining data or model updates from many IoT devices, secure aggregation solutions work to protect user privacy. These methods aggregate the data or model changes while protecting user privacy by using cryptographic mechanisms. Secure aggregation solutions safeguard privacy in IoT ML settings by ensuring that no individual device's data is exposed during the aggregate process.

14.1.4 OVERVIEW OF FEDERATED LEARNING

Recent advances in federated learning eliminate the requirement for centralized data storage by enabling many users to cooperatively train models on various data sources [5], as shown in Figure 14.2. It enables group collaboration on model training without compromising the security of the data being utilized. Participants can avoid providing raw data to a centralized server by locally training their own models. The revised

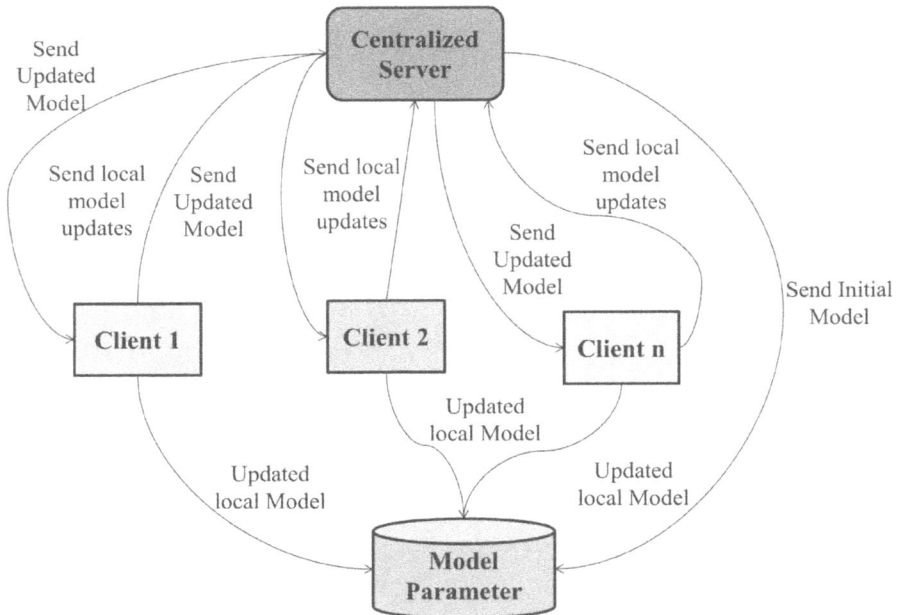

FIGURE 14.2 Federated learning approach.

local models are exchanged and integrated to create the finalized global model [6]. The global model is then made available to the participants. This technique minimizes the danger of data leaking by ensuring that sensitive data stays on local devices.

14.1.5 BENEFITS OF USING FEDERATED LEARNING ON NON-CO-LOCATED DATASETS

Using federated learning on datasets that are not geographically close to one another has a number of advantages. First, it does away with the requirement for physical data centralization, which lowers the cost of data transport and storage and enables more effective data integration and analysis. Federated learning generates more precise and applicable findings by training models on a variety of different data sources [7]. The models are developed to take each dataset's unique characteristics into consideration. Federated learning also protects data privacy by keeping private information locally. This helps to protect privacy and lessen the possibility of data breaches.

Federated learning encourages collaboration across companies or organizations even when their data is stored in different physical locations. By enabling users to train models simultaneously without releasing their data, it promotes knowledge sharing and resource sharing. This collaborative approach may enhance data-driven decision making in several industries, including healthcare, banking, and transportation [8].

The introduction discusses the difficulties of integrating and analyzing datasets that are not physically located in the same location. It advocates using federated learning to address these problems while preserving the data's security and privacy [9]. Federated learning on remote datasets provides a number of benefits, which are all underlined, including improved model performance, data integration, privacy protection, and cooperation [10]. The technical elements, difficulties, and promise of federated learning on non-co-located datasets will be covered in greater detail in the next sections of the chapter, and this section serves as the foundation for that debate [11].

14.2 FEDERATED LEARNING: PRINCIPLES AND APPROACHES

"Federated learning," a new approach to machine learning (ML), enables multiple users to train a single model utilizing data from several dispersed sources without having to share any of the raw data. This section explores the fundamentals of federated learning as well as several methods for tackling the problem, such as horizontal and vertical federated learning [12].

14.2.1 PRINCIPLES OF FEDERATED LEARNING

Federated learning is based on distributed model training and prioritizes protecting users' private information [13, 14]. The following are some of the key ideas:

  i. *Local model training*: As part of this process, each device or organization trains its own local model using its own data. There are several ways to achieve this. The raw data is never exposed to the outside world because this training is performed internally on the device.

ii. *Model aggregation*: We pool and aggregate the modifications made to the local models to produce the global one. This process is known as "model aggregation." The process of aggregation utilizes a variety of methods, such as secure multi-party computation and homomorphic encryption, while still maintaining the privacy of locally made updates.

iii. *Iterative learning*: The model aggregation procedure is carried out iteratively so that the global model can get better over time. Each step entails trading model updates in a round, collecting those updates, and then distributing the revised global model to the iteration's participants.

14.2.2 HORIZONTAL FEDERATED LEARNING

When several entities share some qualities but have distinct data samples or distributions, horizontal federated learning is most useful. Without requesting participants' raw data, the objective is to jointly train a model that can generalize across all these factors [15]. This is shown in Figure 14.3. The following is a summary of the essential stages of horizontal federated learning:

i. *Data partitioning*: Each entity splits its local data into sets while simultaneously ensuring that no raw data is sent outside of the device. This partitioning can be done using any technique that produces subgroups that are easily discernible.

ii. *Local model training*: Each organization uses its own particular subset of data to train its own local model. This training can be done using common ML techniques or specialized privacy protection strategies like differential privacy.

iii. *Model aggregation*: Model changes are frequently transmitted by the entities to a coordinating server as model weights or gradients. This process is

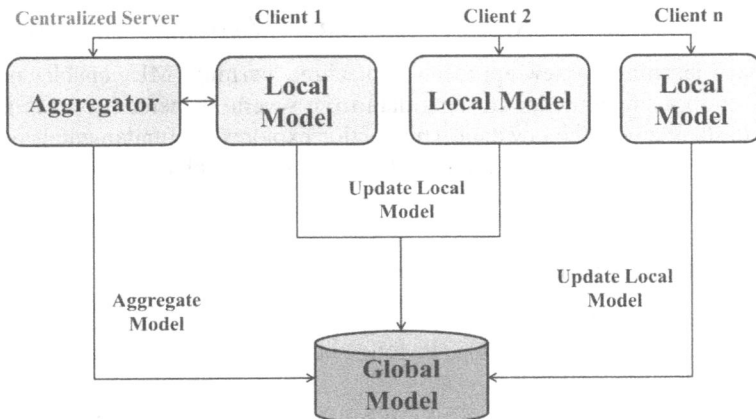

FIGURE 14.3 Horizontal federated learning.

referred to as "model aggregation." In order to give a current global model, the server integrates the changes after performing encrypted aggregation algorithms on them.

iv. *Global model distribution*: Until convergence is attained, the cycle of local model training, model aggregation, and global model distribution is repeated. The updated global model is then disseminated to all entities.

14.2.3 VERTICAL FEDERATED LEARNING

When several entities want to jointly train a model without disclosing their own data and possess features or attributes that are complimentary to one another, vertical federated learning is a viable technique to use [16]. The following list summarizes the main steps of vertical federated learning, as shown in Figure 14.4:

i. *Data preprocessing*: Data is originally transformed from each entity's native format into a common feature space or representation. Even if the raw data is not given, this preparation assures compatibility and enables cooperative training to take place.

ii. *Local model training*: Each entity trains its own local model utilizing its own preprocessed data and focused on its own distinct attributes. The training may make use of several machine learning methods or strategies to handle vertically partitioned data.

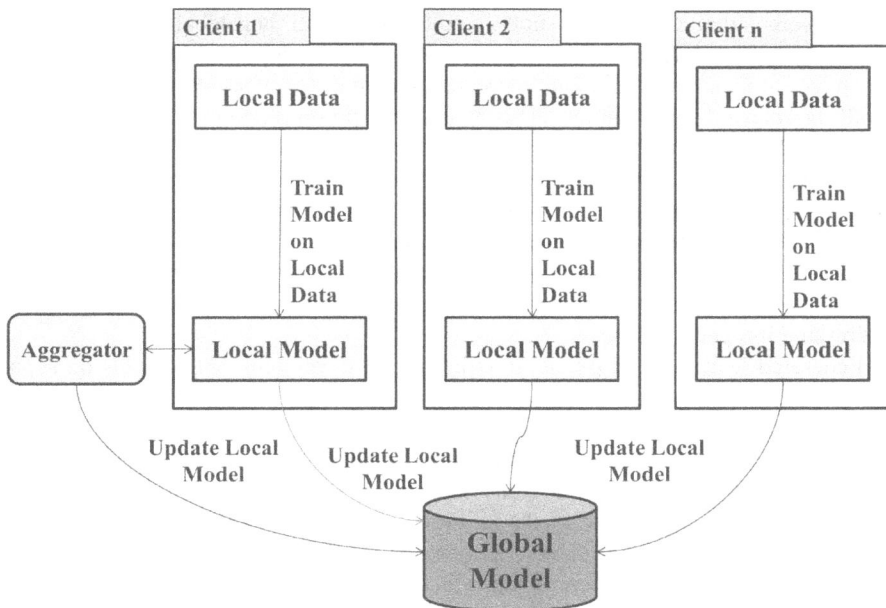

FIGURE 14.4 Vertical federated learning.

iii. *Model aggregation*: Updates to the model, such as model weights or gradients, are sent between the local models and a coordinating server. This process is known as "model aggregation." In order to make a new global model available, the server safely aggregates the modifications.

iv. *Global model distribution*: Until convergence is attained, the cycle of local model training, model aggregation, and global model distribution is repeated. The updated global model is then disseminated to all entities.

14.2.4 OTHER APPROACHES TO FEDERATED LEARNING

There are many different methods and federated learning variations in addition to horizontal and vertical federated learning. These include federated reinforcement learning and federated transfer learning, which train models through interactions with remote environments and employ previously taught models to improve performance [17]. The applications of federated learning might be considerably expanded by using these strategies in a variety of settings.

Transfer learning advantages are brought to dispersed settings using federated transfer learning. With this configuration, models are initially trained on a sizable common dataset before being fine-tuned using data gathered locally by collaborating businesses. Federated transfer learning is a further development of transfer learning [18]. The pre-trained models' insights may be used by entities with less data using this method, improving overall performance and generalization.

Federated reinforcement learning is a technique that combines reinforcement learning (RL) strategies with the philosophy and practices of federated learning. Organizations that employ this strategy collaborate with their local communities to acquire data and shape policies. The policies are then compared to one another and compiled into a single document to create a global policy that reflects the expertise of all the relevant organizations. This enables the preservation of private data when group work is being done in a decentralized environment.

As the discipline develops, federated learning continues to take on new forms and application techniques. Skewed data distributions can be a problem, although techniques like model compression and adaptive learning rate scheduling can assist. These strategies all share the desire to make communication easier. Each strategy aims at enhancing federated learning's performance on non-co-located datasets and addressing its specific issues.

Federated learning is a potential approach for developing ML models on separate datasets while preserving the confidentiality and privacy of the data. Federated learning in this situation might be vertical or horizontal. Each method is designed to be effective with a particular set of facts and circumstances. Federated learning has been expanded into different versions using a variety of methodologies, including federated transfer learning and federated reinforcement learning, in order to further improve performance and enable collaborative learning to occur in decentralized scenarios [19]. The application of these methodologies will enable fascinating new developments in ML and data analytics as federated learning progresses, without the necessity of using physically co-located datasets.

14.2.5   ROLE OF FEDERATED LEARNING IN IoT AND WSN

In the context of the Internet of Things (IoT) and Wireless Sensor Networks (WSNs), federated learning is important because it enables distributed learning and decision-making while protecting data privacy and minimizing connection costs. The following are some crucial functions of federated learning in IoT and WSN:

i.   *Decentralized learning*: In IoT and WSN contexts, a huge amount of data is produced by many connected devices. These devices can learn models cooperatively via federated learning without sending data to a centralized server. This decentralized learning strategy is suited for resource-constrained IoT and WSN devices because it minimizes latency and helps overcome bandwidth restrictions.

ii. *Data privacy preservation*: IoT and WSN systems frequently gather delicate and private information. By retaining the data locally on the devices and only exchanging model changes with the central server, federated learning addresses privacy concerns. As a result, the danger of data breaches or invasions of privacy associated with transmitting sensitive information across the network is decreased.

iii. *Reduced communication overhead*: The battery life, processing power, and network bandwidth of IoT and WSN devices are often constrained. By delivering only model updates rather than raw data, federated learning decreases communication overhead. The operational lifespan of IoT and WSN devices is increased thanks to this effective communication strategy, which also optimizes bandwidth use and conserves energy.

iv. *Distributed intelligence*: Federated learning makes it possible for intelligence to be distributed among IoT and WSN devices. Without relying on a centralized entity, each device can learn from its local data and add to the body of knowledge. This distributed intelligence improves the autonomy and responsiveness of IoT and WSN systems by enabling localized decision-making, real-time analytics, and context-aware processing.

v.   *Scalability and adaptability*: These are important since IoT and WSN environments are very dynamic and constantly experience device additions and deletions. The learning process is unaffected by the addition or removal of devices thanks to federated learning's intrinsic adaptability and scalability. IoT and WSN systems can scale easily and adapt to shifting network topologies and device availability thanks to their flexibility.

vi. *Learning collaboration*: Federated learning encourages communication between IoT and WSN devices. Through model aggregation, devices can exchange knowledge and insights as opposed to learning alone. By fostering collective intelligence through collaborative learning, devices are able to take use of a wider variety of inputs and viewpoints, ultimately improving model accuracy and robustness.

As we have seen, privacy protection, reduced communication overhead, distributed intelligence, adaptability, scalability, and collaborative learning are all advantages of federated learning for IoT and WSN systems. With the use of these capabilities, IoT and WSN data may be used effectively and securely, facilitating the development of smart decision-making processes and a range of industries including smart cities, industrial automation, environmental monitoring, healthcare, and more.

14.3 CHALLENGES OF FEDERATED LEARNING ON NON-CO-LOCATED DATASETS

Federated learning presents several challenges when applied to non-co-located datasets. These challenges include data heterogeneity, data privacy concerns, data security considerations, and other associated obstacles. Understanding and addressing these challenges is crucial for the successful implementation of federated learning on non-co-located datasets.

14.3.1 DATA HETEROGENEITY

Data heterogeneity refers to the fact that diverse datasets that are not co-located may have varied quality, formats, structures, and semantics of their contents derived from different sources. Integrating and interpreting many datasets is difficult and time-consuming. Data heterogeneity can be categorized as follows:

   i. *Data format variations*: Different formats like CSV, JSON, or other proprietary database formats can hold non-co-located datasets. These discrepancies make data harmonization for collaborative analysis difficult.
   ii. *Quality discrepancies*: Data quality can vary across different datasets, with differences in missing values, outliers, or inconsistencies. Addressing these quality discrepancies is essential to ensure accurate and reliable model training.
   iii. *Data heterogeneity challenges*: To overcome these, preprocessing techniques like data normalization, feature engineering, and data cleaning are employed. Data integration strategies such as schema mapping, data alignment, and data transformation are also utilized to ensure compatibility and coherence across non-co-located datasets.

14.3.2 DATA PRIVACY

Data privacy is a paramount concern in federated learning, particularly when dealing with non-co-located datasets. Preserving the privacy of sensitive data while allowing collaborative model training is a significant challenge. The following privacy considerations arise:

   i. *Local data privacy*: Each entity possesses its own data and wants to keep it private. Sharing raw data with other entities could lead to privacy breaches and unauthorized access. The goal is to minimize the exposure of raw data while still contributing to the global model.

ii. *Privacy-preserving model updates*: When participating entities share their local model updates with a central server or coordinator, privacy-preserving techniques must be employed. Encryption methods, such as SMPC and differential privacy, can be utilized to protect the privacy of model updates during aggregation.

iii. *Information leakage*: Model updates may inadvertently reveal sensitive information about the local datasets. Adversaries could analyze the updates to infer details about the local data, compromising privacy. Techniques like noise injection and secure aggregation can mitigate the risk of information leakage.

14.3.3   DATA SECURITY

While using federated learning on disconnected data sources, it is equally crucial to not overlook data security. To maintain its integrity, data must be shielded against tampering, alteration, and assaults. These are the main security concerns:

i. *Secure communication*: To prevent listening in or data interception, secure communication methods must be employed for both the global distribution of models and the dissemination of model changes. When sending sensitive data through encrypted connections, two secure protocols that may be utilized are Secure Sockets Layer (SSL) and Transport Layer Security (TLS).

ii. *Malicious participants*: In a federated setup, there is always the possibility that bad actors would try to gain an advantage for themselves by accessing private data or influencing the training process. We may use safe aggregation methods like homomorphic encryption and SMPC to fight off such assaults.

iii. *Secure model distribution*: A major security challenge is how to distribute the global model to the participating entities while maintaining the authenticity and integrity of the model. To ensure the safe propagation of the global model, mechanisms like digital signatures and cryptography can be utilized.

Data is protected during transmission using encryption techniques and secure communication protocols, which helps address data security concerns. Only approved businesses may participate in the federated learning process thanks to access control methods, authentication procedures, and authorization processes. Utilizing secure aggregation methods and anomaly detection techniques may also help in identifying hostile actors and reducing related risks.

14.3.4   OTHER CHALLENGES

When trying federated learning on datasets that are not co-located, issues can occur such as the heterogeneity of the data, concerns about data privacy, and fears about data security, to name just a few. These issues include the following, as examples:

i. *Resource constraints*: Some datasets could be stored on devices with limited resources rather than being physically arranged together. The federated learning algorithms must be optimized due to the potential effects of resource limitations on the training process.

ii. *Communication overhead*: A further expense of federated learning is the increased communication that is necessary across the many organizations engaged in it in order to convey model changes and aggregate them. Particularly when working with big, high-dimensional datasets, this communication overhead can be a significant factor. Methods like model compression, quantization, and selective participation can be applied to address this issue.

iii. *Ethical and legal considerations*: Due to the fact that federated learning necessitates collaboration between several parties, there are moral and legal issues. Before putting into practice any procedures that are not only legal but also ethical, data ownership, data use rights, and compliance with laws like the General Data Protection Regulation (GDPR) and the Health Insurance Portability and Accountability Act (HIPAA) must all be carefully considered.

iv. *Federated infrastructure management*: Managing the federated infrastructure necessary for federated learning on several datasets that are not physically present in the same location can be challenging. Organizing the involvement of several organizations, establishing secure communication routes, managing updates and synchronization, and other tasks are required to maintain the infrastructure for the collection and dissemination of model information.

To deal with resource limitations and decrease communication overhead, optimization strategies can be used, such as federated optimization algorithms that use optimizing techniques. The ethical and legal difficulties may be resolved in several ways, including by establishing explicit data sharing agreements, ensuring compliance with applicable laws, and adopting ethical standards. Federated learning on scattered datasets could be made easier by federated server designs and other effective infrastructure management and coordination techniques.

The obstacles that federated learning on non-co-located datasets encounters include data heterogeneity, data privacy, data security, resource limitations, communication overhead, ethical dilemmas, and infrastructure management, to name just a few. A range of preprocessing approaches, privacy-preserving mechanisms, secure communication protocols, optimization methodologies, and lawful practices are needed to get over these obstacles. Several issues must be resolved in order to build and implement federated learning on non-co-located datasets. Cooperative ML will be safe, private, and trustworthy if this is done.

14.4 TECHNIQUES AND BEST PRACTICES FOR FEDERATED LEARNING ON NON-CO-LOCATED DATASETS

To use federated learning on datasets that are not co-located, features like data pretreatment, encryption, secure aggregation, and others must be introduced. In this section, we will examine several federated learning strategies and offer recommendations on the best practices for producing accurate and secure results on geographically scattered datasets, as shown in Figure 14.5.

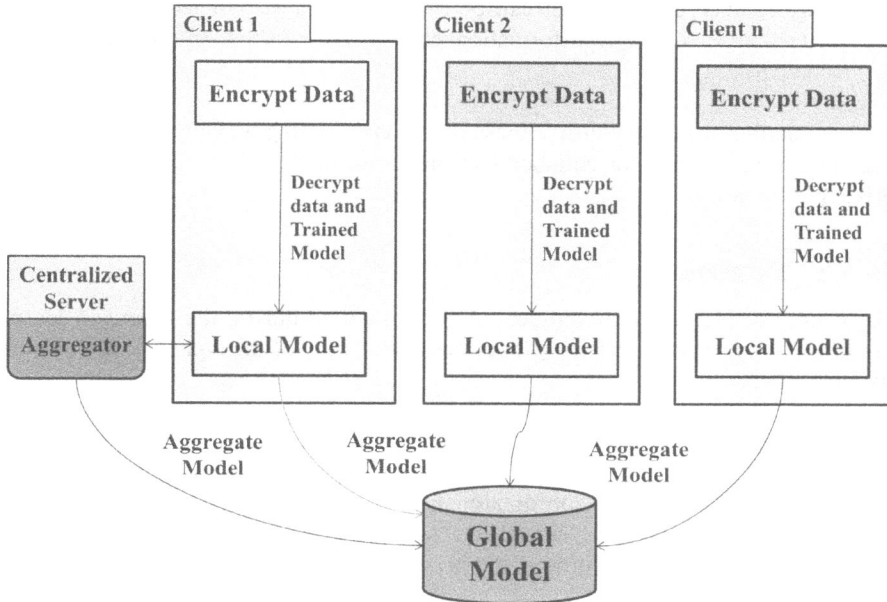

FIGURE 14.5   Federated learning on non-co-located datasets.

14.4.1   DATA PREPROCESSING

In the context of federated learning, data preprocessing is essential, especially when working with disparate datasets. The preprocessing of data can be streamlined using the following methods:

i. *Feature engineering*: Raw data must be transformed into a format that is appropriate for engineering features in order to train a ML model. Feature extraction, dimensionality reduction, and feature scaling are some of the processes included. Each business is free to independently do feature engineering on its own data while keeping that knowledge private.

ii. *Data normalization*: This is essential to ensuring that data from various institutions are measured in accordance with the same standards. The data may be normalized and made more comparable using techniques like min-max scaling and z-score normalization.

iii. *Data imputation*: In real-world datasets, missing data is a frequent problem. Imputation methods like mean imputation and regression imputation can be employed when missing data has to be inferred. Care must be taken to guarantee that imputation is carried out safely and without compromising any personal data.

iv. *Data transformation*: This may be used to spread collections of data that are not physically close to one another in a number of different ways. By converting the data to a standard distribution, the convergence and overall performance of the model can be improved. The use of techniques like Log and Box-Cox transformation can provide data normalization and compatibility.

Participating firms may utilize these data preprocessing techniques to guarantee that their data is ready for collaborative model training without jeopardizing data privacy or security.

14.4.2 ENCRYPTION

To secure the privacy and confidentiality of the data during the federated learning process, encryption technologies are essential. Any of the ciphering algorithms listed here can be used:

i. *Homomorphic encryption*: The encryption eliminates the need to first decode the data in order to do computation on it. Before sending local model updates to the coordinator or central server, entities can encrypt them using their own private key. The server can implement encrypted model aggregation to protect the privacy of the data by avoiding examining the raw data.

ii. *Secure multi-party computation (SMPC)*: A method known as SMPC enables several participants to compute a function jointly while preserving the confidentiality of each party's unique inputs. Federated learning can employ SMPC to securely aggregate model changes. The server executes calculations on the encrypted model updates to create an aggregated model update without revealing the specific contributions of the entities.

iii. *Differential privacy*: This fortifies the model-updating process against privacy breaches by introducing a tiny quantity of random noise. Participating firms may "noise up" their data with extra information before sending local updates to the network in order to safeguard the privacy of their users. In order to obtain differential privacy, aggregation techniques like the Laplace or Gaussian processes might be used.

iv. *Secure communication*: Employing technologies like SSL and TLS is crucial since data and model changes delivered from one organization to another must be transmitted securely. These protocols include SSL and TLS, for instance. Data integrity and privacy can be safeguarded via secure communication, which can prevent data theft or manipulation while it is being transmitted over a network.

By using encryption mechanisms, participating firms may ensure the privacy of data and model changes stored locally during the federated learning process.

14.4.3 SECURE AGGREGATION

Use secure aggregation if you want to maintain the privacy of the changes you make to your models while still reaping the benefits of federated learning, as

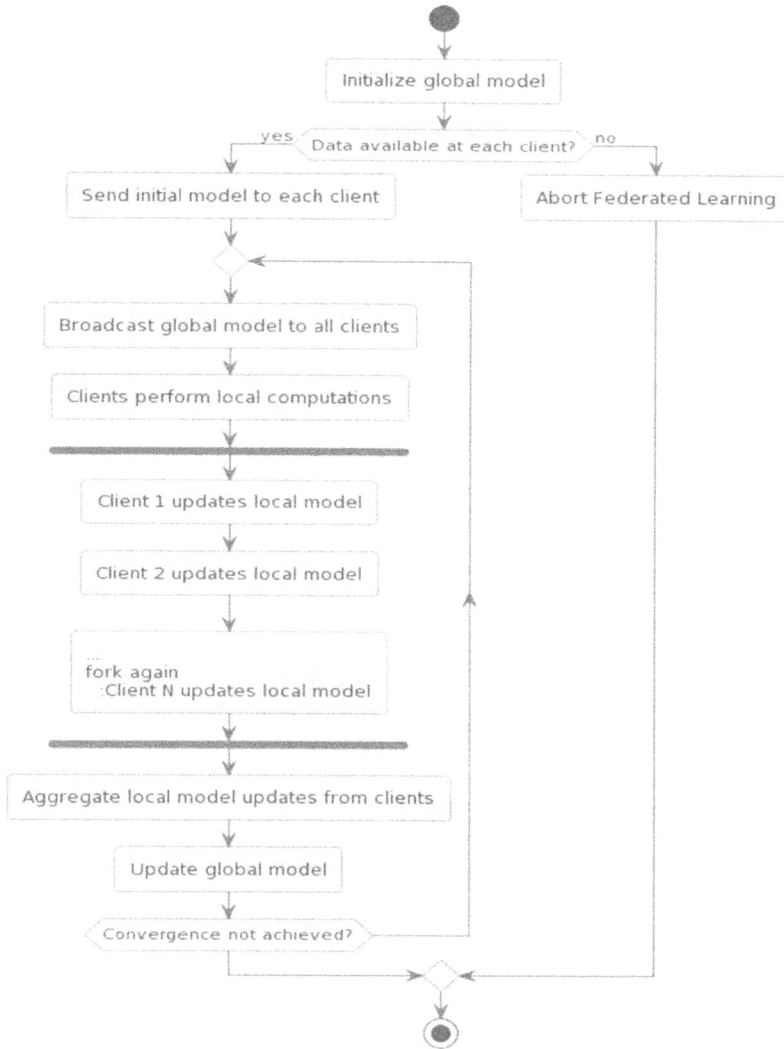

FIGURE 14.6   Flow diagram for federated learning with secure aggregation.

shown in Figure 14.6. The following techniques can be applied to ensure safe aggregation:

    i. *Federated averaging*: It is a standard method for securely aggregating models. Participating entities submit updated local model versions to a central server, which then combines all the uploaded versions into a single global model. To provide the maximum level of data security, the aggregate is conducted on the server without the server having access to the raw data or individual model changes. The model updates it gets are always differentially private or encrypted to safeguard the privacy of the data.

ii. *Secure weighted aggregation*: Depending on variables like data quality or entity knowledge, it can be essential to give the model updates varying weights. A group of methods called secure weighted aggregation can be used to combine updates while maintaining the privacy of the weights. This guarantees that the collective knowledge is accurately and fairly represented while maintaining its confidentiality.

iii. *Secure federated learning with trusted execution environments (TEEs)*: With the aid of TEEs like Intel SGX and ARM TrustZone, computing in a safe environment is now feasible. This opens the door for a method called "federated learning." To enable the secure aggregation of model revisions, TEEs may be utilized by all parties engaged in the process. This protects the confidentiality and reliability of the data being saved.

iv. *Privacy-preserving aggregation protocols*: While gathering data by using one of many privacy-preserving aggregation methods, such as secure sum or secure maximum, it is feasible to combine model changes without disclosing the unique contributions of individuals. These methods allow the server to compute the total while preserving the privacy of each individual update.

The federated learning approach ensures that the pooled knowledge is gathered by using safe aggregation algorithms. These techniques safeguard the privacy of sensitive data, including user information and model updates.

14.4.4 OTHER TECHNIQUES AND BEST PRACTICES

For a successful application of federated learning on non-co-located datasets, the following methodologies and best practices are required in addition to data preprocessing, encryption, and safe aggregation:

i. *Model compression*: Pruning, quantization, and knowledge distillation are some of the techniques used to compress the models in order to make them smaller. As a result, less communication is needed during model updates and aggregation, which makes the federated learning process more efficient.

ii. *Adaptive learning rate scheduling*: It is crucial to plan learning rates adaptively since different non-co-located datasets may have different data distributions and features. Utilizing adaptive learning rate scheduling alternatives like federated adaptive learning rate or client-driven learning rate adaptation, the learning rate may be dynamically changed in accordance with the local data properties. These techniques demonstrate the idea of flexible learning rates. This makes sure that during model training, convergence occurs across all the various datasets.

iii. *Participant selection strategies*: The proper individuals must participate for federated learning to be successful. It is possible to use participant selection techniques like actively picking participants or weighting participants according to the caliber of their data or level of expertise to get the best levels of collaboration and representation across the many datasets.

iv. *Regularization techniques*: Regularization techniques used throughout the federated learning process help reduce overfitting and generalization. These techniques help to manage the models' complexity and produce reliable results on various geographically distributed datasets.

Federated learning on non-co-located datasets may be improved in terms of effectiveness, model performance, and privacy protection if particular methodologies and best practices are applied. To ensure the compatibility and coherence of diverse data sources, preprocessing measures might be employed. Data confidentiality is guaranteed by encryption-based techniques during both data aggregation and transmission. Global models can be built using secure aggregation methods without sacrificing people's privacy [20]. The federated learning process may be further optimized by using additional approaches including model compression, adaptive learning rate scheduling, participant selection strategies, and regularization techniques. Businesses may realize the benefits of federated learning on scattered datasets by employing certain strategies and best practices, enabling cooperative ML without sacrificing accuracy, security, or privacy. Datasets that are not co-localized can nevertheless be used for federated learning.

14.5 OPPORTUNITIES FOR FEDERATED LEARNING ON NON-CO-LOCATED DATASETS

The use of federated learning on datasets that are not co-located offers up significant prospects for unleashing the potential of distant data sources, facilitating cooperation across organizations and sectors, and enhancing decision-making in a variety of different domains. In this section, we investigate the options afforded by federated learning on datasets that are not co-located.

14.5.1 UNLOCKING THE POTENTIAL OF DISTRIBUTED DATA SOURCES

Use real-time data updates to eliminate data silos and increase model generalization from remote data sources. By providing continuous and synchronized access to disparate datasets, companies may eliminate silos and unify data. It improves data availability and makes models more applicable. Real-time updates help companies remain current by training models on the latest and most relevant data for more accurate inferences [21].

i. *Enhanced data availability*: Because non-co-located datasets are frequently dispersed among several entities or organizations, it can be difficult to access and analyze the collective data. This issue is addressed by improving data availability. Federated learning offers a solution to this problem by facilitating collaboration amongst businesses without requiring them to provide their raw data. When this is done, the potential of dispersed data sources is unlocked, and it becomes possible to gain access to a dataset that is both larger and more diverse for the purpose of training machine learning models [22].

ii. *Improved model generalization*: The training of ML models on a varied collection of data results in improved generalization of the models. Variations in data distributions, demographics, and local contexts are characteristic of datasets that are not co-located with one another. Models are able to acquire a deeper grasp of the underlying data and enhance their capacity to generalize to data they have not before seen when they combine the knowledge, they obtain from a variety of sources using federated learning.

iii. *Overcoming data silos*: Datasets that are not co-located are frequently stored in separate silos inside several organizations or entities owing to privacy concerns or regulatory limits. Federated learning is a collaborative framework that enables entities to pool their data resources while still protecting data privacy. This eliminates data silos and encourages information exchange and cooperation among the many stakeholders.

iv. *Real-time data updates*: Non-co-located datasets may contain data that is constantly changing or streaming. Some examples of this type of data are sensor data in IoT environments or health monitoring data. By capitalizing on the dispersed nature of the data sources, federated learning makes it possible to perform model upgrades in real time. Without having to centralize the data, entities can continually update their local models based on the most recent information and contribute this information as timely updates to the global model.

Federated learning on non-co-located datasets allows organizations to use a plethora of different data while also respecting data privacy and legal limitations. This is made possible by tapping into the possibilities of remote data sources.

14.5.2 ENABLING COLLABORATION ACROSS ORGANIZATIONS AND INDUSTRIES

Enabling collaboration across organizations and industries involves fostering cross-domain insights through collaborative research and development. By facilitating data sharing in resource-constrained environments, organizations can pool their strengths for collective progress. Industry collaboration enhances services by leveraging shared knowledge and expertise. This collaborative approach not only accelerates innovation but also addresses challenges more effectively, leading to the creation of robust solutions that benefit multiple domains and industries [23].

i. *Cross-domain insights*: Datasets that are not co-located might originate from a variety of organizations or fields, each of which brings its own knowledge and point of view to the table. Federated learning makes it easier for entities to work together by enabling them to collaboratively train models on the datasets that are specific to each entity. This makes it possible to get insights from other domains and transfer information, which in turn encourages creativity and new discoveries in a variety of disciplines.

ii. *Collaborative research and development*: The use of federated learning on non-co-located datasets makes it possible for organizations to engage in research and development activities in collaboration with one another. It is possible for research institutions, healthcare providers, technology businesses, and other stakeholders to pool their data and knowledge in order to tackle complicated issues and create improvements in a variety of fields, including but not limited to healthcare, finance, and transportation.

iii. *Data sharing in resource-constrained environments*: Entities may have restricted access to data infrastructure or computing resources in contexts that are resource-constrained, such as underdeveloped nations or distant places. These organizations are now able to take part in collaborative model training thanks to federated learning, which eliminates the requirement for data centralization. This encourages the sharing of data and makes it possible for entities to profit from common knowledge without putting a strain on the limited resources they have.

iv. *Industry collaboration for improved services*: Federated learning on non-co-located datasets encourage collaboration among organizations working in the same sector, which ultimately leads to improved services. For instance, healthcare practitioners can work together to train models using patient data. This can result in improved diagnosis, more individualized therapy, or more disease surveillance. In a similar vein, financial institutions may work together to detect fraudulent activity or to construct models for risk assessment. This type of collaboration improves the efficiency as well as the quality of the services that are supplied within the sector.

Federated learning on non-co-located datasets encourage the sharing of knowledge, speeds up research and development, and improves the overall quality of goods and services. It does this by making it possible for individuals and businesses from different organizations and sectors to work together.

14.5.3 IMPROVING DECISION-MAKING IN VARIOUS DOMAINS

The use of federated learning techniques on healthcare datasets that are not co-located has the potential to completely transform patient treatment as well as medical research. In order to enhance illness diagnosis, treatment planning, and the ability to predict patient outcomes, models may be trained by combining data from a variety of healthcare providers while still protecting patient data confidentiality. Federated learning makes it possible to integrate a wide variety of data sources, such as electronic health records, medical imaging, genomics, and wearable devices, which ultimately results in healthcare solutions that are more accurate and individualized.

i. *Finance*: Transaction records, credit histories, and market data are examples of non-co-located datasets that are used in the financial sector. These datasets provide significant insights that may be used for risk assessment, the detection of fraud, and the development of investment strategies.

Federated learning makes it possible for financial organizations to work together and draw insights from communal data while also maintaining the secrecy of sensitive financial information. This makes it much easier to construct strong models that can improve decision-making, identify abnormalities, and reduce risks.

ii. *Transportation*: The transportation sector is responsible for the generation of massive volumes of data thanks to the proliferation of connected automobiles, traffic sensors, and public transit computer networks. The use of federated learning on datasets that are not co-located can improve traffic management, optimize routing algorithms, and increase safety measures. Intelligent transportation systems may be built to minimize congestion, reduce the number of accidents, and increase the overall efficiency of transportation networks. This can be accomplished by utilizing the collective wisdom that can be gleaned from a variety of data sources.

iii. *Energy and utilities*: In the energy and utilities industry, non-co-located datasets, such as smart meters, meteorological data, and patterns of energy usage, give significant insights that may be used for energy optimization, demand forecasting, and grid management. Federated learning makes it possible for various energy suppliers and grid operators to work together to construct models that improve the efficiency of energy distribution, pinpoint opportunities to save energy, and increase the system's resilience.

The decision-making processes may be improved with more accurate models if cooperation and the exchange of knowledge are made possible. This will result in improved results, increased services, and better utilization of available resources.

Federated learning on non-co-located datasets gives exciting prospects for unlocking the potential of distant data sources, increasing cooperation across organizations and sectors, and enhancing decision-making in a variety of disciplines. Organizations may benefit from a bigger and more representative datasets, cross-domain insights, and better services by utilizing varied datasets while respecting privacy and security concerns. In addition to this, federated learning makes it possible for organizations with limited resources to take part in collaborative model training and reap the benefits of pooled knowledge. Organizations can promote innovation, find solutions to complicated challenges, and make educated decisions in order to progress their respective sectors by effectively utilizing federated learning on datasets that are not co-located with one another.

14.6 CONCLUSION

14.6.1 Summary of Key Points

In this chapter, the idea of federated learning was examined along with how it relates to the provision of secure healthcare. We discussed how difficult it is to combine and assess datasets that are geographically separated. Then, we presented federated learning, a collaborative model training approach that uses decentralized data sharing. There was discussion of the benefits of employing federated learning on distributed datasets, including better model generalization, more accessible data, the dissolution of data silos, and real-time data updates.

We delved deeper into federated learning's theories and practices, talking about ideas like horizontal and vertical federated learning as well as many more that are immediately useful. We investigated a number of problems that might occur when federated learning is applied to non-co-located datasets, including data heterogeneity, concerns about data privacy and security, and others. Potential solutions included data preprocessing, encryption, secure aggregation, model compression, adaptive learning rate scheduling, participant selection strategies, and regularization techniques. Several approaches were suggested as remedies for the problems mentioned.

We discussed how to maximize the usage of datasets that aren't geographically adjacent to one another using federated learning. One of these opportunities was the possibility to improve decision-making by having access to previously unavailable distributed data sources in industries as various as healthcare, banking, transportation, and energy. It was underlined that it could be possible to access scattered data sources.

14.6.2   FUTURE DIRECTIONS FOR RESEARCH AND PRACTICE

Although federated learning on non-co-located datasets has a promising future, much more research and application must be done before it can fully realize its promise. One way to address data privacy and security issues is by creating more sophisticated methods of data preservation. Researchers must investigate differential privacy techniques, homomorphism encryption, and SMPC to safeguard sensitive data during model training and aggregation.

Research on the scalability and effectiveness of federated learning should also be prioritized. To allow effective model training, transmission, and aggregation, new algorithms and optimization approaches will be needed as the size and complexity of non-co-located datasets continue to increase.

The use of methods for standardization and interoperability is also important for effective communication and information exchange across the various institutions. Common data formats, protocols, and application programming interfaces (APIs) may make it easier to integrate data and interoperate models, which may lead to a rise in the use of federated learning in real-world contexts.

14.6.3   IMPLICATIONS FOR MACHINE LEARNING AND DATA ANALYTICS IN GENERAL

The federated learning paradigm has broad implications for ML and data analytics when used with non-co-located datasets. Decentralized and collaborative techniques, which enable organizations to capitalize on the value of dispersed data while simultaneously preserving user privacy and data security, are a sign of the trend away from centralized data processing.

Users may now operate beyond the constraints of traditional data silos and centralized models thanks to federated learning. The final models utilized are of higher quality because of collaboration, knowledge exchange, and collective intelligence. For distributed learning, federated learning developed techniques and best practices that may be used to a variety of edge computing and multi-party cooperation scenarios, including as secure aggregation, encryption, and adaptive learning. This is because federated learning was considered right from the initial stages of developing the procedures and recommendations.

The devotion to user privacy by federated learning is also congruent with the rising awareness of moral issues and laws pertaining to personal information. By offering a method for using data while protecting individual privacy, it fosters trust between data suppliers and customers.

Finally, by applying federated learning to divergent datasets that are geographically dispersed, the fields of ML and data analytics stand to gain significantly. Businesses may leverage the value of dispersed data sources, encourage innovation, and make better decisions by incorporating collaborative efforts, safeguarding data privacy, and adopting decentralized methodologies. Future research, novel approaches to securing students' private information, and attempts to standardize the field will be necessary to increase the uptake and effectiveness of federated learning.

REFERENCES

1. M. Abadi, P. Barham, J. Chen, Z. Chen, A. Davis, J. Dean, et al., "Tensorflow: A system for large-scale machine learning," in Proceedings of the 12th USENIX Symposium on Operating Systems Design and Implementation (OSDI 16), pp. 265–283, 2016.
2. C. Vaidya, P. Khobragade, and A. Golghate, "Data leakage detection and security in cloud computing," in GRD Journals: Global Research Development Journal for Engineering, vol. 1, no. 12, pp. 137–140, 2016.
3. S. Ajani, and M. Wanjari, "An efficient approach for clustering uncertain data mining based on hash indexing and Voronoi clustering," 2013 5th International Conference and Computational Intelligence and Communication Networks, 2013, pp. 486–490.
4. R. Asati, H. R. Turkar, A. V. Anjikar, C. Vaizdya, and P. Khobragade, "A survey on spatial based image segmentation techniques" in International Journal of Innovative Research in Computer and Communication Engineering, vol. 3, no. 10, October 2015.
5. J. Deng, W. Dong, R. Socher, L.-J. Li, K. Li, and L. Fei-Fei, "ImageNet: A large-scale hierarchical image database," in 2009 IEEE Conference on Computer Vision and Pattern Recognition, pp. 248–255, 2009.
6. D. Duplyakin, R. Ricci, A. Maricq, G. Wong, J. Duerig, E. Eide, et al., "The Design and Operation of CloudLab," in Proceedings of the 2019 USENIX Annual Technical Conference (USENIX ATC 19), pp. 1–14, 2019.
7. M. Gharibi, S. Bhagavan, and P. Rao, "Federatedtree: A secure serverless algorithm for federated learning to reduce data leakage," in 2021 IEEE International Conference on Big Data (Big Data), pp. 4078–4083, 2021.
8. M. Gharibi, and P. Rao, "Refinedfed: A refining algorithm for federated learning," in 2020 IEEE Applied Imagery Pattern Recognition Workshop (AIPR), pp. 1–5, 2020.
9. K. He, X. Zhang, S. Ren, and J. Sun, "Deep residual learning for image recognition," in Proceedings of the IEEE conference on computer vision and pattern recognition, pp. 770–778, 2016.
10. A. G. Howard, M. Zhu, B. Chen, D. Kalenichenko, W. Wang, T. Weyand et al., "Mobilenets: Efficient convolutional neural networks for mobile vision applications," arXiv:1704.04861, 2017.
11. V. Khetani, Y. Gandhi, S. Bhattacharya, S. N. Ajani, and S. Limkar, "Cross-domain analysis of ML and DL: Evaluating their impact in diverse domains," in International Journal of Intelligent Systems and Applications in Engineering, vol. 11, no. 7s, pp. 253–262, 2023.

12. T.-M. H. Hsu, H. Qi, and M. Brown, "Measuring the effects of nonidentical data distribution for federated visual classification," arXiv:1909.06335, 2019.
13. G. Huang, Z. Liu, L. Van Der Maaten, and K. Q. Weinberger, "Densely connected convolutional networks," in Proceedings of the IEEE conference on computer vision and pattern recognition, pp. 4700–4708, 2017.
14. K. Huang, and G. Zhu, "Broadband analog aggregation for low-latency federated edge learning," in IEEE Transactions on Wireless Communications, 2020.
15. L. Wang, W. Wang, and B. Li, "CMFL: Mitigating communication overhead for federated learning," 2019 IEEE 39th International Conference on Distributed Computing Systems (ICDCS), Dallas, TX, USA, 2019, pp. 954–964, doi: 10.1109/ICDCS.2019.00099.
16. S. N. Ajani, P. Khobragade, M. Dhone, B. Ganguly, N. Shelke, and N. Parati, "Advancements in computing: Emerging trends in computational science with next-generation computing," in International Journal of Intelligent Systems and Applications in Engineering, vol. 12, no. 7s, pp. 546–559, 2023.
17. S. Ajani, and M. Wanjari, "An efficient approach for clustering uncertain data mining based on Hash indexing and Voronoi clustering," *2013 5th International Conference and Computational Intelligence and Communication Networks*, Mathura, India, 2013, pp. 486–490, doi: 10.1109/CICN.2013.106.
18. B. Kumar, S. Singh, R. Grover, K. R. Isabels, A. Garg, and B. Charudatta Dattatraya, "Analysis of mathematical modelling deterministic and stochastic problems in federated learning," 2023 3rd International Conference on Advance Computing and Innovative Technologies in Engineering (ICACITE), Greater Noida, India, 2023, pp. 1700–1704, doi: 10.1109/ICACITE57410.2023.10183114.
19. X. He, H. Zhu, and Q. Ling, "Byzantine-robust and communication-efficient distributed non-convex learning over non-IID data," ICASSP 2022 - 2022 IEEE International Conference on Acoustics, Speech and Signal Processing (ICASSP), Singapore, Singapore, 2022, pp. 5223–5227, doi: 10.1109/ICASSP43922.2022.974709.
20. S. Li, E. Ngai, and T. Voigt, "Byzantine-robust aggregation in federated learning empowered industrial IoT," in IEEE Transactions on Industrial Informatics, vol. 19, no. 2, pp. 1165–1175, Feb. 2023, doi: 10.1109/TII.2021.3128164.
21. J. Wang, H. Liang, and G. Joshi, "Overlap local- SGD: An algorithmic approach to hide communication delays in distributed SGD," in ICASSP 2020 - 2020 IEEE International Conference on Acoustics Speech and Signal Processing (ICASSP), pp. 8871–8875, 2020.
22. S. Wang, T. Tuor, T. Salonidis, K. K. Leung, C. Makaya, T. He, et al., "Adaptive federated learning in resource-constrained edge computing systems," in IEEE Journal on Selected Areas in Communications, vol. 37, no. 6, pp. 1205–1221, 2019.
23. H. T. Nguyen, N. C. Luong, J. Zhao, C. Yuen, and D. T. Niyato, "Resource allocation in mobility-aware federated learning networks: A deep reinforcement learning approach," in Proceedings of the 2020 IEEE 6th World Forum on Internet of Things (WF-IoT), pp. 1–6, 2020.

15 Deep Learning Method for Hyperspectral Image Classification

Parul Bhanarkar and Salim Y. Amdani

15.1 INTRODUCTION TO HYPERSPECTRAL IMAGING

The collection and analysis of spectrum data from scenes or objects at a finer scale than is achievable with conventional imaging techniques is made possible by advanced remote sensing techniques like hyperspectral imaging (HSI) [1]. Contrary to traditional imaging methods, which only record a small number of distinct spectral bands, hyperspectral imaging captures a far larger number of bands, as illustrated in Figure 15.1. The upshot is that each and every pixel in the image will have a distinct spectral fingerprint [2]. The electromagnetic spectrum is only partially captured by conventional imaging methods. The incredible power of hyperspectral photography has transformed a number of industries, including agriculture, environmental monitoring, mineral extraction, urban planning, and medical diagnostics [3].

15.1.1 ADVANTAGES OF HYPERSPECTRAL IMAGING

For extensive analysis and categorization jobs, hyperspectral imaging is a helpful tool since it offers a number of benefits over conventional imaging approaches [4]. These are a few of the most important advantages:

i. *Spectral discrimination*: With the expanded number of spectral bands offered by hyperspectral imaging, it is considerably easier to discern various materials with identical outward appearances. This is possible because of spectrum analysis. With this improved spectral resolution, objects may be precisely identified and categorized according to their distinctive spectral fingerprints.

ii. *Material identification*: The identification and characterization of certain substances and chemicals in a scene are made possible by the capacity of hyperspectral imaging to offer comprehensive spectral information. Applications such as geological surveys, agricultural analysis, and environmental monitoring can all benefit greatly from this capacity.

iii. *Sub-pixel analysis*: Hyperspectral imaging enables the study of sub-pixel data in order to detect and define the many materials included within a single image pixel. Multispectral photography, which possesses this capability, is needed for applications that require accurate information of the distribution of different materials inside a pixel, such as land cover mapping.

DOI: 10.1201/9781003437079-15

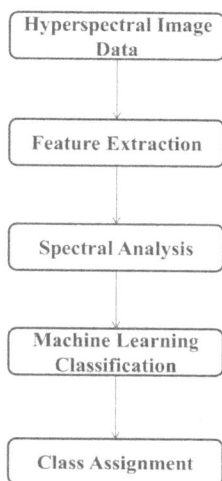

FIGURE 15.1   Hyperspectral image classification.

iv. *Enhanced feature extraction*: The high-dimensional spectrum data produced by hyperspectral imaging allows for the application of more sophisticated feature extraction methods. It is feasible to extract discriminative features that are not easily detectable using traditional imaging techniques by making use of the tiny spectrum changes that occur between various types of materials.

15.1.2   LIMITATIONS OF HYPERSPECTRAL IMAGING

While using hyperspectral imaging has numerous benefits, there are certain drawbacks that should be considered as well [5]. Some of the most significant constraints include the following:

i. *Data volume and complexity*: complicated data sets The vast number of spectral bands that define hyperspectral data and contribute to its high dimensionality. The already significant data volume and complexity is increased by this. Data processing, analysis, and storage become problematic as a result. To manage such enormous amounts of data, efficient and effective methods and algorithms are needed.

ii. *Cost and acquisition time*: The cost of purchasing and maintaining hyperspectral imaging equipment might be significant. Typically, hyperspectral data collecting uses specialized sensors that need careful calibration and maintenance. Additionally, the time required to acquire hyperspectral data is longer than the time required for traditional imaging techniques, which restricts the use of hyperspectral data in real-time in some circumstances.

iii. *Atmospheric interference*: The accuracy of the analysis and categorization of hyperspectral data may be hampered by atmospheric distortions and artefacts. One of the numerous factors that must be considered and accounted for during the data processing stage are scattering, absorption, and atmospheric conditions.

15.1.3 Applications of Hyperspectral Imaging

The broad spectrum data from hyperspectral imaging (HSI) may be used for analysis and decision-making in a variety of circumstances, opening up new applications for the technology [6]. Here are a few of the most significant uses:

i. *Agriculture and crop monitoring*: HSI is a technology for farming and agricultural monitoring that allows for accurate disease identification, nutrient analysis, and yield forecasting. Farmers may make more educated judgments about irrigation, fertilization, and pest control by examining the spectral fingerprints of each variety of plant.

ii. *Environmental monitoring*: Hyperspectral photography has grown to be a crucial component of environmental monitoring because of its value in determining plant health, tracking land cover, and identifying environmental trends. In lakes, rivers, and seas, it may be used to monitor changes in water quality, locate contaminants, and identify dangerous algal blooms.

iii. *Mineral exploration*: Geologists can locate possible mineral resources and map the distribution of those resources by using spectral analysis, which includes examining the spectral fingerprints of rocks and minerals. For effective resource management and efficient mining operations, this information is essential.

iv. *Urban planning and infrastructure assessment*: The assessment of urban infrastructure and land use is made easier by the use of HSI, which also provides useful information on plant cover and land use. Finding urban heat islands and determining whether there are any issues with a structure's materials or design are two examples of how it might be useful.

v. *Forestry and deforestation monitoring*: HSI may be used to track forest health, identify tree species, and locate regions that have experienced deforestation or degradation of the forest. It is of utmost importance for forest management, furthering conservation initiatives, and monitoring the effects of climate change on forest ecosystems.

vi. *Medical diagnostics*: Potential applications for HSI in medical diagnosis include tissue analysis. It helps in the designation of particular biomarkers, the identification of malignant tissue, and the description of skin lesions. Spectrum analysis aids medical professionals in better diagnosis and provides direction when selecting a course of treatment.

vii. *Remote sensing and Earth observation*: HSI is frequently utilized in these fields. It offers useful information that may be used to investigate atmospheric composition, keep an eye out for natural disasters like floods and wildfires, and evaluate how climate change is affecting ecosystems.

viii. *Archaeology and cultural heritage*: Numerous industries, including archaeology and the preservation of cultural objects, use hyperspectral photography. Finding hidden characteristics, defining material composition, and guiding restoration efforts are all made possible by analyzing the spectral fingerprints of artifacts, works of art, and historical sites. There might be a few benefits to this.

ix. *Remote sensing in astronomy*: Astronomy can benefit from using remote sensing techniques, such as HSI, to analyze celestial objects and occurrences. This makes it possible to examine cosmic radiation, identify chemical elements in far-off stars and galaxies, and investigate stellar spectra.

The variety and importance of HSI are demonstrated by the numerous applications that have been developed. Because of its capacity to include the entire spectrum, it opens the door to new approaches to analysis, categorization, and decision-making that advance a variety of industries, including science, technology, and environmental management. These developments are made possible by the fact that this potential exists.

15.2   HYPERSPECTRAL IMAGE CLASSIFICATION TECHNIQUES

15.2.1   OVERVIEW

Hyperspectral image (HSI) classification is the process of assigning a specific class or label to each pixel in a HSI based on its spectral signature. It plays a vital role in extracting meaningful information and understanding the composition of the observed scene [7, 8]. Over the years, various classification techniques have been developed to effectively analyze hyperspectral data.

15.2.2   TRADITIONAL CLASSIFICATION METHODS

Traditional hyperspectral image classification methods [9–11], form the foundation of hyperspectral data analysis and some of the commonly used traditional classification methods are:

i. *Spectral angle mapper*: Its main function is to determine the angle between the spectral signature of a pixel and a reference spectral signature that corresponds to a certain class. The lowest-valued category is given to the pixel after measuring its spectral angle.

ii. *Spectral information divergence (SID) metric*: The amount by which each pixel's spectral distribution deviates from the reference spectra for the class is measured using the SID metric. Multiply the SID by the spectral distribution of each pixel to obtain this separation. Two statistical methods are employed to assess the data: the Kullback–Leibler divergence and the Bhattacharyya distance.

iii. *Support vector machines (SVM)*: SVM is a most commonly used machine learning (ML) algorithm that finds an optimal hyperplane to separate different classes based on a training dataset. It maps the input spectral features into a high-dimensional space to achieve better separability.

iv. *Maximum likelihood classifier (MLC)*: MLC assumes that each class follows a multivariate Gaussian distribution and estimates the class parameters from the training data. The classification decision is made based on the maximum likelihood estimation.

v. *Pixel Purity Index (PPI)*: PPI identifies pure pixels within the image by analyzing their spectral characteristics. These pure pixels can then be used as training samples for classification algorithms.

15.3 INTEGRATION OF HYPERSPECTRAL IMAGING WITH IoT AND WSN

The Internet of Things (IoT) and Wireless Sensor Networks (WSN), as shown in Figure 15.2, can be used in conjunction with HSI to create new opportunities for real-time monitoring and analysis of hyperspectral data across a range of application domains [12–17].

i. *Improved sensing capabilities*: The scalable and distributed sensing infra-structure that the IoT and WSN provide makes it possible to deploy hyper-spectral sensors in a range of scenarios. By integrating HSI devices into IoT and WSN networks, a large number of geographically scattered sensors can simultaneously capture hyperspectral data, offering thorough and accurate monitoring of an area of interest.

ii. *Real-time data capture*: Real-time data transmission and capture are made possible by the integration of HSI with IoT and WSN. Spectral data can be continuously captured by hyperspectral sensors placed throughout the network, and the gathered information can be sent to a central or edge

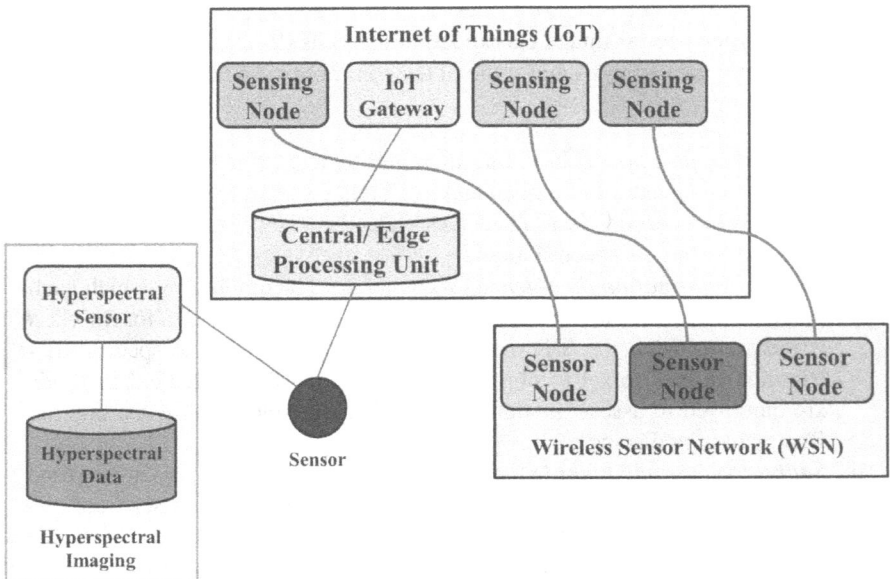

FIGURE 15.2 Integration of hyperspectral imaging (HSI) with IoT and WSN.

processing unit for additional analysis. For applications like environmental monitoring or precision agriculture that demand quick decision-making, this real-time data collecting capacity is essential.

iii. *Effective data transmission*: Due to the great spectral and spatial resolution, hyperspectral data is frequently large in size. By utilizing data compression methods and intelligent data aggregation, IoT and WSN provide mechanisms for effective data transport. In networks with limited resources, this permits the reduction of data traffic, optimization of bandwidth use, and minimization of energy consumption.

iv. *Distributed processing*: Hyperspectral data can be processed in a distributed manner thanks to the combination of IoT and WSN with HSI. To avoid sending massive volumes of raw data to a centralized processing unit, the acquired data can be processed locally at the sensor nodes. By utilizing the computational power of the sensor nodes, distributed processing algorithms may be created to distribute tasks like target detection, feature extraction, and spectral analysis.

v. *Context-aware decision-making*: A comprehensive picture of the monitored environment can be obtained by fusing hyperspectral data with additional sensor measurements from IoT and WSN. By linking hyperspectral data with environmental characteristics like temperature, humidity, or pollution levels, this integration enables context-aware decision-making. The interpretability and dependability of the hyperspectral data are improved by this contextual information, allowing for more precise and well-informed decision-making.

New opportunities for applications like precision agriculture, environmental monitoring, disaster management, and surveillance are made possible by the integration of hyperspectral photography with IoT and WSN. To fully realize the potential of this integration, however, issues with data management, network scalability, energy efficiency, and security must be resolved. To further improve the integration of hyperspectral imaging with IoT and WSN, future research efforts should concentrate on building effective and scalable algorithms, optimizing resource utilization, and investigating new sensing technologies.

15.4   DEEP LEARNING TECHNIQUES

15.4.1   Overview of Deep Learning Techniques

Deep learning (DL) has recently received a lot of interest in the field of HSI classification due to its capacity to automatically learn and extract complicated properties from high-dimensional data. DL's recent stratospheric development can be attributed in part to this. A high-level overview of popular DL strategies for hyperspectral picture classification is given in the next section and a summary is discussed in Table 15.2.

The hierarchical structure of the architectures of DL, which are made up of numerous layers of interconnected artificial neurons, is one of its defining characteristics. These architectures have the ability to autonomously learn hierarchical

TABLE 15.1
Publically Available Hyperspectral Image Datasets

Dataset	Spectral Bands	Spatial Resolution	Ground Truth Information	Applications
Indian Pines	145	20 m	Detailed land cover classes	Crop classification, land cover mapping, anomaly detection
Pavia University	103	1.3 m	Detailed land cover classes	Agriculture monitoring, vegetation analysis, urban planning
Salinas	224	3.7 m	Detailed land cover classes	Crop yield estimation, disease detection, precision agriculture
University of Houston	144	1 m	Urban land cover classes	Urban land use classification, infrastructure monitoring, disaster management
Urban 3D	364	0.25 m	Urban land cover classes	Urban planning, 3D reconstruction, object detection
Cuprite	188	10 m	Mineral composition	Geological mapping, mineral exploration, remote sensing applications
Hyperion	242	Varies	Various land cover classes	Environmental monitoring, disaster assessment, vegetation mapping

representations of data, capturing both granular and abstract properties, the description of dataset is shown in Table 15.1. Convolutional neural networks (CNNs), recurrent neural networks (RNNs), and deep belief networks (DBNs) are often used as deep learning algorithms for categorizing hyperspectral images [18].

15.4.2 PUBLICALLY AVAILABLE HYPERSPECTRAL IMAGE DATASETS

Choosing the right datasets is one of the most important parts of making hyperspectral images more useful. Table 15.1 shows a wide range of different datasets, each with its own special features that make it useful for different kinds of analysis. The famous Indian Pines dataset has 145 spectral bands and a spatial resolution of 20 m, which makes it possible to classify land cover in great detail for tasks like crop classification and finding anomalies. The complex Urban 3D dataset has 364 spectral bands and an impressive 0.25-m spatial resolution, which makes it possible to plan cities accurately, reconstruct 3D models, and find objects. As we look through datasets like Pavia University, Salinas, the University of Houston, Cuprite, and Hyperion, we find a lot of different spectral bands, spatial resolutions, and ground truth information. Each of these is designed for a different purpose, from monitoring agriculture to geological mapping. This collection is a great resource for students, practitioners, and fans who are interested in the wide field of hyperspectral remote sensing.

15.4.3 CONVOLUTIONAL NEURAL NETWORKS (CNNS)

In recent years, CNNs have been increasingly popular as effective DL techniques for image processing. Another use for CNNs that has shown success is the categorization of hyperspectral images. CNNs are exceptionally good at collecting spectral and spatial correlations when working with hyperspectral data.

In HSI, each spectral band is treated as a separate channel, similar to the RGB channels in traditional image analysis. Convolutional layers consist of filters that slide over the input data, extracting local features by performing convolutions [18]. The filters capture spatial and spectral patterns, learning important features from the hyperspectral data. Pooling layers downsample the learned features, reducing the dimensionality of the data, and provide translational invariance. Fully connected layers aggregate the learned features and make final classification decisions.

CNNs have several advantages for hyperspectral image classification. They can automatically learn discriminative spatial–spectral features, capture hierarchical representations, and handle high-dimensional data, as shown in Figure 15.3. Transfer learning, where pretrained CNN models on large-scale image datasets are fine-tuned for hyperspectral data, can also be employed to leverage the learned representations and improve classification performance.

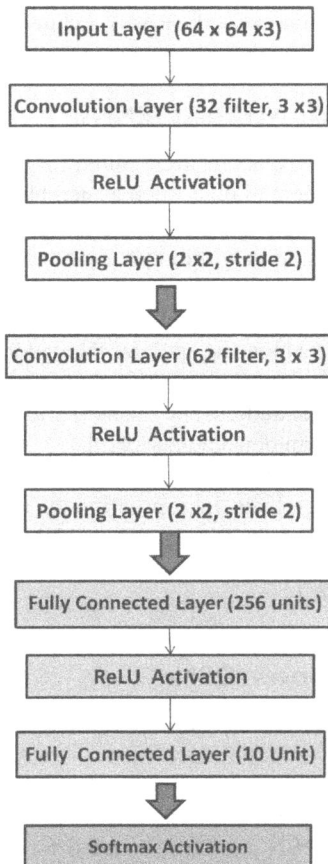

Input Layer (64 x 64 x3)

Convolution Layer (32 filter, 3 x3)

ReLU Activation

Pooling Layer (2 x2, stride 2)

Convolution Layer (62 filter, 3 x 3)

ReLU Activation

Pooling Layer (2 x2, stride 2)

Fully Connected Layer (256 units)

ReLU Activation

Fully Connected Layer (10 Unit)

Softmax Activation

FIGURE 15.3 Convolutional neural network (CNN) architecture for hyperspectral image classification.

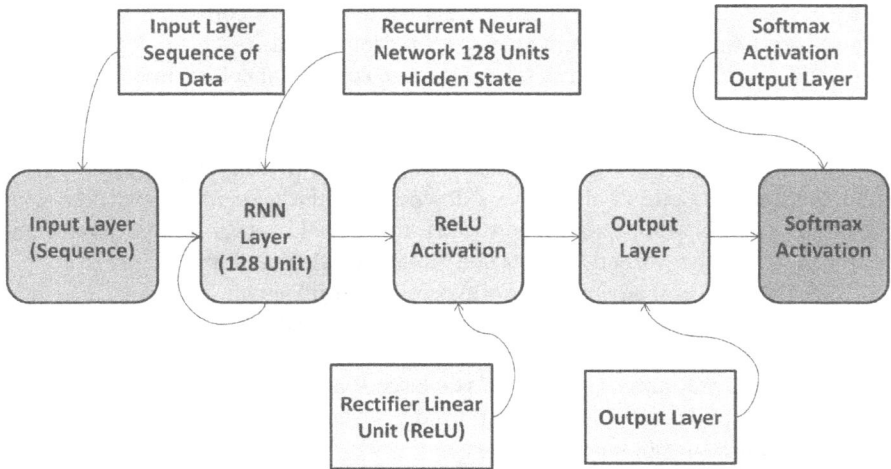

FIGURE 15.4   Recurrent neural network (RNN) architecture for hyperspectral image classification.

15.4.4   RECURRENT NEURAL NETWORKS (RNNs)

Recurrent neural networks (RNNs) are well-suited for hyperspectral image classification tasks that involve sequential data, such as neighboring pixels in a hyperspectral image. RNNs can model the temporal dependencies between adjacent pixels and capture the contextual information within hyperspectral data.

The key feature of RNNs is the presence of recurrent connections, allowing information to be propagated not only forward but also backward in the network. This enables RNNs to learn and exploit the spectral dependencies present in hyperspectral data. Long short-term memory (LSTM) and gated recurrent unit (GRU) are popular variants of RNNs that address the vanishing gradient problem and improve the ability to capture long-term dependencies [19].

In hyperspectral image classification, RNNs can effectively model the spectral variations across neighboring pixels, incorporating the spatial–spectral information. By considering the context of neighboring pixels, RNNs can improve the classification accuracy and handle the challenges posed by spectral variability and spatial–spectral coupling (Figure 15.4).

15.4.5   DEEP BELIEF NETWORKS (DBNs)

DBNs can learn hierarchical representations and extract latent variables from the hyperspectral data. Each layer of the DBN learns increasingly abstract features, capturing different levels of abstraction within the data. The unsupervised pretraining of DBNs allows them to learn effective representations without the need for labeled data. Once pretrained, the DBN can be fine-tuned for supervised classification. A systematic illustration is shown in Figure 15.5. DBNs have been applied to hyperspectral image classification by treating the spectral bands as visible variables and

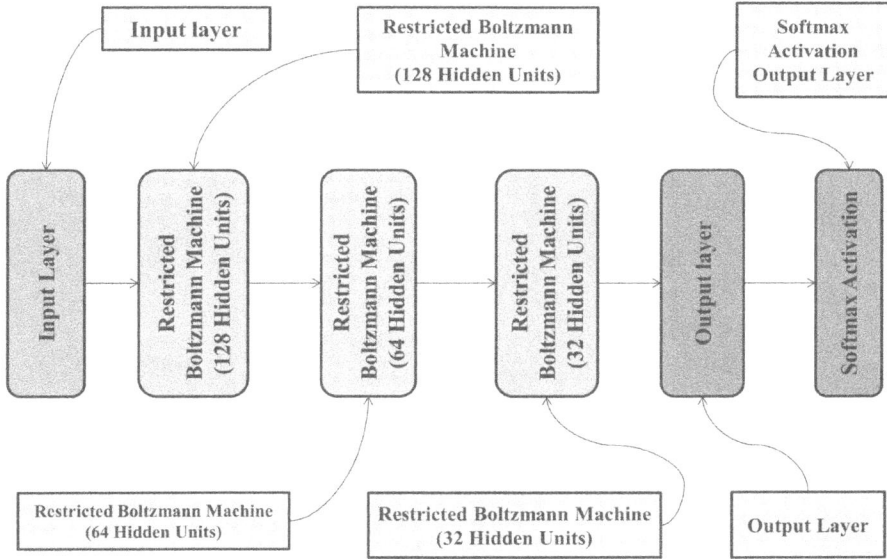

FIGURE 15.5 Deep belief network (DBN) architecture for hyperspectral image classification.

learning hidden representations that capture the underlying structure of the data. The learned features can then be used for classification tasks, improving the accuracy and robustness of the model [20].

15.4.6 COMPARISON OF DEEP LEARNING TECHNIQUES FOR HYPERSPECTRAL IMAGE CLASSIFICATION

There are three main model designs used to analyze hyperspectral data [21]: Convolutional Neural Networks (CNNs), Recurrent Neural Networks (RNNs), and Deep Belief Networks (DBNs) compared in the Table 15.2. Each one has its own way of figuring out how complicated hyperspectral video is. The main parts of each design are different, as are the ways that they handle raw data and their pros and cons.

15.5 IMPLEMENTATION OF HYPERSPECTRAL IMAGE CLASSIFICATION USING CNNS

15.5.1 PREPROCESSING WITH THE DATASET

Using the WSN and IoT based sensor, hyperspectral data was gathered over an agricultural area close to Pavia, Italy, for the Pavia University dataset. This dataset is often utilized by scientists. The HSI has a spatial resolution of 1.3 m and 103 spectral bands. The dataset has been extensively used in a variety of remote sensing

TABLE 15.2

Comparison of Deep Learning Techniques for Hyperspectral Image Classification

Model Architecture	CNNs	RNNs	DBNs
Key Components	Convolutional layers, pooling layers, fully connected layers	Recurrent connections (LSTM/GRU units), dense layers	Stacked Restricted Boltzmann Machines (RBMs)
Input Data	Hyperspectral data with each spectral band treated as a separate channel	Sequential hyperspectral data, considering the neighboring pixel relationships	Hyperspectral data with spectral bands as visible variables
Spatial–Spectral Analysis	Learns spatial and spectral relationships, capturing local and global features	Focuses on temporal dependencies, modeling contextual information within hyperspectral data	Captures hierarchical representations, learns latent variables that capture the underlying structure of the hyperspectral data
Advantages	• Effective in capturing spatial–spectral patterns • Handles high-dimensional data • Enables transfer learning from pretrained models	• Handles sequential hyperspectral data with temporal dependencies • Robust to spectral variability and spatial–spectral coupling • Improves classification accuracy through contextual modeling	• Learns hierarchical representations and latent variables • Effective in handling high-dimensional data • Captures underlying structure of hyperspectral data
Limitations	• Computationally intensive, requiring substantial resources • May require large amounts of labeled training data for optimal performance	• Longer training times compared to CNNs • May struggle with long-range dependencies	• Computationally demanding during training • Requires careful parameter tuning
Applications	Hyperspectral imagery	Land cover classification, anomaly detection, and change detection	Hyperspectral data analysis, feature extraction, and dimensionality reduction
Parameters	• Number of convolutional layers: 4 • Number of pooling layers: 2 • Number of fully connected layers: 1 • Filter size in convolutional layers: 3×3 • Pooling size in pooling layers: 2×2	• Number of recurrent layers: 2 • Number of dense layers: 1 • LSTM/GRU units per layer: 128 • Dropout rate: 0.2 • Optimizer: Adam	• Number of RBM layers: 3 • Number of hidden units per RBM layer: [100, 75, 50] • Number of output units: Number of classes in the classification task • Learning rate: 0.001 • Batch size: 32

TABLE 15.3
Pavia University Hyperspectral Dataset

Dataset	Pavia University
Labeled Samples	42,776
Spatial Resolution	1.3 m
Image Dimensions	610 × 340 pixels
Sensor	ROSIS
Spectral Range	0.43–0.86 μm
Wavelengths	115 (12 excluded due to noise)
Bands per Pixel	103
Objects	9
Location	Pavia University, Italy
Application	Versatile – Agriculture, Urban Analysis, Environmental Study

applications, such as crop monitoring, vegetation study, and urban planning, as discussed in Table 15.3.

To guarantee high-quality data and algorithm compatibility, hyperspectral image classification with CNNs on the Pavia University dataset has to be preprocessed. The following preprocessing methods can be applied if necessary:

i. *Data format conversion*: Datasets from Pavia University are frequently made available in raw or other uncommon file types. The data must be transformed into a recognized format before employing a DL framework or library. The data may be converted into numpy arrays or tensors in TensorFlow.

ii. *Spectral band selection*: The nature of the classification task at hand, as well as the availability of computational resources, may necessitate the selection of a subset of spectral bands for processing. The spectrum's band selection preserves critical information while reducing computing load and the risk of noise from irrelevant bands.

iii. *Spatial resampling*: The spatial resolution of hyperspectral imaging varies depending on the spectral band being studied. It is possible to align the spatial resolution of all spectral bands using resampling methods like k-nearest-neighbors or bilinear interpolation. These methods can be applied to guarantee uniformity.

iv. *Data normalization*: It is normal to observe significant fluctuation in the intensity levels of the various bands while analyzing hyperspectral data. By bringing all of the bands to a similar scale, data normalization approaches like min-max scaling and z-score normalization can make it easier to train the model.

v. *Data augmentation*: Increasing the variety of training samples and improving the model's generalization performance are two goals of data augmentation. The images are cropped, given a noisy, grainy look, then randomly rotated or flipped.

TABLE 15.4
Experimental Results

Methods	Epochs	Precision	Recall	Accuracy	F1-Score
CNN	200	0.96	0.97	0.96	0.97
RNN	200	0.93	0.94	0.94	0.93
DBN	200	0.91	0.91	0.90	0.91

15.5.2 Experimental Results and Analysis

Here we show the outcomes of our research on the Pavia University dataset, where we used CNNs, RNNs, and DBNs to classify hyperspectral pictures. These networks were trained using the image library at Pavia University. The models' performance measures like precision, recall, accuracy, and F1-score were evaluated after they had been trained for a total of 200 epochs, shown in Table 15.4.

Figure 15.6 shows the precision of each approach deep learning approach, emphasizing how well it performs in hyperspectral image categorization tasks. The outcomes demonstrate that the CNN model outperformed all other metrics utilized in the research. After being assessed for accuracy (0.96), recall (0.97), and precision (0.96), it received an F1-score of 0.97. According to these results, the CNN model did a great job identifying the hyperspectral data obtained from the Pavia University dataset.

Figure 15.7 shows the recall of each approach deep learning approach, emphasizing how well it performs in hyperspectral image categorization tasks. With scores of 0.93 for precision, 0.94 for recall, 0.94 for accuracy, and 0.93 for F1, the RNN model, in comparison, performed just marginally better than the other models. The RNN model fared worse than the CNN model, although produced decent results. The emphasis on spatial–spectral patterns in the Pavia University dataset for classification may have led to the performance discrepancy between the two schools. Convolutional layers in CNNs, which enable the extraction of spatial features at various scales, are especially created to capture these patterns successfully. On the other hand, because their recurrent connections, RNNs are more suited to capture temporal associations in sequential data as opposed to spatial information.

Figure 15.8 shows the F1-Score of each approach deep learning approach, emphasizing how well it performs in hyperspectral image categorization tasks. The DBN model did the least well of the three, scoring 0.91 for precision, 0.91 for recall, 0.90 for accuracy, and 0.91 for F1. Although DBNs are widely renowned for their capacity to uncover underlying structures and build hierarchical representations, it is likely that more training or hyperparameter tweaking will be required to improve performance on the Pavia University dataset.

Figure 15.9 shows the accuracy of each approach deep learning approach, emphasizing how well it performs in hyperspectral image categorization tasks. Overall, the results demonstrate that CNNs successfully classify hyperspectral images on the Pavia University dataset. The CNN model's outstanding precision, recall, accuracy, and F1-score show how well it learned and generalized the spatial–spectral patterns

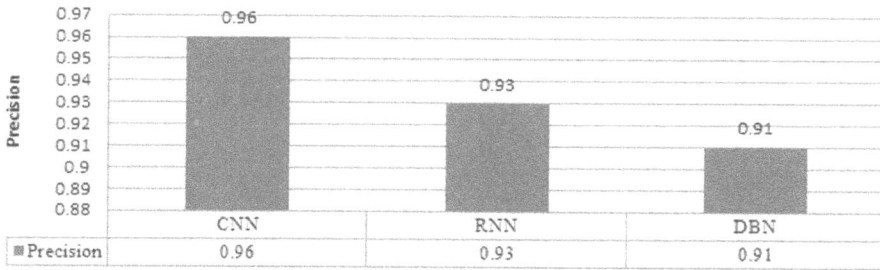

FIGURE 15.6   Precision.

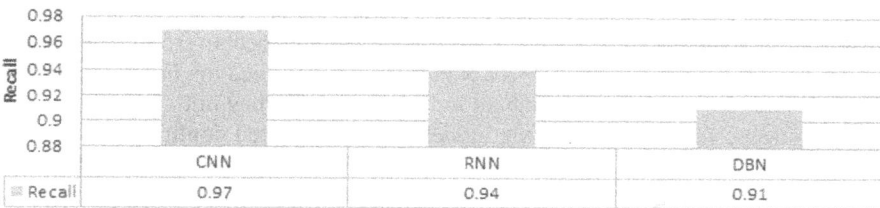

FIGURE 15.7   Recall.

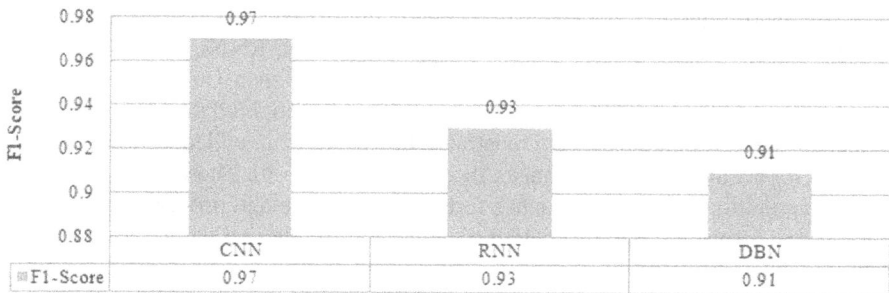

FIGURE 15.8   F1-Score.

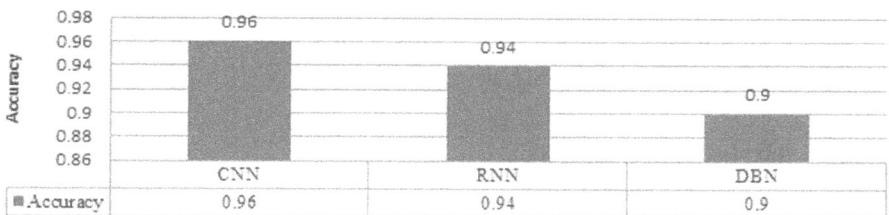

FIGURE 15.9   Accuracy.

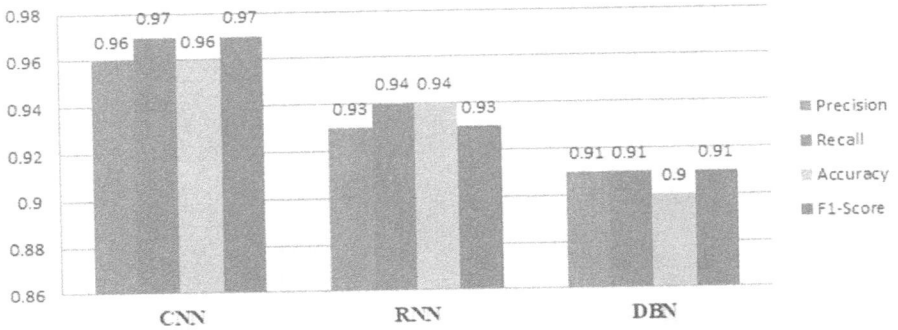

FIGURE 15.10 Comparative analysis.

visible in the hyperspectral images. The DBN model may benefit from more optimization or other techniques, but the RNN model's performance implies that temporal dependencies may not be as significant for this dataset as they initially appear.

It should be highlighted that the models and their tested configurations on the Pavia University dataset are the only ones that can provide the findings that have been given. Performance may be impacted by several factors, including hyperparameter selection, network design, and dataset properties. Only after thorough testing and experimenting with several configurations can the optimum model for a certain hyperspectral image classification assignment be discovered.

Figure 15.10 compares the performance of convolutional neural networks (CNNs), recurrent neural networks (RNNs), and deep belief networks (DBNs) in classifying hyperspectral images. Our discussion of applying CNNs, RNNs, and DBNs to the challenge of hyperspectral image classification on the Pavia University dataset is concluded in this section. The experiment's findings highlighted the superior performance of the CNN model and demonstrated how well it can collect spatial–spectral patterns for accurate classification, as shown in Figure 15.10. These findings deepen our understanding of DL approaches to hyperspectral image processing and show how these approaches may be applied to datasets from the real world, such as those kept by Pavia University.

15.6 USE-CASE: OBJECT DETECTION ON HYPERSPECTRAL IMAGES IN WIRELESS NETWORKS

Hyperspectral image object identification in wireless networks is shown in Figure 15.11. In order to identify objects, wireless sensor nodes collect, classify, and localize features from the hyperspectral sensor's exact spectral data.

15.6.1 Hyperspectral Imaging

Hyperspectral sensor: This element is a representation of the hyperspectral imaging sensor. It gathers data over a broad spectrum of wavelengths, obtaining precise spectral data for each pixel in the image.

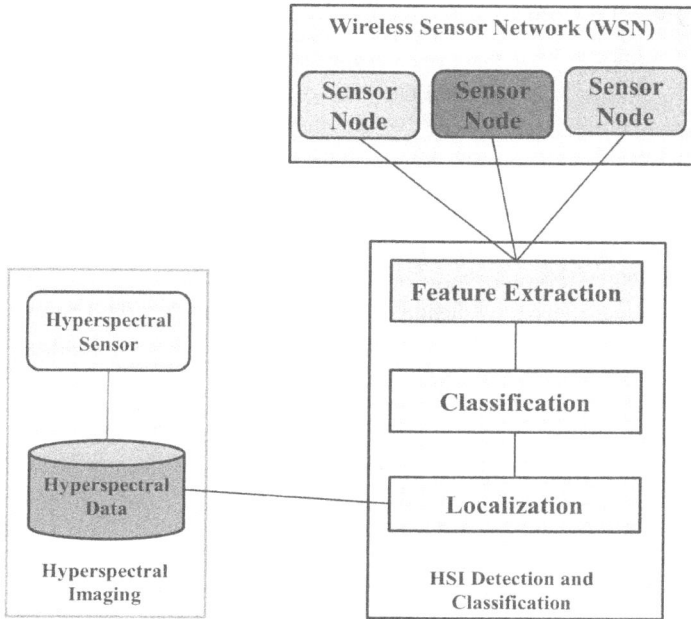

FIGURE 15.11   Object detection on hyperspectral images in wireless networks.

Hyperspectral data: This database serves as a repository for the hyperspectral image data that the sensor has collected. It preserves the spectral details of the scene's objects.

15.6.2   WIRELESS SENSOR NETWORKS (WSN)

Sensor node: These nodes are situated in the monitoring region and are a part of the wireless sensor network. They receive hyperspectral data from the hyperspectral sensor and provide it to the algorithm's later stages for item detection. Prior to transmission, the nodes may also do basic preprocessing or filtering on the data.

15.6.3   OBJECT DETECTION AND CLASSIFICATION

Feature extraction: In this step, pertinent features are extracted from the hyperspectral data to reflect the properties of the scene's objects. Spectral signatures, texture patterns, and spatial data are examples of features.

Classification: The classification stage divides the extracted features into various object classes or categories using ML or pattern recognition techniques. In this stage, the kind or class of the objects found in the hyperspectral image is determined.

Localization: In a hyperspectral image, localization refers to locating or defining the boundaries of the items that have been detected. The location of the objects within the scene is revealed during this step.

15.7 CONCLUSION

Remote sensing specialists have been trying to classify hyperspectral photos for a long time. In our analysis, we used DL, including CNNs, RNNs, and DBNs, to categorize hyperspectral photos. The Pavia University dataset was collected from a nearby agricultural region using popular hyperspectral data. We researched hyperspectral Image Analysis. According to our Pavia University dataset analysis, CNNs were better at detecting hyperspectral pictures than RNNs and DBNs.

DBNs, known for learning hierarchical representations, fared poorly compared to CNNs and RNNs. It's important to remember that DBN analysis and optimization, such as evaluating alternative designs or changing hyperparameters, may enhance output. DBNs have shown promise in other domains, therefore categorizing hyperspectral pictures with them warrants more research. For training and assessment, the Pavia University dataset was large and diversified. It has 103 spectral bands and a 1.3-metre spatial resolution. Remember that the dataset is one approach to representing hyperspectral data and may have its own limitations. Extrapolating to other datasets or situations requires significant care. CNN had the greatest outcomes across all factors. It scored 0.97 for accuracy, recall, and precision. These findings show that the CNN model identified Pavia University hyperspectral data well. The RNN model fared slightly better than the others with precision, recall, accuracy, and F1-scores of 0.93, 0.94, 0.94, and 0.93, respectively. The CNN model outperformed the RNN model. The performance gap between the two institutions may be due to Pavia University's categorization focus on spatial–spectral patterns. CNN convolutional layers, which extract spatial data at multiple sizes, are designed to catch these patterns. The RNN model is better at capturing temporal relationships in sequential data than spatial information because of their recurrent connections. RNNs tend to have more former than latter. The DBN model has the worst precision, recall, accuracy, and F1 scores. DBNs are known for uncovering underlying structures and building hierarchical representations; however they may need more training or hyperparameter tweaking to perform better on the Pavia University dataset.

CNNs categorize hyperspectral pictures well on the Pavia University dataset. The CNN model's precision, recall, accuracy, and F1-score demonstrate how effectively it learnt and generalized hyperspectral spatial–spectral patterns. The RNN model's performance suggests that temporal dependencies may not be as important for this dataset as first thought. The DBN model may benefit from greater optimization or alternative methods.

These findings are solely available from the models and their tested settings on the Pavia University dataset. Hyperparameter selection, network architecture, and dataset attributes affect performance. The best model for hyperspectral picture categorization can only be found after extensive testing and experimentation. This section ends our discussion of using CNNs, RNNs, and DBNs to classify hyperspectral images on the Pavia University dataset. The experiment showed how well the CNN model can gather spatial–spectral patterns for appropriate categorization. The model's emphasis on trial outcomes supported this. These findings improve our knowledge of DL techniques in hyperspectral image processing and demonstrate how they may be applied to real-world datasets like those of Pavia University.

REFERENCES

1. B. Borasca, L. Bruzzone, L. Carlin and M. Zusi, "A Fuzzy-Input Fuzzy-Output SVM Technique for Classification of Hyperspectral Remote Sensing Images," Proceedings of the 7th Nordic Signal Processing Symposium - NORSIG 2006, 2006, pp. 2–5. doi:10.1109/NORSIG.2006.275261.
2. F. Melgani and L. Bruzzone, "Classification of Hyperspectral Remote Sensing Images with Support Vector Machines," IEEE Transactions on Geoscience and Remote Sensing, vol. 42, no. 8, 2004, pp. 1778–1790. doi:10.1109/TGRS.2004.831865.
3. X. Wang and Y. Feng, "New Method Based on Support Vector Machine in Classification for Hyperspectral Data," 2008 International Symposium on Computational Intelligence and Design, 2008, pp. 76–80. doi:10.1109/ISCID.2008.61.
4. N. Alajlan, Y. Bazi, H. AlHichri and E. Othman, "Robust classification of hyperspectral images based on the combination of supervised and unsupervised learning paradigms," 2012 IEEE International Geoscience and Remote Sensing Symposium, 2012, pp. 1417–1420. doi:10.1109/IGARSS.2012.6351270.
5. S. N. Ajani, P. K. Ingole and A. V. Sakhare "Modality of Multi-Attribute Decision Making for Network Selection in Heterogeneous Wireless Networks", Ambient Science, vol. 9, no. 2, 2022, pp. 26–31, ISSN 2348-5191.
6. A. Samat, P. Du, S. Liu, J. Li and L. Cheng, "E2LMs: Ensemble Extreme Learning Machines for Hyperspectral Image Classification," IEEE Journal of Selected Topics in Applied Earth Observations and Remote Sensing, vol. 7, no. 4, 2014, pp. 1060–1069. doi:10.1109/JSTARS.2014.2301775.
7. Y. Gu and H. Liu, "Sample-Screening MKL Method via Boosting Strategy for Hyperspectral Image Classification", Neurocomputing, vol. 173, Part 3, 2016, pp. 1630–1639, ISSN 0925-2312, doi:10.1016/j.neucom.2015.09.035.
8. Y. Guo, S. Han, H. Cao, Y. Zhang and Q. Wang, "Guided Filter Based Deep Recurrent Neural Networks for Hyperspectral Image Classification", Procedia Computer Science, vol. 129, 2018, pp. 219–223, ISSN 1877-0509, doi:10.1016/j.procs.2018.03.048.
9. S. A. Medjahed and M. Ouali, "Band Selection Based on Optimization Approach for Hyperspectral Image Classification", The Egyptian Journal of Remote Sensing and Space Science, vol. 21, no. 3, 2018, pp. 413–418, ISSN 1110-9823, doi:10.1016/j.ejrs.2018.01.003.
10. S. Ajani and M. Wanjari, "An Efficient Approach for Clustering Uncertain Data Mining Based on Hash Indexing and Voronoi Clustering," 2013 5th International Conference and Computational Intelligence and Communication Networks, 2013, pp. 486–490.
11. M. Imani and H. Ghassemian, "An Overview on Spectral and Spatial Information Fusion for Hyperspectral Image Classification: Current Trends and Challenges", Information Fusion, vol. 59, 2020, pp. 59–83, ISSN 1566-2535, doi:10.1016/j.inffus.2020.01.007.
12. G. Ortac and G. Ozcan, "Comparative Study of Hyperspectral Image Classification by Multidimensional Convolutional Neural Network Approaches to Improve Accuracy", Expert Systems with Applications, vol. 182, 2021, p. 115280, ISSN 0957-4174, doi: 10.1016/j.eswa.2021.115280.
13. S. Kutluk, K. Kayabol and A. Akan, "A New CNN Training Approach with Application to Hyperspectral Image Classification", Digital Signal Processing, vol. 113, 2021, p. 103016, ISSN 1051-2004, doi:10.1016/j.dsp.2021.103016.
14. S. N. Ajani, P. V. Potnurwar, V. K. Bongirwar, A. V. Potnurwar, A. Joshi and N. Parati, "Dynamic RRT* Algorithm for Probabilistic Path Prediction in Dynamic Environment," International Journal of Intelligent Systems and Applications in Engineering, vol. 11, no. 7s, 2023, pp. 263–271.
15. B. Zhang, W. Yang, L. Gao et al. Real-Time Target Detection in Hyperspectral Images Based on Spatial-Spectral Information Extraction," EURASIP Journal on Advances in Signal Processing, vol. 2012, 2012, p. 142. doi:10.1186/1687-6180-2012-142.

16. R. Allauddin Mulla, M. Eknath Pawar, S. S. Banait, S. N. Ajani, M. Pravin Borawake and S. Hundekari, "Design and Implementation of Deep Learning Method for Disease Identification in Plant Leaf," International Journal on Recent and Innovation Trends in Computing and Communication, vol. 11, no. 2s, 2023, pp. 278–285. doi:10.17762/ijritcc. v11i2s.6147.

17. V. Kumar and J. K. Ghosh, "Objects Detection in Hyperspectral Images Using Spectral Derivative," Journal of the Indian Society of Remote Sensing, vol. 45, 2017, pp. 603–610. doi:10.1007/s12524-016-0627-9.

18. J. Liu, Z. Wu, Z. Xiao and J. Yang, "Region-Based Relaxed Multiple Kernel Collaborative Representation for Hyperspectral Image Classification," IEEE Access, vol. 5, 2017, pp. 20921–20933. doi:10.1109/ACCESS.2017.2758168.

19. N. Imamoglu, Y. Oishi, X. Zhang et al., "Hyperspectral Image Dataset for Benchmarking on Salient Object Detection," 2018 Tenth International Conference on Quality of Multimedia Experience (QoMEX), 2018, pp. 1–3, doi:10.1109/QoMEX.2018.8463428.

20. L. Ma, G. Lu, D. Wang et al. Adaptive Deep Learning for Head and Neck Cancer Detection Using Hyperspectral Imaging. Visual Computing for Industry, Biomedicine, and Art, vol. 2, 2019, p. 18. doi:10.1186/s42492-019-0023-8.

21. X. Sun, H. Zhang, F. Xu, Y. Zhu and X. Fu, "Constrained-Target Band Selection with Subspace Partition for Hyperspectral Target Detection," IEEE Journal of Selected Topics in Applied Earth Observations and Remote Sensing, vol. 14, pp. 9147–9161, 2021 doi:10.1109/JSTARS.2021.3109455.

16 Toward Smarter Industries
Security Framework Using Attribute-Based Encryption and Systematic Solutions

Shruti and Shalli Rani

16.1 INTRODUCTION TO INDUSTRY 4.0

Industry 4.0, the current trend in the manufacturing industry, is also known as the Fourth Industrial Revolution. It is an integration of physical and digital systems characterized by using technologies like the Internet of Things (IoT) [1, 2], artificial intelligence (AI), cloud computing, and big data analysis to optimize and automate manufacturing processes. This era of manufacturing promises to deliver improved efficiency, agility, and flexibility, as machines become smarter, more connected, and more autonomous, increasing the productivity of the production processes [3]. Healthcare, automotive, aerospace, and consumer products (Figure 16.1) are just a few examples where Industry 4.0 is expected to completely transform manufacturing. The shift toward Industry 4.0 also presents some challenges, such as the need for new skills and training, cybersecurity risks, and ethical concerns. As we enter this new era of manufacturing, it is important to make sure that its advantages are shared equally and any associated threats are also addressed.

To automate a production process, the First Industrial Revolution used water and steam power, while in the Second Industrial Revolution, electricity allowed the introduction of mass production. The Third Industrial Revolution gave rise to the modern digital age where computerization and automation happened. And now, we are in the Fourth Industrial Revolution (or Industry 4.0), which is ready to transform manufacturing once again. Industry 4.0 centers around merging the physical and digital experience by integrating cyber-physical systems (CPS) that combine sensors, software, and communication technologies with physical objects, machines, and systems. This makes real-time monitoring and controlling physical processes easy, resulting in enhanced production efficiency and takes automation to a new level. Industry 4.0 signifies a revolutionary concept that represents a fundamental shift in how we think about manufacturing. Its effects will reverberate throughout diverse sectors, such as healthcare, automotive, aerospace, and consumer goods. As we embrace this new manufacturing era, it is fascinating to consider the potential opportunities that await

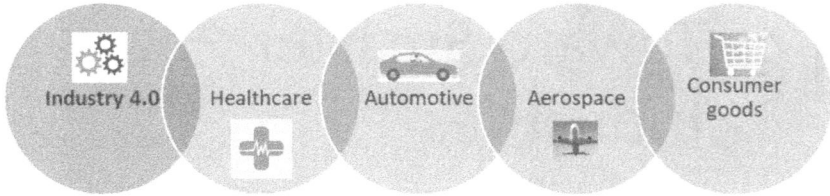

FIGURE 16.1   Industry 4.0's various manufacturing sectors.

us. Industry 4.0 has specific features that differentiate it from previous industrial revolutions. These features include [4]:

1. *Cyber-physical systems (CPS)*: An integration of physical systems with digital technology is known as CPS. It is a key feature of Industry 4.0 that combines hardware, software, and data processing making real-time monitoring and control of physical processes easy.
2. *Internet of Things (IoT)*: IoT is a network of devices that do not require humans for communication and data exchange. It plays an important role in Industry 4.0 as it makes communication much easier between machines, sensors, and other devices.
3. *Big data analytics*: It allows analysis and decision-making using large amounts of data produced by IoT and other sources, that helps to increase efficiency and quality.
4. *Advanced technology*: Technologies like machine learning (ML) and AI can be used to automate the process of decision-making that will enhance production efficiency.
5. *Cloud computing*: In Industry 4.0, cloud computing makes it possible to store and share data and other software resources online, eliminating the need for any physical infrastructure, and it also provides processing and storage capability.
6. *Decentralized decision-making*: Industry 4.0 supports a decentralized decision-making process, so that machines and systems can make decisions independently based on real-time data.
7. *Customization and personalization*: Industry 4.0 supports flexibility and enables the production of personalized and customized products according to the demands of the industry.
8. *Interoperability*: In Industry 4.0, all machines, devices, sensors, and people are connected and can communicate with each other.
9. Transparency in information: To contextualize information, the information systems in Industry 4.0 use sensor data to build a virtual representation of the real world.
10. *Flexibility*: Industry 4.0 uses technology to make operations more adaptive, in contrast to conventional automation which focuses on maximizing efficiency and throughput.

16.1.1 ADVANTAGES OF INDUSTRY 4.0

Industry 4.0 is driving manufacturing toward greater agility, efficiency, and flexibility. In order to respond quickly to changing conditions and demands in real-time, production systems and machines are becoming more intelligent, more autonomous, and better connected to each other. Significant improvements in terms of efficiency, quality, sustainability, and customer satisfaction are expected to be made in this new era of manufacturing. It offers several advantages, including:

1. *Enhanced productivity*: Using Industry 4.0 leads to the automation of various tasks, which can result in more rapid and efficient manufacturing processes.
2. *Better quality*: By using cyber-physical systems, real-time monitoring and regulation of production processes can be done, leading to improved product quality and consistency.
3. *Flexibility*: Industry 4.0 can produce customized goods to meet customer needs and meet the changing trends of the market.
4. *Cost-effectiveness*: Industry 4.0 technologies also result in cost savings by reducing labor costs, optimizing resource utilization, and minimizing waste.
5. *Improved safety*: Industry 4.0 technologies can improve worker safety by automating dangerous or repetitive tasks, providing real-time monitoring of equipment and processes, and identifying potential safety hazards before they occur.
6. *Environmental sustainability*: Industry 4.0 technologies promote environmental sustainability by optimizing resource usage, reducing waste, and improving energy efficiency.

16.1.2 CHALLENGES IN INDUSTRY 4.0

Nonetheless, Industry 4.0 also poses some challenges, like the need for new skills, risks related to cybersecurity, and ethical considerations. As we head toward a more connected and automated future, it is important to ensure that the benefits of Industry 4.0 are distributed equally, and its potential threats are mitigated. Some of the challenges faced by Industry 4.0 include:

1. *Data privacy and security*: Industry 4.0 generates and exchanges vast amounts of data, raising concerns about data privacy and security. Unauthorized access or cyber-attacks could cause significant damage to businesses and their customers.
2. *Skills gap*: Implementing Industry 4.0 requires highly skilled employees, which can be challenging for some companies especially those that lack the resources to provide training or hire skilled workers.
3. *Compatibility and standardization issues*: Integrating different technologies and systems can be challenging due to compatibility issues between devices from different manufacturers, leading to inefficiencies, errors, and increased costs.

4. *Infrastructure and connectivity*: Industry 4.0 technologies require reliable and fast Internet connectivity, which may not be available in all locations. Moreover, implementing and maintaining such connections is costly and can be a barrier for some companies.

5. *Regulatory challenges*: The fast pace of Industry 4.0 development has created a lack of regulatory frameworks to ensure safe and ethical use, creating uncertainty for businesses and consumers.

6. *Ethical and social issues*: In Industry 4.0, ethical and social issues are being raised due to the growing use of automation and AI, including their potential impact on employment and concerns about the fairness of decision-making processes that may exhibit bias.

7. *Technical integration*: The use of technologies that aren't equipped to handle digitalization could result in the production of poor-quality products. Also, more effort is needed to integrate modern IT technology [5].

8. *Safety and quality issues*: Another challenge faced by the companies is the growth in system complexity and the adoption of technologies that need more power. This creates a threat to the safety and quality of the product [6].

16.1.3 SECURITY IN INDUSTRY 4.0

The security and privacy in Industry 4.0 can be provided using one of the latest security algorithms, ABE. It is an encryption scheme that enables access control policies to be defined in terms of characteristics or attributes [7], such as age, job title, or role, instead of individual user identities.

In ABE, information is encrypted based on a collection of attributes, and only users who have those matching attributes can decrypt and obtain access to the data, as shown in Figure 16.2. This makes managing access control in large-scale systems simpler as there is no need for a unique access control list for each user. ABE is becoming more dominant in scenarios where accurate access control is needed,

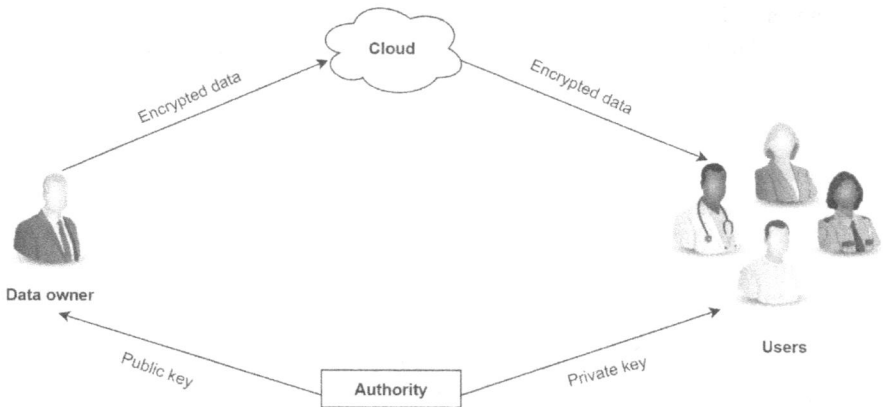

FIGURE 16.2 Architecture of the attribute-based encryption (ABE) scheme.

such as in healthcare systems, cloud computing, and IoT as it is a type of encryption method that provides an enhanced level of security and privacy to data in Industry 4.0. In this era of smart factories and linked devices, it is a necessity to protect confidential information and maintain secure communication between devices and machines. ABE provides a solution by enabling data owners to encrypt their data with specific attributes, such as job title or department, rather than with a specific recipient's public key. This approach allows only authorized users with matching attributes to decrypt and access the data, while others are denied access. Due to its features like adaptability, scalability, and the ability to provide granular data security, ABE has become popular in Industry 4.0, allowing industries to satisfy their security needs. However, there are some difficulties with using ABE techniques in industrial data. Due to ABE's lack of a standardized policy language, the design and maintenance of policies in particular are a significant problem [8].

16.1.4 LIMITATIONS OF INDUSTRY 4.0

Industry 4.0 also has certain limitations, which are discussed below:

1. *Technology dependent*: Industry 4.0 heavily relies on technology, which means that any disruption to technology can significantly impact production and supply chains.
2. *High initial investment*: Implementing Industry 4.0 technologies requires a large investment in hardware, software, and training, which can be challenging for small and medium-sized industries.
3. *Limited human interaction*: The increased use of automation and AI in Industry 4.0 can reduce human interaction, potentially impacting creativity and innovation.
4. *Ethical and social concerns*: Using AI and automation can lead to ethical and social concerns leading to job displacement, privacy issues, and bias.
5. *Environmental impact*: While Industry 4.0 can improve efficiency and reduce waste, the increased use of technology can have a negative environmental impact, such as increased energy consumption and electronic waste.

16.2 TOWARD INDUSTRY 5.0

Industry 5.0 is a new revolution in manufacturing and production, which emphasizes the collaboration between humans and machines to order to improve manufacturing processes. Unlike the previous trend of Industry 4.0, where machines and automation dominated, the goal of Industry 5.0 is to increase the importance of humans in the manufacturing process. By integrating digital tools, robotics, and automation technologies, workers can perform tasks more efficiently and safely. The main emphasis of Industry 5.0 is on technologies like sensors, AI, and ML, to build a more interconnected and intelligent manufacturing system. In turn, this will lead to a flexible, adaptable, and sustainable manufacturing environment that uses less energy, and produce less waste and emissions. Overall, Industry 5.0 represents a shift toward more socially responsible and environmentally conscious manufacturing [9]. It can be defined as:

> A new revolutionary approach that prioritizes collaboration between humans and machines in order to build a more sustainable and human-centric manufacturing environment.

The need to overcome the limitations and challenges of Industry 4.0, and to increase the advantages it provides, is what motivates the shift from Industry 4.0 to Industry 5.0. In contrast to Industry 4.0, which places a strong emphasis on automating production process, Industry 5.0 aims to reintroduce the humans back into the manufacturing process while retaining the advantages of automation and digitalization. It focuses on creating an environment where humans and machines work in collaboration to achieve common goals. This shift is thought to be necessary to overcome Industry 4.0's limitations like lack of flexibility, skills gap, and the ethical and social considerations that arise from the increasing use of automation and AI. The difference between Industry 4.0 and Industry 5.0 is discussed in Table 16.1.

16.2.1 Characteristics of Industry 5.0

The main characteristics of Industry 5.0, as shown in Figure 16.3, include:

1. *Human-centric manufacturing*: Industry 5.0 gives priority to human workers and their empowerment along with advanced digital tools and automation technologies.

TABLE 16.1
Differences between Industry 4.0 and Industry 5.0

Aspects	Industry 4.0	Industry 5.0
Focus	Automation, data exchange, efficiency	Human-centric, social responsibility, sustainability
Key Technologies	IoT, AI, big data and cloud computing	CPS, AI, augmented reality, virtual reality, and blockchain
Decentralized Decision-Making	Yes	More centralized with human oversight
Customization	On-demand, highly customizable	Personalized, cocreated with customers
Workforce	Highly skilled technicians and engineers	Collaborative, cross-functional, diverse
Supply Chain	Digitalized and connected	Emphasis on sustainability and ethical sourcing
Sustainability	Emphasis on energy efficiency	Sustainable and environmentally conscious
Social Responsibility	Limited focus on social responsibility	Focus on social and ethical responsibility
Innovation	Incremental improvements	Disruptive and innovative

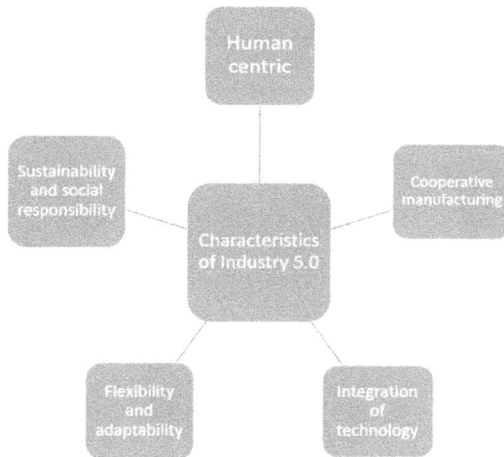

FIGURE 16.3 Characteristics of Industry 5.0.

2. *Cooperative manufacturing*: Industry 5.0 encourages cooperation between humans and machines to accomplish manufacturing objectives, with each party contributing their unique strengths.
3. *Integrating digital technologies*: A more intelligent and connected industrial system is created by Industry 5.0 by integrating digital technologies like AI, sensors, ML, and augmented reality.
4. *Flexibility and adaptability*: Industry 5.0 fosters a manufacturing environment that is flexible and adaptable, allowing for quick responses to changing market demands and customer preferences.
5. *Sustainability and social responsibility*: Industry 5.0 places a focus on reducing environmental impact and promoting social responsibility through the implementation of sustainable practices like renewable energy, waste reduction, and employee welfare.

16.2.2 OPPORTUNITIES OF INDUSTRY 5.0

Industry 5.0 provides various opportunities, as shown in Figure 16.4, and discussed below:

1. Automation will have favorable impact on employment across various sectors in industries.
2. Customization will be possible according to the need of the customer using automated manufacturing system.
3. Industry 5.0 will increase chances for creative people to work, which promotes the improvement of labor productivity.
4. Industry 5.0 will assist the customer digitally in handling frequent follow-up tasks by giving machines the flexibility needed to generate a high-level option based on employee needs [10].

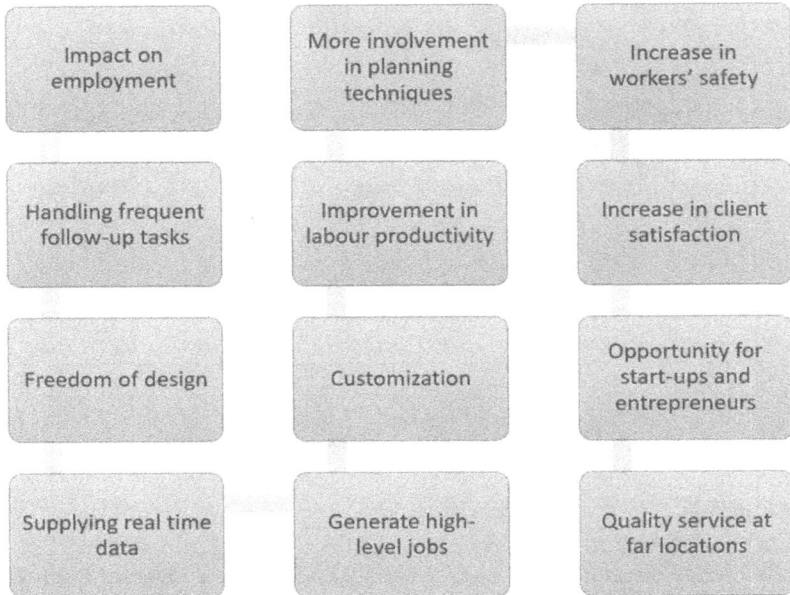

FIGURE 16.4 Opportunities of Industry 5.0.

5. Industry 5.0 will generate higher-value jobs, as it will give people control of its construction processes [11].
6. In industry 5.0, the operator in the production cell will become more involved in the planning technique.
7. More custom-made products will be made as a result of the freedom of design.
8. Industry 5.0 will make automation process easier by supplying real-time data from the industry.
9. Increase in worker safety as cobots (collaborative robots) will perform the more hazardous tasks.
10. More customized goods and services will boost client satisfaction and loyalty, and draw in new clients, which boosts revenue and market share for the businesses.
11. If appropriate money and infrastructure are available, it will offer great opportunities for start-ups and entrepreneurs to develop new goods and services related to Industry 5.0.
12. Human–machine interaction will be given more weight in Industry 5.0, and a bigger platform will be provided for its research and development.
13. With Industry 5.0, quality services will be delivered in remote locations, particularly in the healthcare sector where robots can perform surgery.

16.2.3 APPLICATIONS OF INDUSTRY 5.0

The applications of Industry 5.0 are diverse and far-reaching, and it has the potential to transform various sectors, including consumer products, automotive, aerospace,

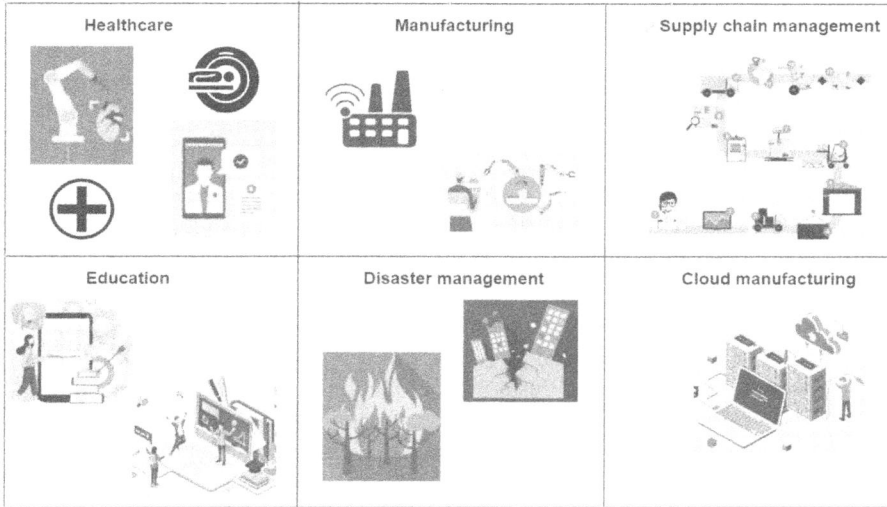

FIGURE 16.5   Various applications of Industry 5.0.

and healthcare [12]. For example, Industry 5.0 in the automotive industry can be used to manufacture customized and eco-friendly vehicles, and in healthcare it can be used to develop more efficient and personalized medical devices. Both large and small industries can benefit from Industry 5.0 as it has the capability to change according to the market needs. Some of the applications of Industry 5.0 are shown in Figure 16.5 and are discussed next [13]:

Healthcare: One of the key applications of Industry 5.0 in real-time is in hospitals, also known as "smart hospitals". It is essential for improving the working practices of medical professionals. As with the COVID-19 pandemic, clinicians can remotely inspect patients and provide vital data for better medication by utilizing smart healthcare technologies [14, 15]. AI, natural language processing (NLP) and ML can be used to assess a range of things, including levels of glucose. The disease detection process allows Industry 5.0 to build customized smart implants accurately in accordance with changing client requests [16, 17].

Manufacturing: Industry 5.0 focuses mainly on human–machine communication. Its major goal is to enhance the interoperability of human creativity with more precise machinery. To make the production sustainable, Industry 5.0 establishes mechanisms for resource recycling and reuse [18, 19]. Moreover, manufacturing must have a less harmful impact on the environment to minimize waste and maximize resource efficiency, and personalization must be increased through additive manufacturing. Industry 5.0 is transforming all industrial processes by relieving human workers of repetitive tasks. Production facilities can be located near sources of cheap raw materials and in regions where the cost of manufacturing is low [20]. Cloud manufacturing will be used to manage operations connected to the production and plant machinery control. The production efficiency, value addition, and market share of the manufacturing industries can be increased by including service components in the production process. The cost-effective virtualized platform can be used to handle the manufacturing services [21].

Supply chain management: According to [22], Industry 5.0 emphasizes on the importance of collaboration between people and cobots. For supply chain management, common manufacturing processes require the use of robots, which is a challenging task as robots require appropriate instructions [23]. According to [24], products can be personalized and customized without human involvement. However, the supply chain needs to run smoothly from beginning to end, including the choice of raw materials, after taking into consideration the unique customization and modification requirements of every customer. Industry 5.0 aims to combine automated, intelligent digital ecosystems with social interaction. The development of efficient operations and the personalization of end-user experiences are made possible by the incorporation of human factors in this process. Industry 5.0 aims to combine automated, intelligent digital ecosystems with social interaction. Human interaction plays an important role in such processes as it allows for customized end-user experiences and creates efficient operations [25].

Education: Education is seen as a necessity for every country's technological reformation. The goal of Industry 4.0 training was to minimize human involvement and give priority to machines; however, the goal of Industry 5.0 training is to develop a collaboration between autonomous machines and humans. As an example, a Lead Robotics Officer is an expert on the interface between machines and human operators and has prior training in robotics and AI. His or her job is to make decisions based on this training, and he or she can only do this with a combination of technology, leadership, and communication skills.

Disaster management: Disasters are sudden, catastrophic events that cause damage to people's life and property, and their prevention and management measures can help us to lessen their effects. Disaster assistance is a crucial component for any organization, but it focuses only on short term goals. A lot of disaster recovery plans have been updated as a result of the COVID-19 pandemic, which could mean that long-term resilience will take the place of disaster recovery methods in future plans. The author of [26] proposes how Industry 5.0 can be used in disaster management, especially after the 7.0 magnitude disaster that happened in Indonesia in 2018. The study revealed that Industry 4.0 has issues with disaster recovery and management, and that including human intervention in AI and IoT systems can assist in resolving problems related to disaster. Any type of disaster, including earthquakes and pandemics, can be managed with Industry 5.0.

Cloud manufacturing: In the cloud manufacturing process, multinational stakeholders can work together to run an effective and affordable manufacturing process. It includes features like reliability, high quality, cost effectiveness, and flexibility. Cloud manufacturing also benefits the environment by removing the need for long-distance raw material deliveries during the manufacturing process. Designers can also protect their designs by storing them on a cloud server and having robust access control [27]. This way designers can establish their manufacturing plants close to the place of the raw materials where the cost is low. Here, the cloud is responsible for controlling machines and managing the production cycle, including service composition and scheduling, as well as machine control in the facility [28, 29]. In [30, 31] the authors showed that cloud manufacturing can be used as a service-oriented manufacturing model. With Industry 5.0, cloud manufacturing systems are expected to meet the various needs of engineering, production, and logistics.

16.2.4 SECURITY IN INDUSTRY 5.0

Trust plays an important role in any industrial data transmission. In an operation technology and information technology environment where businesses struggle to combat the current widespread ransomware and other cybercrimes, trust is hard to find. This lack of security needs to be addressed in Industry 5.0. For example, in order to trust any industrial AI application, and ensure that it won't harm anyone, workers need to be confident that the data being provided is reliable and hasn't been fabricated or sourced incorrectly. Furthermore, these systems have the ability to gather sensitive data on workers and citizens; therefore, trust in the security of the data handling will be essential. Circular economy and decentralized industry applications also face the same problems. Data is essential in each of these scenarios, but our current data infrastructures are not sufficiently trustworthy.

Manufacturers now have to ensure that there is a secure and seamless interface between operational technology systems and information technology systems in order to take full advantage of digital innovation and stay competitive. However, operational technology network components like automated robots, food processing systems, and industrial control systems are typically not categorized as "secure-by-design", as they weren't made to be connected to enterprise networks. This lack of security has made manufacturing prone to cyber threats. The highest risk of a cyber security threat to Industry 5.0 is from the following:

1. Phishing attack
2. Theft of IP
3. Attack on supply chain
4. Ransomware
5. Outside interference/interruption to equipment

The standard supply chain is quite complicated; therefore, industry frequently outsources, and depends more on sub-suppliers, which can lead to security concerns.

Most of the data in Industry 5.0 is retrieved from sensors and IoT devices and the techniques being used to ensure trust and security of this data are decades old. Therefore, to prevent attacks like spoofing, it is necessary to securely identify the devices that send data. For this, PKI (public key infrastructure), which powers today's Internet security, can be crucial in providing distinctive device IDs for authentication. To further protect the software running on the devices, they can use a combination of hardware security and software tamper resistant solutions. Digital signatures and encryption techniques can be used for data integrity as data moves through the different networks.

In addition, ABE has the ability to play a significant role in Industry 5.0; it secures data by controlling access to it based on attributes like job roles, departments, or locations. For example, in a manufacturing environment, different employees may need access to different data based on their job roles, and ABE ensures that only users who are authorized can access the data they need. ABE can also be used to secure data in the supply chain and logistics sectors. With ABE, data can be encrypted based on attributes such as product type, delivery location, or shipment date and only authorized users are allowed to access and decrypt it, improving the security and privacy of the supply chain.

In healthcare computer systems, ABE can also provide security by encrypting patient data based on attributes such as medical conditions, patient age, or treatment requirements. Only authorized medical personnel who have the corresponding attributes would be able to access the patient data, ensuring that patient privacy is protected while still enabling medical professionals to provide effective treatment. In these ways, ABE can play a crucial role in securing data and managing access control in Industry 5.0 applications.

16.3 DIFFERENT CAPABILITIES OF INDUSTRY 4.0 VS INDUSTRY 5.0

Industry 4.0 and Industry 5.0 represent two different phases of the manufacturing industry's evolution. Industry 4.0, also known as the Fourth Industrial Revolution, has made the manufacturing process increasingly digital and interconnected. It focuses mainly on automation, connectivity, and data-driven decision-making resulting in improved efficiency and reduction in the costs through the use of technologies. Industry 4.0 optimizes the manufacturing process by collecting and analyzing data using robotics, sensors, and cloud computing to find inefficiencies and improvement opportunities. Its aim is to leverage data and automation to improve quality, increase production speed, and promote customization to support changing customer preferences. However, some criticisms of Industry 4.0 include concerns around job displacement and the dehumanization of the manufacturing process as automation can lead to fewer job opportunities for humans.

On the other hand, Industry 5.0 is defined as a more collaborative and human-centric approach, where human workers play an essential part in the manufacturing process, in addition to automation and digital technologies. This approach aims to empower workers by providing them with the resources and tools they need to do their tasks efficiently, while simultaneously addressing issues with job displacement and worker welfare. Industry 5.0 understands that humans have special skills like creativity and critical thinking, that machines simply cannot match. By working together, they can complement each other's strengths and create a more effective and sustainable manufacturing environment.

One example of how Industry 5.0 is being applied is through the use of cobots, or collaborative robots. Cobots are simple to use and do not require much training, therefore workers can quickly incorporate them into their workflows. These are robots can work alongside humans, assisting with tasks such as assembly or material handling, which helps to improve overall efficiency and reduces the chances of mistakes. Another example of Industry 5.0 in action is the use of augmented reality (AR) and virtual reality (VR) technologies that help workers with real-time data and instructions. AR and VR can also be used to train workers, providing a more immersive and engaging learning experience.

The main difference between Industry 4.0 and Industry 5.0 is how each focus on automation. Fully automating the manufacturing process is the goal of Industry 4.0, while Industry 5.0 intends to balance automation and human involvement. Therefore, they represent two different approaches to manufacturing. Industry 5.0 recognizes that some tasks are better suited to machines, while others require the unique capabilities of human workers. It encourages collaboration between humans and machines to achieve better outcomes than either could achieve alone. With the advent of the Industry 5.0 paradigm, the primary research goals shifted from sustainability to human-centricity [32].

ABE can be used in both Industry 4.0 and Industry 5.0 for fine-grained security and privacy but it use differs in some aspects. Table 16.2 summarizes the differences between ABE in Industry 4.0 and Industry 5.0.

While ABE in Industry 4.0 is mainly focused on data security and access control in cloud computing and IoT systems, ABE has a wider range of applications in Industry 5.0 that includes healthcare and preventive maintenance systems. Industry 5.0 also features stronger integration with AI and ML, allowing for more advanced decision-making. However, both Industry 4.0 and 5.0 face challenges such as scalability, complexity in managing attributes, and ensuring data privacy and ethical concerns.

TABLE 16.2
Differences between the use of ABE in Industry 4.0 and Industry 5.0

Aspects	Industry 4.0	Industry 5.0
Focus	Data security and access control	Data security and access control
Scope	Mainly used in cloud computing and IoT systems	Used in a broader range of applications, including healthcare systems
Integration	Limited integration with AI and ML	Strong integration with AI and ML
Applications	Limited to access control and data sharing	Used in various applications, including predictive maintenance
Benefits	Improved data security and access control	Improved data security, access control, and decision-making
Challenges	Scalability and complexity in managing attributes	Ensuring data privacy and addressing ethical concerns

16.4   SYSTEMATIC ANALYTICS OF INDUSTRIAL IoT

Industrial IoT (IIoT) is a technology that connects industrial equipment and devices to the Internet and one another. This connection makes it possible to collect and analyze data to enhance the effectiveness of industrial processes. It includes machine-to-machine and industrial communication technologies [33] and allows real-time monitoring and control of equipment, which helps organizations make informed decisions and respond to changes quickly. The sensors and devices present in IIoT collect data such as temperature, pressure, and vibration, and then send it to the cloud where algorithms are used to store, process, and analyze it in real-time. The analyzed data can then be used to optimize industrial processes and improve operational efficiency. For example, if a machine starts behaving in an abnormal way, IIoT sensors will detect this behavior and send an alert message to the system. The system will shut down the machine to prevent any damage. IIoT can also be used to keep an eye on logistics, supply chain, and environmental factors that have an impact on industrial processes. Another example is the use of sensors attached to a delivery truck where its location, speed, and temperature can be tracked to monitor its status and delivery time. Overall, IIoT makes decision-making and process optimization easy for industrial organizations, resulting in increased production and efficiency.

The architecture of IIoT is divided into layers (Figure 16.6), each having its own functions and components. The hardware layer is at the bottom and contains sensors, actuators, and other devices that gather information about industrial equipment and processes. These devices can be connected wired or wirelessly to the Internet. Next is the connectivity layer that provides the framework for data transmission between devices and cloud. This layer includes gateways and routers that can handle the flow of data between the devices and the cloud. The third layer is the data layer, that has cloud-based systems that store, handle, and examine the data gathered by the sensors and devices. This layer contains a variety of data processing techniques, including AI and ML. The application layer is the top layer that includes applications and the user interfaces to interact with the data. It also includes some tools for real-time visualization of the industrial processes.

The systematic analytics of IIoT allow the use of data analysis techniques to make decisions according to the industrial demands. In order to get meaningful results, complex data is broken down into smaller sets using statistical techniques. One example of the systematic analytics of IIoT is predictive maintenance where algorithms are capable of predicting when a machine is likely to fail by analyzing data from sensors and other linked equipment before any malfunction occurs. This may save maintenance expenses, reduce idle time, and increase overall equipment effectiveness. Another example is quality control. By analyzing data from sensors and cameras, ML algorithms can identify defects in products as they are being manufactured. This enables the production process to be modified in real-time, minimizing waste and enhancing product quality.

The systematic analytics of IIoT also involve the use of data visualization tools, such as dashboards and reports, to help users make sense of the data by identifying patterns and unusual events. This will help to boost production processes, cut costs, and enhance efficiency in industrial settings. Organizations can receive operational

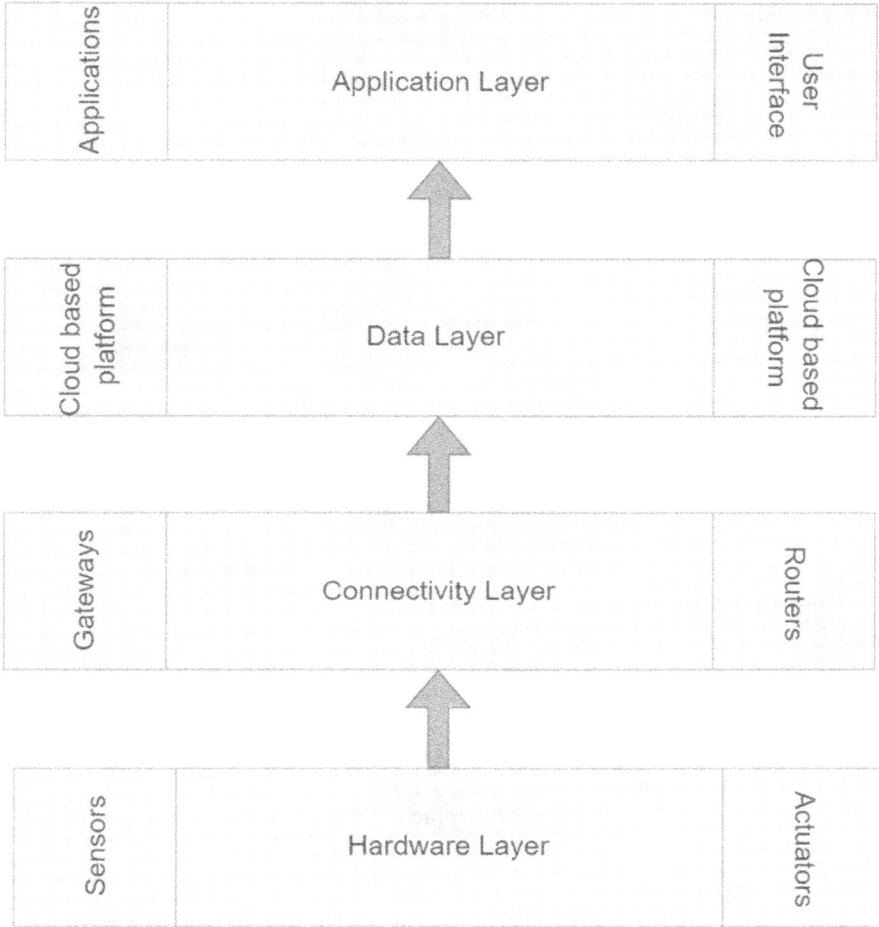

FIGURE 16.6   Architecture of the Industrial Internet of Things (IIoT).

insights that were previously difficult to access by analyzing enormous amounts of data, enabling them to make wise decisions and optimize their operations.

16.5   THE SINGULARITY AND CONCERNS FOR THE FUTURE

The term "singularity" in relation to IIoT refers to the point at which machine intelligence overtakes human intellect, causing a significant change in the nature of industrial operations and the economy as a whole. Taking AI as an example, there are fears about its potential hazards and unfavorable effects.

One concern is the possibility of widespread job losses as machines and algorithms replace human workers in industrial settings, which could cause social and economic disruption if workers are not enough prepared for the shift. Another issue is the possibility that machines could start taking decisions that have adverse effects on humans. There are concerns that the IIoT could worsen existing power imbalance and inequities, especially if the technology is monopolized by powerful governments or corporations.

IIoT also raises ethical questions around surveillance, privacy, and environmental effect. Achieving interoperability, scalability, and reliability seamlessly in these systems is crucial. Moreover, security and cybersecurity threats are also high in IIoT as these systems are interconnected making them more susceptible to cyberattacks. To address these concerns and ensure that IIoT benefits society as a whole, researchers and policymakers are exploring various approaches. In order to share the advantages of the technology fairly and equitably, certain regulations and guidelines are currently being developed that will provide a suitable framework.

Encryption makes sure that the privacy of the data is maintained against the cloud [34], therefore to address the issue of security in IIoT we can use the ABE method that is based on attributes and processes the data according to the encryption keys [35]. The ABE model for IIoT would involve defining the attributes and access control policies needed to secure data and devices within the IoT network. Here is a proposed model that includes attributes like device type (e.g. sensor, actuator, gateway), device manufacturer, device model, device firmware version, device location, device owner, data type (e.g. temperature, humidity, pressure), data source (e.g. sensor, actuator, external system), data owner and access level (e.g. read-only, read-write, admin). The access control policies are:

1. Only devices with a certain manufacturer or firmware version can access specific data sources.
2. Devices located in a certain geographic region can access data with specific attributes.
3. Only users with certain job titles or roles can access data at a certain access level.
4. Data owners can grant access to specific individuals or groups based on their attributes.
5. Devices must be authenticated before being granted access to data.

The ABE model will first assign attributes to the users and devices in the system, attributes can be anything from type of the device to location or date of manufacturing. Then the access policies will be defined for these attributes that will allow only specific users to access the data. When a device generates a data, it will be encrypted using the ABE model where the access policies for this data is defined. The encrypted data will then be stored on the cloud with its access policies. Whenever a user/device wants to access this data, they need to authorize themselves by verifying their attributes. Once the verification is done, the system will decrypt the data. The outline of the proposed IIoT framework using ABE is shown in Figure 16.7. The model is also capable of allowing the administrator to track the access and usage of the data and take care of key management mechanisms to maintain security and integrity of the system.

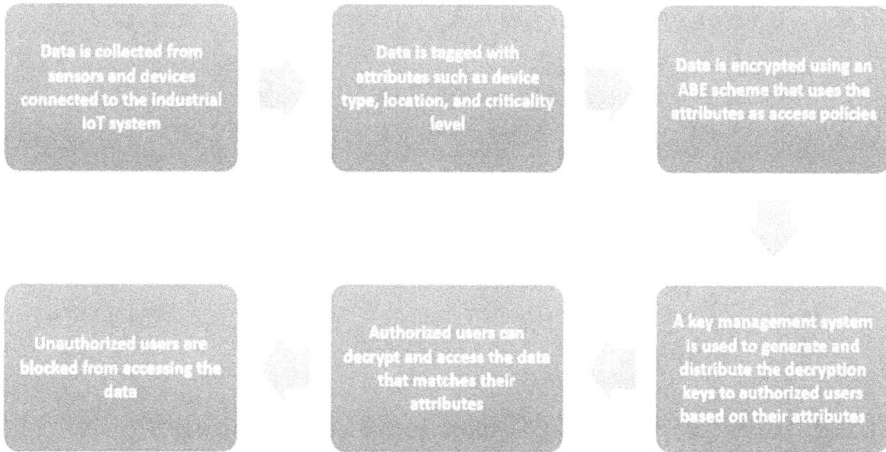

FIGURE 16.7 Outline of the proposed IIoT framework using the ABE model.

According to the proposed ABE model, access policies are framed to make access control easier in large IoT systems. This model will help the industries to improve their data security and access control.

16.6 SUSTAINABLE DEVELOPMENT UNDER INDUSTRY 5.0

Unlike traditional industrial transformations, the main priority of Industry 5.0 is environmental conservation. It focuses on finding environmentally friendly solutions to fight climate change. Sustainability is the most important feature of Industry 5.0, and this makes it very different from Industry 4.0. Industry 5.0's main focus is to reduce the harmful impact of fossil fuels on the environment by using renewable energy resources, and reducing any kind of wastage and carbon emissions. One way in which Industry 5.0 can promote sustainability is through the use of IoT and advanced analytics to improve the use of resources and reduce any kind of wastage. For example, energy and water usage in manufacturing process can be monitored using sensors that will be helpful in reducing consumption and costs. Another way in which Industry 5.0 can support sustainability is by adopting the principles of the circular economy that involves designing goods and processes to maximize the length of time materials can be put to use before being thrown away.

Industry 5.0 can also promote social responsibility by focusing on improving working conditions and fostering greater collaboration between humans and machines. It ensures that when machines are designed and manufactured that the needs and safety of the workers are considered, and there is an emphasis on giving the training and education needed to adapt to new jobs and technologies. Industry 5.0 offers a framework for encouraging sustainable development by combining digital technologies and human-centered production methods that will minimize environmental impact and promote social responsibility.

16.6.1   Benefits of Industry 5.0 for Sustainable Manufacturing

1. Cutting production costs can help manufacturers become more competitive.
2. Manufacturers, society, and workers will benefit.
3. By upgrading their skills, workers can empower their workforce.
4. While maintaining their profitability, manufacturers can concentrate on issues like resource conservation, tackling climate change, and supporting social stability.
5. Advanced technologies that efficiently employ resources are used to support circular production models, which promote sustainable manufacturing.
6. By concentrating on sustainability, industry becomes more resistant to outside factors like pandemics, and financial crises.

16.6.2   Steps to Obtaining a Sustainable Model in Industry 5.0

A sustainable model in Industry 5.0 can be obtained by following these steps, which are also shown in Figure 16.8:

1. *Evaluate and set sustainability objectives*: First, analyze your present activities and find areas that require improvement in sustainability. Establish clear objectives and targets to improve resource efficiency, social responsibility, and reduces environmental impact.
2. *Use energy-saving measures*: Find ways to make manufacturing processes, machines, and buildings more energy efficient. This involves developing energy-management systems, optimizing manufacturing schedules, and introducing energy-efficient technologies.
3. *Adopt circular economy principles*: Reducing waste, encouraging recycling and reuse, and designing products with longevity and recyclability are the steps toward circular economy strategy. Implementing strategies like waste reduction programs, product lifecycle extensions, and remanufacturing can also help.
4. *Use renewable energy and clean technologies*: A clean technology can be integrated using energy-efficient machines and renewable energy sources which will help to reduce dependence on fossil fuels and greenhouse gas emissions. Alternate energy resources like solar or wind energy can also be used.
5. *Improve supply chain sustainability*: Suppliers should work together to advance sustainability along the supply chain. Practices like choosing eco-friendly products and lowering emissions caused by transportation need to be implemented.
6. *Involve the workers and stakeholders*: Sustainable work environments can be promoted among workers by giving them training, increasing their understanding of sustainability issues, and involving them in sustainable projects. To align with the goal of sustainable vision, all clients, suppliers, and communities need to be involved.
7. *Monitor, measure, and report*: Monitoring programs need to be established to keep an eye on progress toward sustainability objectives. Identify and track key performance indicators for energy use, waste generation, carbon

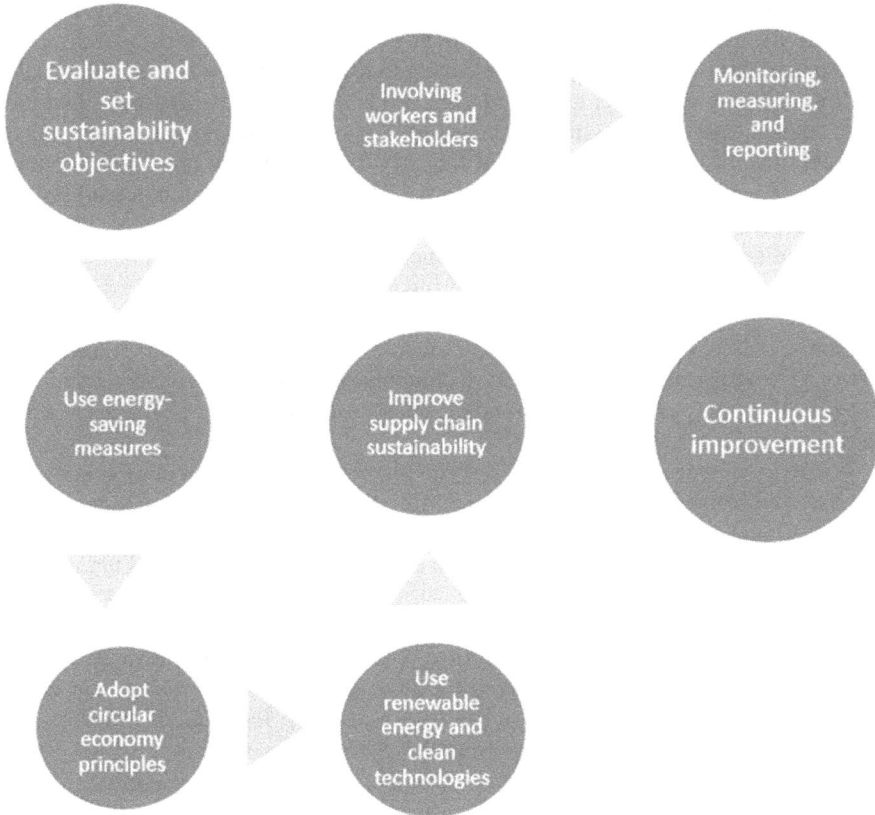

FIGURE 16.8   Steps for obtaining a sustainable model in Industry 5.0.

emissions, and social impact. Reports on sustainability achievements should be regularly maintained.

8. *Continuous improvement*: Regular reviewing and assessing the sustainable objectives are steps for continuous improvement. Feedbacks from workers, customers, and stakeholders also helps innovation and further improvement.

By taking the steps discussed above, we can contribute to the development of a sustainable Industry 5.0 model that will minimize any negative environmental effects while prioritizing social responsibility and resource efficiency.

16.7   WHO IS GETTING SMARTER?

Industry 5.0 is a new way of manufacturing that merges the skills of humans with cutting-edge digital technologies to establish a smarter, more effective, eco-friendly, and human-oriented production ecosystem. This change will impact various sectors like manufacturing, supply chain, energy management, healthcare, and agriculture.

For the Industry 5.0 revolution to succeed, the IIoT must increase efficiency, increase productivity, and improve asset management [36].

In manufacturing, Industry 5.0 is expected to bring smartness by enhancing automation, reducing idle time, and increasing productivity, and it will make factories more adaptable to the changing customer needs. By eliminating many tedious, dirty, and repetitive tasks from human labor, Industry 5.0 will change production systems all over the world [16]. Furthermore, Industry 5.0 will empower human workers and enhance their role in the manufacturing process, enabling them to work alongside the latest digital tools and automation technologies.

In the supply chain, Industry 5.0 can bring greater transparency and efficiency by incorporating IoT sensors and data analytics to track and manage goods from production to delivery. It can be helpful in reducing waste and improve customer satisfaction. The integration of Industry 5.0 technologies can also lead to more responsive and flexible supply chains that can adapt quickly to changes in demand or supply increasing its smartness.

In energy management, Industry 5.0 can lead to the development of smart grids that use data from connected devices to optimize energy consumption, reduce waste and associated costs. These require real-time data so that they can respond to the user demand and makes process costs efficient [37]. In collaboration with renewable energy sources, Industry 5.0 can reduce carbon emissions and promote sustainable development.

In healthcare, Industry 5.0 can improve patient care using IoT devices, AI, and data analytics to develop personalized treatment plans and monitor patient health in real-time ensuring more effective healthcare delivery, fewer medical errors, and improved patient recovery results making this industry smart.

In agriculture, Industry 5.0 can transform the industry by using sensors and data analytics to optimize crop yields, reduce waste, and minimize the environmental impact of farming making it smart. Increases in productivity and sustainability can be achieved by farmers using real-time data to optimize water consumption, fertilizer application, and pest control. Industry 5.0 can also help in the development of precise agriculture, where farmers can use data to make decisions on when and where to plant crops that would lead to higher yields and reduced wastage.

16.8   SECURITY STANDARDS AND REGULATIONS

Security standards and regulations are crucial for ensuring the safety and integrity of IIoT systems. The weakness of these systems against cyberattacks and other security threats can have a serious impact on the safety of workers, the reliability of manufacturing processes, and the confidentiality of sensitive information.

To address these issues, several standards and regulations have been implemented. One such example is the Industrial Internet Security Framework (IISF), developed by the Industrial Internet Consortium (IIC), that offers guidelines on how to secure IIoT systems. Additional security-related topics covered by the IISF are secure architectures, identity and access management, data protection, and incident response. Other standards and laws that are related to IIoT security are ISO/IEC 27001, which offers a framework for information security management systems, and

NIST Cybersecurity Framework, that offers recommendations and best practices for enhancing cybersecurity.

In addition to these standards and regulations, there are also various industry-specific regulations that apply to IIoT systems. For example, in the USA the National Institute for Occupational Safety and Health (NIOSH) and the Occupational Safety and Health Administration (OSHA) have approved guidelines regarding the safe use of machines and robots in manufacturing processes, respectively. Security standards and regulations are essential for ensuring the security and reliability of IIoT systems as well as the privacy of sensitive data and intellectual property. Further, to stay informed about the most recent security standards and laws and to make sure that their IIoT systems are compliant, companies should maintain regular contact with industry and regulatory bodies.

16.9 CONCLUSION

After all this discussion, we have concluded that Industry 5.0 plays a more important role in the future development of manufacturing processes than the previously used Industry 4.0, and obtaining a sustainable model using Industry 5.0 with IIoT will be an additional benefit. Moreover, its security can be enhanced using the latest ABE scheme which will protect it from any kind of attack. However, in the future, more work needs to be done to make Industry 5.0 a success in every field. In our view, more trust and security-based algorithms need to be incorporated into this system.

REFERENCES

1. Wang, L., Törngren, M., Onori, M., *Current status and advancement of cyber-physical systems in manufacturing*, Journal of Manufacturing Systems, 37, pp. 517–527, 2015.
2. Jeschke, S., Brecher, C., Meisen, T., Özdemir, D., Eschert, T., *Industrial Internet of Things and Cyber Manufacturing Systems*, In: Jeschke, S., Brecher, C., Song, H., Rawat, D. (eds) Industrial Internet of Things. Springer Series in Wireless Technology. Springer, Cham, 2017.
3. Ghobakhloo, M., *Industry 4.0, digitization, and opportunities for sustainability*, Journal of Cleaner Production, 252, p. 119869, 2020.
4. Perales, D.P., Valero, F.A., García, A.B., *Industry 4.0: A Classification Scheme*, In: Viles, E., Ormazábal, M., Lleó, A. (eds) Closing the Gap Between Practice and Research in Industrial Engineering. Lecture Notes in Management and Industrial Engineering. Springer, Cham, 2018.
5. Kiel, D., Müller, J.M., Arnold, C., Voigt, K.I., *Sustainable industrial value creation: Benefits and challenges of industry 4.0*, International Journal of Innovation Management, 21(08), p. 1740015, 2017.
6. Khan, M., Haleem, A., Javaid, M., *Changes and improvements in Industry 5.0: A strategic approach to overcome the challenges of Industry 4.0*, Green Technologies and Sustainability, 1(2), p. 100020, 2023.
7. Rasori, M., Manna, M.L., Perazzo, P., Dini, G., *A survey on attribute-based encryption schemes suitable for the internet of things*, IEEE Internet of Things Journal, 9(11), pp. 8269–8290, 2022.
8. Chiquito, A., Bodin, U., Schelén, O., *Attribute-based approaches for secure data sharing in industrial contexts*, IEEE Access, 11, pp. 10180–10195, 2023.

9. Xu, X., Lu, Y., Vogel-Heuser, B., Wang, L., *Industry 4.0 and Industry 5.0—Inception, conception and perception*, Journal of Manufacturing Systems, 61, pp. 530–535, 2021.
10. Javaid, M., Haleem, A., Singh, R.P., Haq, M., Raina, I.U., Suman, A., *Industry 5.0: Potential applications in COVID-19*, Journal of Industrial Integration and Management, 5(04), pp. 507–530, 2020.
11. Rossi, B., Manufacturing Gets Personal in Industry 5.0., Raconteur, 2018.
12. Leng, J., Sha, W., Wang, B., Zheng, P., Zhuang, C., Liu, Q., Wuest, T., Mourtzis, D., Wang, L., *Industry 5.0: Prospect and retrospect*, Journal of Manufacturing Systems, 65, pp. 279–295, 2022.
13. Maddikunta, P.K.R., Pham, Q.V., Prabadevi, B., Deepa, N., Dev, K., Gadekallu, T.R., Ruby, R., Liyanage, M., *Industry 5.0: A survey on enabling technologies and potential applications*, Journal of Industrial Information Integration, 26, p. 100257, 2022.
14. Longo, F., Padovano, A., Umbrella, S., *Value oriented and ethical technology engineering in Industry 5.0: A human-centric perspective for the design of the future*, Applied Sciences, 10(10), p. 4182, 2020.
15. Wu, H., Nguyen, G.T., Chorppath, A.K., Fitzek, F., 2017, *Network slicing for conditional monitoring in the industrial internet of things*, Transport, 2018.
16. Nahavandi, S., *Industry 5.0 - a human - centric solution*, Sustainability, 11, p. 4371, 2019.
17. Lutz, J., Memmert, D., Raabe, D., Dornberger, R., Donath, L., *Wearable for integrative performance and tactic analyses: Opportunities, challenges and future directions*, International Journal of Research in Public Health, 7(1), pp. 59, 2020.
18. Aslam, F., Aimin, W., Li, M., Ur Rehman, K., *Innovation in the era of IoT and industry 5.0: Absolute innovation management (AIM) framework*, Information, 11(2), p. 124, 2020.
19. Alhassan, A.B., Zhang, X., Shen, H., Xu, H., *Power transmission line inspection robots: A review, trends and challenges for future research*, International Journal of Electrical Power & Energy Systems, 118, p. 105862, 2020.
20. Ghobakhloo, M., Fathi, M., Iranmanesh, M., Maroufkhani, P., Morales, M.E., *Industry 4.0 ten years on: A bibliometric and systematic review of concepts, sustainability value drivers, and success determinants*, Journal of Cleaner Production, 302, p. 127052, 2021.
21. Deepa, N., Pham, Q.V., Nguyen, D.C., Bhattacharya, S., Prabadevi, B., Gadekallu, T.R., Maddikunta, P.K.R., Fang, F., Pathirana, P.N., *A survey on blockchain for big data: Approaches, opportunities, and future directions*, Future Generation Computer Systems, 131, pp. 209–226, 2022.
22. Nguyen, T.N., Ebrahim, F.M., Stylianou, K.C., *Photoluminescent, upconversion luminescent and nonlinear optical metal-organic frameworks: From fundamental photophysics to potential applications*, Coordination Chemistry Reviews, 377, pp. 259–306, 2018.
23. Wang, S., Wang, H., Li, J., Wang, H., Chaudhry, J., Alazab, M., Song, H., *A fast CP-ABE system for cyber-physical security and privacy in mobile healthcare network*, IEEE Transactions on Industry Applications, 56(4), pp. 4467–4477, 2020.
24. Babamiri, B., Bahari, D., Salimi, A., *Highly sensitive bioaffinity electrochemiluminescence sensors: Recent advances and future directions*, Biosensors and Bioelectronics, 142, p. 111530, 2019.
25. Adel, A., *A conceptual framework to improve cyber forensic Administration in Industry 5.0: Qualitative study approach*, Forensic Sciences, 2(1), pp. 111–129, 2022.
26. Sukmono, F.G., Junaedi, F., *Towards industry 5.0 in disaster mitigation in Lombok Island, Indonesia*, Jurnal Studi Komunikasi (Indones J Commun Stud), 4(3), pp. 553–564, 2020.
27. Akbaripour, H., Houshmand, M., Van Woensel, T., Mutlu, N., *Cloud manufacturing service selection optimization and scheduling with transportation considerations: Mixed-integer programming models*, The International Journal of Advanced Manufacturing Technology, 95, pp. 43–70, 2018.

28. Liu, Y., Xu, X., Zhang, L., Tao, F., *An extensible model for multitask-oriented service composition and scheduling in cloud manufacturing*, Journal of Computing and Information Science in Engineering, 16(4), pp. 041009, 2016.

29. Helo, P., Phuong, D., Hao, Y., *Cloud manufacturing–scheduling as a service for sheet metal manufacturing*, Computers & Operations Research, 110, pp. 208–219, 2019.

30. Bo-Hu, L.I., Lin, Z., Shi-Long, W., Fei, T., Jun-wei, C.A.O., Xiao-dan, J., Xiao, S., Xu-dong, C., *Cloud manufacturing: A new service-oriented networked manufacturing model*, Computer Integrated Manufacturing System, 16(01), 2010.

31. Tao, F., Zhang, L., Venkatesh, V.C., Luo, Y., Cheng, Y., *Cloud manufacturing: A computing and service-oriented manufacturing model*, Proceedings of the Institution of Mechanical Engineers, Part B: Journal of Engineering Manufacture, 225(10), pp. 1969–1976, 2011.

32. Zizic, M.C., Mladineo, M., Gjeldum, N., Celent, L., *From industry 4.0 towards industry 5.0: A review and analysis of paradigm shift for the people, organization and technology*, Energies, 15(14), pp. 5221, 2022.

33. Sisinni, E., Saifullah, A., Han, S., Jennehag, U., Gidlund, M., *Industrial internet of things: Challenges, opportunities, and directions*, IEEE Transactions on Industrial Informatics, 14(11), pp. 4724–4734, 2018.

34. Verma, S., Ahuja, S., *A hybrid two-layer attribute-based encryption for privacy preserving in public cloud*, In 2016 International Conference on Inventive computation Technologies (ICICT), IEEE, 2, pp. 1–5, 2016.

35. Kumar, A., Kumar, S.A., Dutt, V., Kumar Dubey, A., Narang, S., *A hybrid secure cloud platform maintenance based on improved attribute-based encryption strategies*, International Journal of Interactive Multimedia and Artificial Intelligence, 8(2), pp. 150–157, 2021.

36. Kumar, N., Sharma, B., Narang, S., *Emerging Communication Technologies for Industrial Internet of Things: Industry 5.0 Perspective*, In Proceedings of Third International Conference on Computing, Communications, and Cyber-Security: IC4S 2021, pp. 107–122, Singapore: Springer Nature Singapore, 2022.

37. Babbar, H., Rani, S., *Security architecture and its methodology for fog computing*, ECS Transactions, 107(1), p. 4549, 2022.

Index

A

Actuator nodes, 298, 299
Application programming interfaces (APIs), 298–299, 333
Ant colony optimization (ACO), 154
Anti-money laundering (AML), 137
Artificial intelligence (AI), 3, 26, 38, 77, 91–92, 97
applications, 103–106
principles, 90–91
Artificial neural networks (ANN), 89, 151, 195, 212, 243
Attribute-based encryption (ABE), 355, 358
Augmented reality (AR), 76, 198, 267, 277, 360–361, 367

B

Blockchain, 26–28, 106–107, 114–116
core principles of, 116–117
cryptography, 114, 117
decentralization, 116
immutability, 117
privacy and identity protection, 117
smart contracts, 117
transparency, 117
components of, 117–120
consensus mechanisms, 27–28, 117–122, 130, 132, 137, 139
delegated proof of stake (DPoS), 119, 132
practical Byzantine fault tolerance (PBFT), 27, 117, 119
proof of stake (PoS), 27, 117, 119, 304
proof of work (PoW), 27, 117, 119
types of, 120–121
consortium, 121
private, 121
public, 121
Blockchain-based, 132–134, 137, 141
communication framework, 132–134, 137, 141
benefits and challenges, 133–134
design considerations, 132–133
Bluetooth, 3, 17–18, 22–23, 36, 66, 72–73, 76, 83, 112, 148, 207, 296, 304
Brain tumor, 243, 245–256, 258

C

Carbon capture utilization and storage (CCUS), 45
Carbon footprint, 38–39, 43–47, 49, 162
Clinical decision support systems (CDSS), 231

Cloud computing, 2, 80, 84, 102, 144, 146, 148, 155, 162, 264–265, 268, 270, 272, 277–279, 282–284, 300, 355–356, 359–360, 366–367
Cloud data centers (CDCs), 265, 269, 271–273, 278–280
Computer vision, 89, 92–96, 106
Constrained application protocol (CoAP), 13, 26, 30, 82
Content centric networking (CCN), 172, 181–182
Convergence, 12–14, 103, 319–320, 325, 328
Convolutional neural networks (CNN), 195, 214, 219, 228, 250, 342, 345, 350
Counter-terrorism financing (CTF), 137
COVID-19, 35, 190, 213–214, 263–264
Cyber-physical systems (CPS), 355–357

D

Data granularity, 68–69
Data heterogeneity, 314, 322, 324, 333
Data integrity, 26–28, 122, 126–127, 129, 133–136, 138–141, 189, 296, 326, 366
Dedicated short-range communication (DSRC), 112, 147–148
Deep belief network (DBN), 181, 213, 219, 221, 342, 345, 350
Deep learning, 81, 95, 217
classification algorithms, 95
auto encoders (AE), 221
convolution neural networks (CNN), 218
long short term memory (LSTM), 95, 153, 220, 222, 228, 251, 341, 346
restricted Boltzmann machine (RBM), 95, 181, 220, 346
applications in healthcare, 213–214
Distributed computing, 20, 82–83, 268–269, 273–274, 276–277
for IoT applications, 274, 276
technologies, 276–278, 286
cloud computing, 278
container-based computing, 281
fog computing, 278
serverless computing, 280

E

E-commerce, 209–210
Edge computing, 15, 20–23, 50, 75, 77, 82–84, 98, 101, 134
edge analytics, 22

379

For Product Safety Concerns and Information please contact our EU
representative GPSR@taylorandfrancis.com
Taylor & Francis Verlag GmbH, Kaufingerstraße 24, 80331 München, Germany

www.ingramcontent.com/pod-product-compliance
Lightning Source LLC
Chambersburg PA
CBHW060751220326

41598CB00022B/2402